FLORA OF DUDHWA NATIONAL PARK

(KHERI DISTT. U.P.)

By

DR. K.K. SINGH

National Botanical Research Institute

Lucknow - 226 001

BISHEN SINGH MAHENDRA PAL SINGH

23-A New Connaught Place

Dehra Dun - 248001, [INDIA]

1996

FLORA OF DUDHWA NATIONAL PARK

ISBN-81-211-0127-1

Printed by Gajendra Singh Gahlot at Shiva Offset Press,
and Laser Typesetting at Doon Phototype Printers
14, Old Connaught Place, Dehra Dun,
for Bishen Singh Mahendra Pal Singh, Dehra Dun, India.

FOREWORD

Although plant taxonomy has expanded from the herbarium to the field and experimental garden, and from type specimens to living populations and molecular biology, the need for floristic research has not decreased. In the first place, plant taxonomy serves an eminently practical function through floristics. Our knowledge of plant biodiversity is made available in an organised form in Floras. Many developing countries have felt the increased need to assess and revise their Floras for economic utilization of plant wealth, and to conserve rare, endangered and threatened plants. Many Floras, especially of flowering plants, are being prepared and published under international collaboration. They provide users with information on specific, subspecific and varietal diversity, distribution, endemicity, ecology, and rarity or endangerment of plants. They are the products of floristic research and enable the users to identify plants. The detailed knowledge of the flora of an area is acquired by systematic plant exploration, identification and documentation, compilation of floristic inventories and check-lists, preparation of descriptive floras (phytography, mapping), and study of the origin and relationship of the flora. The present work is an outcome of intensive botanical exploration in Dudhwa National Park situated on Indo-Nepal border in Kheri district of Uttar Pradesh, India. Dr. K.K. Singh has spent many years botanizing in the Park and has an intimate field knowledge of the plant life. He has prepared a concise flora with keys and brief descriptions, intended primarily to provide a means of identifying the species. In 1978, he prepared a doctoral thesis on this subject under my supervision, which won him the Ph.D. degree of Agra University. During the last few years, the manuscript has been further revised and improved for publication in book form. The author has spared no pains in making this work as complete as possible, and has provide many field notes based on personal experience. He has also tried to modernize and update the nomenclature of plants in light of the International Code of Botanical Nomenclature.

The protected areas like parks and reserves have a vital role to play in the conservation and sustainable use of biodiversity, because of their potential for the monitoring of populations, ecological processes and the habitats. Unfortunately, many of them lack sound knowledge and understanding of their biota, except some sort of listing of major or dominant flora and fauna. The present protected area network is roughly four per cent of the geographical area of the country, and needs to be expanded to include more areas of high conservation value. Highest priority for protection of biodiversity and basic inventory work should be given to parks and reserves. The importance of Dudhwa National Park is in large part due to its rich biological wealth. It is the main stronghold of the relict population of swamp deer or the barasingha. It is also one of the better known tiger reserves in the country, and is considered as most suitable area for reintroduction of Great Indian rhinoceros. The Park serves as an invaluable refuge for plants driven to the verge of extinction due to habitat loss. The publication of an exhaustive account of 821 species of angiosperms indicates the floristic richness of the Park.

The overriding need in coming decades is to ensure the long term survival of our fascinating plants and animals, and to evolve strategies which integrate conservation and development. Flora and fauna should be regarded as significant socio-economic and environmental resources, since they are of benefit to people and economy. Dr. Singh has made valuable contribution by writing up the flora of the Park. It will be of immense help to students and teachers of botany, foresters, environmentalists, conservationists, and persons concerned with utilization and management of plant resources.

Lucknow, U.P. — J.K. Maheshwari
July 1994 — Ph.D., F.N.E.

PREFACE

The Flora of Dudhwa National Park, Kheri District in Uttar Pradesh is of prime importance because of its interesting flora, fauna and ecological spreadouts. The park has a vital role to play in the conservation and sustainable use of biodiversity. The importance of the park is in large part due to its rich biological wealth. It is the main stronghold of the population of swamp deer. It is also one of the better known tiger reserves in the country and is a suitable additional habitat for Rhinoceros.

The present work which is the outcome of about seven years intensive botanical exploration of the area, provides details of floristic and phytogeographical accounts of 821 Angiospermic taxa. Besides it also provides plant wealth of the area, ethnobotanical uses of plants by the Tharu tribe of the park. It is hoped that the publication will be useful to the students and teachers of botany, foresters, environmentalists, conservationists and persons concerned with utilization and management of the biological resources.

I am thankful to Dr. J.K. Maheshwari, Retired Scientist-incharge, National Botanical Research Institute, Lucknow for his valuable guidance and advice during the course of this study and preparation of this flora. I am grateful to Dr. P.V. Sane, Director, N.B.R.I. for his encouragement and Dr. S.R. Paul, Retd. Scientist for his valuable advice. I am also thankful to Dr. M.M. Dhar, Retd. Director, CDRI for his valuable advice and encouragement. I also record my special thanks to my wife Mrs. Asha Singh, who have been a constant source of encouragement during the entire course of this study and work.

Place : Lucknow (K.K. Singh)
Dated : July, 1994

CONTENTS

Pages

Wadia, D.N. 1953 Geology of India (ed.3). London.

Webster, G.L. 1957 A monographic study of the West Indian species of *Phyllanthus. J. Arn. Arb.* 38:51-80; 170-98-295-73.

Wulff, E.V. 1950 An Introduction to Historical Plant Geography. Waltham.

1. INTRODUCTION

Knowledge of vegetation and flora of any region is essential for the study of biodiversity and environment. This necessitates researchers in the taxonomic field to assess and evaluate the flora and vegetation of an area from time to time. The publication of "The Flora of British India" by Sir J.D. Hooker (1872-1897) gave a stimulus to taxonomic studies in our country and several regional and provincial floras pertaining to different parts of the country were published within a short span of time. With the revival of Botanical Survey of India much emphasis has been laid on an intensive study of local floras with a view to gather precise information on the identity,floristic diversity, distribution, ecological assoication, phenology, medicinal and economic uses, and local names of Indian plants. In the Silver Jubilee Volume of "Botanica", Maheshwari (1975) emphasised the need for trained floristicians in the execution of a number of research projects dealing with the survey and availability of raw materials of vegetable origin, collection of Ayurvedic and Unani drugs, search of new sources of plants of potential economic and medicinal value, biological screening of indigenous flora, establishment of germplasm banks for crop improvement, botanical conservation etc., It is, therefore, of utmost importance that in a vast country like India, local and regional floras must be prepared in the light of modern taxonomic researches. Generally speaking, local floras are much more valuable than those of bigger areas because explorations can be carried out intensively for the former minimising the possibilities of plants being left out, and for studying biodiversity utilization and conservation strategies.

The need of working out the local floras has become all the more essential because a large number of additional and exotic species have been described from different parts of the country since the publication of earlier floras. Many of the plants in Indian floras do not possess valid scientific names in the light of the International Code of Botanical Nomenclature. Several genera and families have been revised by taxonomists in recent monographic treatments. Santapau (1958a) suggested that what we want now is not mere more collections but critical observations on the spot with illustrative specimens. This is possible only when limited areas are explored intensively. He has placed great emphasis on preparing a repository of local floras by various universities in India. Similar suggestions and recommendations were also made at the Summer School in Taxonomy held at Kodaikanal (1962) and by Rao (1970) and Sreemadhavan and Subramanyam (1970).

The local herbaria serve as centres of special information to medical, chemical, pharmaceutical, pharmacological, agricultural, educational and other allied research institutes and industrial organizations. Above all, the herbaria will be helpful in the preparation of a vegetation map of India, in the preparation and revision of local and state floras and ultimately leading to the revision of the flora of India.

In light of these suggestions, a detailed floristic and phytogeographical accounts of the forests of Dudhwa National Park and adjacent areas of Kheri district, Uttar Pradesh has been prepared. It is hoped that this work will prove useful to the students, teachers, naturalists, wildlife experts, phytochemists, pharmacologists, medical and ayurvedic practitioners of the region. These studies would also be of great importance in enhancing our knowledge of the indigenous plant resources. Further, such studies would be useful for the economic development of the region and for implementing different schemes on small scale industries, cottage industries etc., for the welfare of the local tribe Tharu and would advance our knowledge of the flora of India.

2. GEOGRAPHICAL FEATURES

Kheri district forms a part of Lucknow Division and is the largest one in Oudh. It occupies north-west position in the State of Uttar Pradesh between 27°41' and 28°42' N. latitude and 80°2' and 81°19' E. of longitude. The district is bounded on the east by the district of Bahraich, from which it is separated by river Kauriala, on the south by Sitapur and Hardoi, on the west by the Shahjahanpur and Pilibhit districts, and on the north by the territory of Nepal, separated

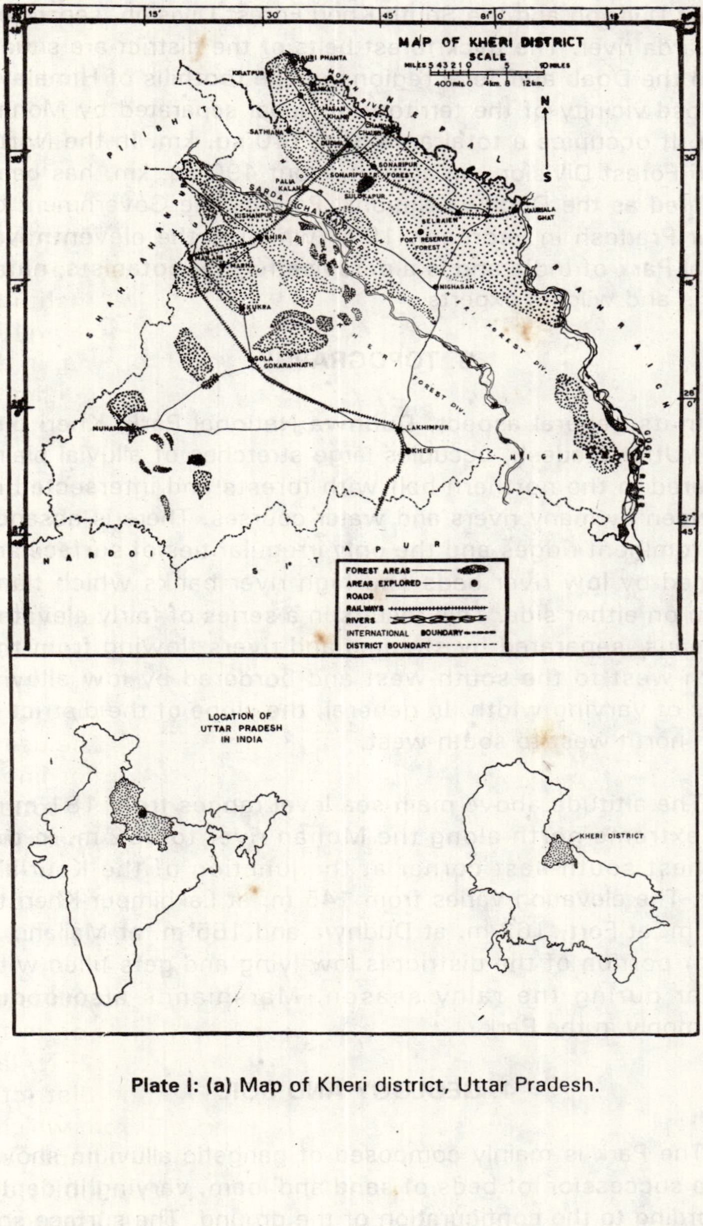

Plate I: (a) Map of Kheri district, Uttar Pradesh.

by Mohan river. **(Plate Map I)** The forests of the district have broadly been divided in two zones, i.e., the North Kheri Forest Division and the South Kheri Forest Division, separated by Sarda river. The thick forest belts of the district are situated in the Doab and Terai regions of the foothills of Himalaya in close vicinity of the territory of Nepal separated by Mohan river. It occupies a total area of 4,740 sq. km. In the North Kheri Forest Division, an area of about 490 sq. km. has been declared as the Dudhwa National Park by the Government of Uttar Pradesh in February, 1977 which is the eleventh National Park of India and is a paradise for the botanists, naturalists and wildlife experts.

3. TOPOGRAPHY

In its general aspect, Dudhwa National Park, Kheri district, Uttar Pradesh, occupies large stretches of alluvial plain, covered in the northern half with forests and intersected in between by many rivers and water courses. There is absence of prominent ridges and the only irregularities of surface are formed by low river beds and high river banks which flank them on either side. This results in a series of fairly elevated plateaus, separated by streams and rivers flowing from the north-west to the south-west and bordered by low alluvial belts of varying width. In general, the slope of the district is from north-west to south-west.

The altitude above main sea level ranges from 181 m in the extreme north along the Mohan river to 152 m. in the furthest south-east corner at the junction of the Kauriala river. The elevation varies from 145 m. at Lakhimpur-Kheri to 155 m. at Fort, 162 m. at Dudhwa and 165 m. at Mailani. A major portion of the district is low-lying and gets filled with water during the rainy season. Marshlands also occur commonly in the Park.

4. GEOLOGY AND SOIL

The Park is mainly composed of gangetic alluvium showing a succession of beds of sand and loam, varying in depth according to the configuration of the ground. The surface soil in the low alluvium is very recent, but that in the high alluvium is mostly a loamy sand varying in depth and composition. The underlying soil of the forest area consists of gangetic alluvium with a succession of beds of sand and loam of varying depth. In areas where clay has been deposited on the surface as in depressions, the ground becomes swampy during the rainy season. The following broad types of soil are met within the area.

1. **Low alluvium:** It consists of pure sand and is found in the beds of river. It is also met with in wastelands situated along the banks of rivers. The recent loamy sand is much coarser and shallower than that found in the high alluvium.

2. **High and middle alluvium:** A rich loamy sand with a variable proportion of clay is best found in the sal forests.

   In addition, the following types of soil are also met with :

(i) A moist sandy loam mixed with fair proportion of decayed vegetable matters. This is met within depression and water courses covered with grasses.

(ii) A stiff sandy loam with a fair proportion of clay and very slight admixture of humus.

(iii) A micaceous sand with little or no clay and marked by an almost complete absence of humus.

(iv) A stiff clay with a large amount of decayed and partly decayed vegetable matter. This soil is found in low-lying area.

Minerals are absent. Kankar is found in several parts of the district.

5. CLIMATE

The climate of the Park is of monsoon type. The year may be broadly divided into rainy, winter and summer seasons. The winter season extends from the last week of October to middle of March and is characterised by cold shining days and cold foggy nights. There is heavy deposition of dew during the winter season when the vegetation remains damp until about the middle of the day. The coldest months of this season are December and January. The mean daily maximum and minimum temperatures are 22.13°C and 8.32°C respectively. Mean relative humidity is 74.2%.

The summer season begins from the middle of March and extends up to the break of monsoon. During this period, the days are hot although the nights are moderately cool. Mean relative humidity is 63.52% and mean daily maximum and minimum temperatures are 35.93°C and 23.52°C respectively. From April onwards, the maximum temperature is seldom below 38°C and during May it rises to as much as 46°C. The prevailing westerly winds begin to increase in force during the month of March and usually blow strongly during the months of April and May.

The rainy season starts from the middle of June or the first week of July and extends up to the first week of October. The average total annual rainfall is about 1159.76 mm. The months of July and August are the wettest. Mean relative humidity is about 78.76% during this period and mean daily maximum and minimum temperatures are 30.43°C and 24.35°C respectively.

6. DUDHWA NATIONAL PARK

The Dudhwa National Park lies approximately between 28° 18'-28 ± 42' N and 80 ± 28' - 80-57' E. It is bounded in the north by the territory of Nepal and remaining sides are contiguous with the Kheri district of Uttar Pradesh. The Park area is a vast alluvial plain with a number of small rivers, ponds, tals, the important among these being Mohan river which more or less forms an international boundary between Nepal and Suheli river which forms southern boundary of the Park and Jauraha, Neora, Nagrola nalas and Kakarha, Nagra, Bhadi, Churela tals. The mean elevation above sea level ranges from 182 m. in the extreme north to 150 m. in the

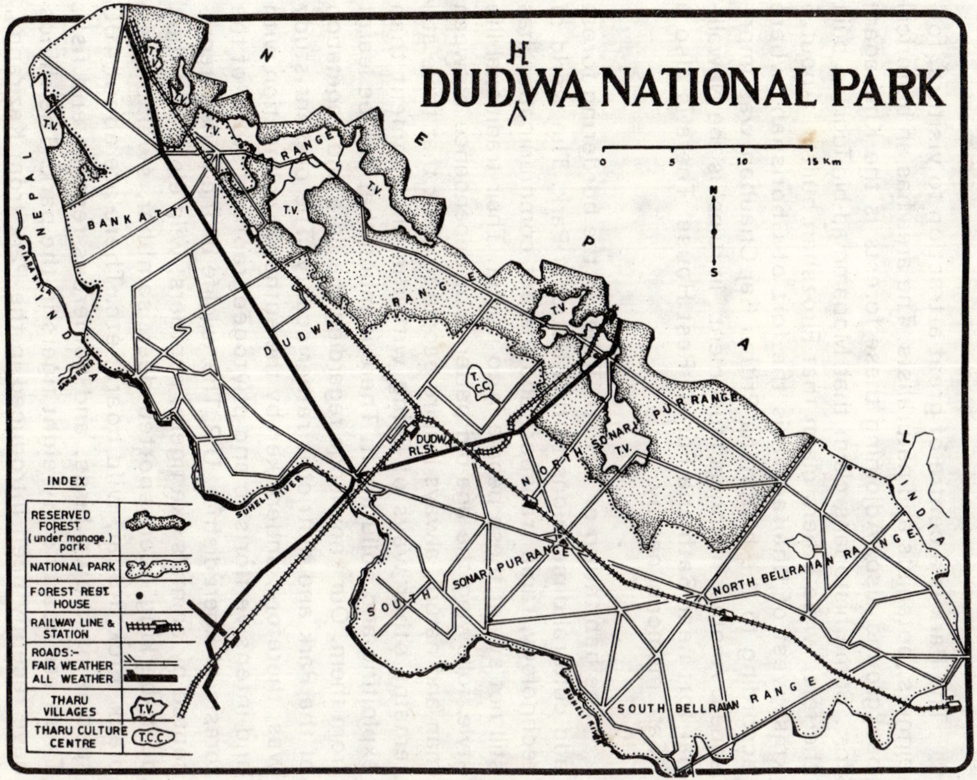

Plate I: (b) Map of Dudhwa National Park, Kheri district

south-east. The total area of the Park is about 490 sq. km. and adjoining area of 123 sq. km. is also under the administrative control of the Park, thus making it a complete block of 613 sq. km. (Map of the Park).

The Park is a centre of great attraction to visitors, foreigners, botanists and naturalists. The area has an historical background also. Adjoining these forests is the Khairagarh fort. Some historians claim that Mohammad-bin-Tughlaq built the fort, while others claim that Firozshah built it. About 5 kms. west of Khairagarh is the fort of Ghorishah, where according to the historians Prithvi Raj Chauhan was imprisoned. Another historical monument `in ruins' is the Nawalkhar fort near Sathiana Forest Rest House. There is also a `Sati' memorial nearby.

The habitation of `Tharu Tribe' in the bordering forest add cultural dimensions to this National Park. The world of technology, rapid transport and instant communication has still not swept these people in to its net. Their traditional life style, following the age-old patterns of dependence between man and nature, always inspire people to think that there still remains other ways of living with the environment than exploiting and polluting it. There is much that can be learnt from them. Our knowledge regarding the flora and vegetation of the Park and district is rather scanty. The present study was, therefore, undertaken by the author in connection with and intensive floristic and phytogeographical study of the forests of Kheri district, U.P. The wildlife of the Park is represented by animals like tigers, panthers, wild elephants, hog deer, barking deer, spotted deer, sambhar, swamp deer, black buck, nilgai, wild boars, etc. There are about 400 species of known birds, and numerous reptiles and fish. There are nearly sixty-eight tigers in the park. The rhinos have recently been introduced in the Park from Kaziranga, Assam.

The Flora of Dudhwa National Park is of prime importance because of its interesting flora, fauna and ecological spreadouts. The flora deals with the floristic, and phytogeographical studies of 821 Angiospermic taxa of the park and adjoining areas of Kheri district U.P. The forest area of the park are of tropical moist deciduous type and dominated by Sal (*Shorea robusta* Gaertn.) trees. The general part deals with the topography, geology, soil, climate, vegetation types, phytogeography and ethnobotanical studies. The descriptive parts deals with the systematic treatment of 821 species of Angiosperms representing 490 genera and 120 families distributed in the park and adjacent areas of the district. Macroscopical dichotomous keys for quick identification of the families, genera and species are provided in the flora. Accurate nomenclature reference for each species and recent monograph, revisions if any have been added. This is followed by basionym and important synonyms relevant to the nomenclature or to the regions. Diagnostic description, habitat range, local name, locality, flowering and fruiting period, frequency, collector's name and field numbers etc., for each species is provided in the book. The flora contains nearly 40 photographs of plants, forests, illustrations and maps. It is useful to the botanists foresters, students, wildlife experts and amateurs interested in the plant and animal life of the park. The park is a natural biosphere reserve and is useful as a data source of genetic resources and conservation of the Flora and Fauna.

7. PLAN FOLLOWED IN THE FLORA

The following plan has been adopted in the presentation of floristic data of Dudhwa National Park and adjacent areas of Kheri district, Uttar Pradesh.

Classification: The families of flowering plants are arranged according to Bentham and Hooker's Genera Plantarum (1862-83). However, in splitting of certain families, Hutchinson's (ed. 1959, 1973) classification has been adopted. A comprehensive dichotomous key to the families of dicotyledons and monocotyledons plants has been provided. Dichotomous keys for quick identification of taxa have been presented, chiefly on the basis of macroscopical characters.

Nomenclature and Literature Citation: The nomenclature of plants has been modernised in the light of the recent *Inter-*

national Code of Botanical Nomenclature. (1983). Every effort has been made to bring the nomenclature up-to-date in light of recent monographic and revisionary works on various taxa. The valid scientific name is followed by the basionym and important synonyms that are commonly cited in important regional floras, especially of adjoining areas. Among the various works that have been cited throughout, mention is made of *The Flora of British India* (Hooker et al., 1872-1897), *Flora of Upper Gangetic Plain and of the Adjacent Siwalik and Sub-Himalayan Tracts"* Vol. I & II (Duthie, rep. ed. 1960), *Forest Flora for the Plains of Uttar Pradesh* (P.C. Kanjilal, 1966) and *Flora Gorakhpurensis* (T.N. Srivastava, 1976). In addition, publications like: *The Flora of Khandala on the Western Ghats of India* (Santapau, 1953), *The Flora of Delhi* (Maheshwari, 1963), *Aquatic Angiosperms* (Subramanyam, 1962), *Supplement to Duthie's Flora of Upper Gangetic Plain* (Raizada, 1976). and *Flora of Hassan District* (Saldanha and Nicolson, 1976), Name changes in Flowering Plants of India and Adjacent regions. (Bennet 1987) have also been cited wherever necessary.

Systematic description: The diagnostic characters of all the species have been provided. The measurements of plant parts are given in the metric system. The description of individual species provides observations on the ecology, habitat, frequency, dominance and features of special botanical interest. The medicinal and economic uses, local names, flowering and fruiting periods, and field numbers of the species studied are provided in detail in this flora.

8. VEGETATION TYPES

The forests of the Park belong to the tropical moist deciduous type. These are mainly found in Dudhwa, Sonaripur, Sathiana, Bellrain, Bankati and Fort Ranges of the North Kheri and Golagokarannath, Mailani, Mohamadi, Kishenpur, Nighasan, Bhira, and Powayan Ranges of South Kheri Forest Division. The permanent vegetation of the forests is represented by trees, shrubs and perennial herbs. The herbaceous flora is mainly seasonal and appears during the rainy season. Based on the floristic composition, the vegetation of the district may be divided into following four forest types:

1. **Sal forest,**
2. **Mixed forest,**
3. **Swamp forest,**
4. **Grasslands.** (Savannah)

1. **Sal forest:** This forest occupies major part of the area. The dominant tree species, *Shorea robusta* Gaertn. f., forms pure belt and can be easily distinguished in various storeys of the forest.

The common associates of the Sal forest are: *Terminalia alata* Heyne ex Roth, *Adina cordifolia* (Roxb.) Hook.f. ex Brandis, *Bridelia squamosa* Gehrm., *Dillenia pentagyna* Roxb., *Lagerstroemia parviflora* Roxb., *Schleichera oleosa* (Lour.) Oken., *Terminalia bellirica* (Gaertn.) Roxb., *Stereospermum suaveolens* DC., *Madhuca longifolia* (Koeng.) Macbr. var., *latifolia* Chev., *Lannea coromandelica* (Houtt.) Merr., *Ougeinia oojeinensis* (Roxb.) Hochr. etc. The middle storey is often dense and consists chiefly of *Aegle marmelos* (Linn.) Corr., *Careya arborea* Roxb., *Cassia fistula* Linn., *Mallotus philippensis* (Lamk.) Muell.-Arg., *Ehretia leavis* Roxb., *Syzygium cumini* (Linn.) Skeels, *Miliusa tomentosa* (Roxb.) Fenet & Gagnep., *Holarrhena pubescens* Wall ex G. Don., *Diospyros exsculpta* Buch.- Ham., *Semecarpus anacardium* Linn.f. etc. The forest undergrowth is composed of *Ardisia solanacea* Roxb., *Azanza lampas* (Cav.) Alef., *Baliospermum montanum* (Willd.) Muell.-Arg., *Flemingia strobilifera* (Linn.) R.Br. ex Aiton, F. chapper Buch.-Ham. ex Benth. *Carissa opaca* Stapf ex Haines, *Callicarpa macrophylla* Vahl, *Clausena pentaphylla* DC., *Clerodendrum indicum* (Linn.) O. Ktze., *C. viscosum* Vent., *Colebrookia oppositifolia* Sm., *Desmodium pulchellum* (Linn.) Benth., *Murraya koenigii* (Linn.) Spreng., *Pogostemon benghalensis* (Burm.f.) O. Ktze., *Leea crispa* Linn., *Desmodium heterocarpon* (Linn.) DC., *Glycosmis mauritiana* (Lamk.) Tanaka etc. The ground flora consists of *Adenostemma lavenia* (Linn.) O. Ktze., *Curculigo orchioides* Gaertn., *Elephantopus scaber* Linn., *Uraria neglecta* Prain, *Uraria alopecuroides* Wight, *Cassia pumila* Lamk., *Hackelochloa granularis* (Linn.) O. Ktze., *Oplismenus compositus* (Linn.) Beauv., *O. burmanii* (Retz.) Beauv., *Cyrtococcum patens* (Linn.) *A. Camus, Acrocephalus indicus* (Burm.f.) O. Ktze. etc.

The common climbers are: *Tiliacora acuminata* (Lamk.) Miers, *Butea parviflora* Roxb., *Dalbergia volubilis* Roxb., *Millettia extensa* (Benth.) Baker, *Abrus precatorius* Linn., *Bauhinia vahlii* W. & A., *Ichnocarpus frutescens* (Linn.) Ait. & Ait.f., *Ziziphus rugosa* Lamk., *Z. oenoplia* (Linn.) Mill., *Scindapsus officinalis* (Roxb.) Schott, *Ventilago denticulata* Willd., *Capparis zeylanica* Linn. *Celastrus paniculatus* Willd. and *Smilax ocreata* A. DC. etc.

The pteridophytic flora of the Sal forest is represented by species like *Lygodium flexuosum* (Linn.) Sw., *Ophioglossum reticulatum* Linn. *Adiantum philippense* Linn., *Christella parasitica* (Linn.) Lev., *Aleuritopteris farinosa* (Forsk.) Fee, *Ceratopteris thalictroides* (Linn.) Brongn., *Helminthostachys zeylanica* (Linn.) Hook., *Equisetum debile* Roxb., *Diplazium esculentum* (Retz.) Sw., and *Menicium proliferum* (Retz.) Sw. etc.

The Sal forest is interspersed here and there with grasslands locally called as `Phantas'. These occur throughout the sal forest and are represented mainly by grasses like *Themeda arundinacea* (Roxb.) Ridley, *Bothriochloa intermedia* (R.Br.) A. Camus, *Saccharum spontaneum* Linn. etc. Some of the important trees and shrubs occurring sparsely here and there are: *Acacia catechu* Willd., *Dalbergia sissoo* Roxb., *Alangium salvifolium* (Linn.f.) Wang., *Bombax ceiba* Linn., *Adina cordifolia* (Roxb.) Hook.f. ex Brandis, *Woodfordia fruticosa* (Linn.) Kurz, *Syzygium cerasoides* (Roxb.) Chatt. & Kanjilal, *Xeromphis spinosa* (Thunb.) Keay, At places, *Tectona grandis* Linn.f. has been planted, especially along roadsides.

2. **Mixed forest:** This type occupies large stretches of both the forest divisions and occurs on higher and more stable alluvial terraces. The main species met with in the upper storey are: *Adina cordifolia* (Roxb.) Hook.f. ex Brandis, *Albizia procera* (Roxb.) Benth., *Ficus semicordata* Buch.-Ham., *F. racemosa* Linn., *Tectona grandis* Linn.f., *Dalbergia sissoo* Roxb., *Wendlandia heynei* (R. & S.) Sant. & Merch., *Bombax ceiba* Linn., *Lannea coromandelica* (Houtt.) Merr. etc. The middle storey consists of trees like *Acacia catechu* Willd., *Hymenodictyon excelsum*

Wall., *Mitragyna prvifolia* Korth, *Drypetes roxburghii* (Wall.) Hurus., *Garuga pinnata* Roxb., *Flacourtia indica* (Burm.f.) Merr., *Ailanthus excelsa* Roxb., *Emblica officinalis* Gaertn., *Ficus hispida* Linn.f., *Bridelia squamosa* Gehrm., *Gmelina arborea* Roxb., *Kydia calycina* Roxb., *Litsea glutinosa* (Lour.) C.B. Robins., *Litsea monopetala* (Roxb.) Pers., *Trewia nudiflora* Linn., *Bauhinia racemosa* Lamk., *Sterculia villosa* Roxb., *Ziziphus mauritiana* Lamk., *Z. xylopyrus* Willd., *Xeromphis uliginosa* (Retz.) Mahesh., *Butea monosperma* (Lamk.) Taub. etc. The common shrubs are: *Carissa opaca* Stapf ex Haines, *Leea crispa* Linn. *Clerodendrum viscosum* Vent., *Calotropis gigantea* (Linn.) R.Br., *Calotropis procera* (Ait.) R.Br. etc. Climbers like *Ampelocissus latifolia* (Roxb.) Planch., *Tinospora cordifolia* (Willd.) Miers ex Hook.f. & Thoms., *Ventilago denticulata* Willd., *Tiliacora acuminata* (Lamk.) Miers etc., are also present.

The common grasses are *Saccharum spontaneum* Linn., *Apluda mutica* Linn., *Heteropogon contortus* (Linn.) Beauv., *Setaria glauca* (Linn.) Beauv. etc. There are also some interesting dwarf species like *Careya herbacea* Roxb., *Pygmaeopremna herbacea* (Roxb.) Moldenke, *Grewia sapida* Roxb., *Combretum nanum* Buch.-Ham. ex D.Don etc., found in areas subject to periodic forest fires.

3. **Swamp forest:** The vegetation occurring in low-lying, water-logged areas, along nalas, river beds and stream beds which remain under water during rains and sometimes many months thereafter, develops into a swamp forest. The aeration of soil is usually poor. The soil is rich in humus and consists of fine clay. This type of forest is chiefly located along Suheli, and Kathana riversides. It belongs to either *Barringtonia* Swamp Forest or *Syzygium* Swamp Forest. The former type is seen along the more swampy areas while the latter on comparatively drier areas. At certain places, these forests get mixed together.

The most common species of the swamp forest are *Syzygium cumini* (Linn.) Skeels, *Barringtonia acutangula*

Gaertn., *Ardisia solanacea* Roxb., *Cephalanthus occidentalis* Linn., *Bischofia javanica* Blume, *Salix tetrasperma* Roxb., *Ficus heterophylla* Linn.f., *F. racemosa* Linn., *F. semicordata* Buch.-Ham., *Glochidion multiloculare* Voigt, *Grewia disperma* Rottl. ex Spreng., *Hyptianthera stricta* W. & A., *Syzygium heyneanum* Wall. ex Duthie, *Kirganelia reticulata* (Poir.) Baill., *Tamarix dioica* Roxb., *Trewia nudiflora* Linn., *Mitragyna parvifolia* (Roxb.) Korth., *Glycosmis mauritiana* (Lamk.) Tanaka, *Phragmites* karka. Trin, *Vetiveria zizanioides* (Linn.) Nash etc. Some of the common herbs are *Bacopa monnieri* (Linn.) Wettest., *Centella asiatica* (Linn.) Urban, *Eclipta prostrata* (Linn.) Linn., *Cnidium monnieri* (Linn.) Cusson. and several species of *Polygonum.* The conspicuous climbers are *Tinospora cordifolia* (Willd.) Miers, *Bridelia stipularis* (Linn.) Bl., *Ipomoea muricata* (Linn.) Jacq., *Ichnocarpus frutescens* (Linn.) R. Br. & Ait. f., *Calamus tenuis* Roxb., *Oxystelma secamone* (Linn.) K. Schum. etc.

Grasslands (Savannah) The grasslands occur in moist Sal savannah i.e. `Phantas' and in riverine flats.

The `Phantas' occupy large stretches and lie scattered in between the Sal forests. The common grasses are *Themeda arundinacea* (Roxb.) Ridley, *Apluda mutica* Linn., *Vetiveria zizanioides* (Linn.) Nash, *Saccharum benghalense* Retz., *S. spontaneum* Linn., *Bothriochloa intermedia* (R.Br.) A. Camus, *Imperata cylindrica* (Linn.) Beauv., *Eulaliopsis binata* (Retz.) Hubb., *Phragmites* karka. Trin., *Heteropogon contortus* (Linn.) Beauv., *Chrysopogon aciculatus* (Retz.) Trin., *Capillipedium assimile* (Steud.) A. Camus, *Dichanthium annulatum* (Forsk.) Stapf, *Desmostachya bipinnata* (Linn.) Stapf, *Eragrostis tenella* (Linn.) Beauv. etc.

The species, *Cymbopogon osmastonii* Parker is endemic to the Park. The riverine flats are liable to become waterlogged during the rains but dry out quickly and support grassland vegetation. The soil is mainly alluvial with scattered clayey patches in the depressions. There is usually a thin surface layer of loamy soil with some amount of humus. The grasslands are interspersed here and there with common trees and shrubs like *Butea monosperma* (Lamk.) Taub.,

Bombax ceiba Linn., *Syzygium cerasioides* (Roxb.) Chatt. & Kanjilal, *Dalbergia sissoo* Roxb., *Acacia catechu* Willd., *Trewia nudiflora* Linn., *Mitragyna parvifolia* Korth., *Gmelina arborea* Roxb., *Calotropis gigantea* (Linn.) R.Br., *Carissa opaca* Stapf ex Haines, *Helicteres isora* Linn., *Woodfordia fruticosa* (Linn.) Kurz, *Xeromphis uliginosa* (Thunb.) Mahesh., *Ziziphus xylopyrus* (Retz.) Willd., *Zizyphus nummularia* (Burm.f.) W. & A., *Grewia sapida* Roxb., *G. sclerophylla* Roxb. ex G.Don, *Leea macrophylla* Roxb., etc. The common herbs are *Trichodesma sedgwickianum* Banerjee, *Heliotropium strigosum* Willd., *Indigofera linifolia* Retz., *Leea alata* Edgew., *Zornia gibbosa* Span., *Tribulus terrestris* Linn., *Convolvulus microphyllus* Sieb. ex Spreng. etc.

A dwarf handsome species, *Erythrina resupinata* Roxb. is also seen in some grasslands of the Park. Some of the common sedges associated with grasses are *Carex fedia* Nees, *Cyperus exaltatus* Retz., *C. compactus* Retz., *Fimbristylis dichotoma* (Linn.) Vahl, *F. ovata* (Burm.f.) Kern, *F. podocarpa* Nees etc.

The general surveys of these forests revealed four dominant types of plant communities. They are *Shorea - Mallotus-Adina, Bridelia* community, *Shorea - Terminalia* community, *Dalbergia - Acacia* community, *Syzygium - Barringtonia - Kirganelia - Trewia* community.

SEASONAL VEGETATION

The seasonal vegetation comprises herbs and undershrubs growing on a variety of habitats during the rainy, winter and summer seasons. The common species found during different seasons are mentioned below:

The rainy season plants are represented by species like *Cassia occidentalis* Linn., *C. tora* Linn., *C. absus* Linn., *Cleome viscosa* Linn., *C. gynandra* Linn., *Trianthema portulacastrum* Linn., *Hybanthus enneaspermus* (Linn.) F. Muell., *Amiscophacelus axillaris* (Linn.) Rolla Rao & Kam., *Commelina benghalensis* Linn., *Cyanotis cristata* (Linn.) D.Don, *Caesulia axillaris* Roxb., *Corchorus aestuans* Linn., *Martynia annua*

Linn., *Bacopa monnieri* (Linn.) Wettest. *Evolvulus alsinoides* Linn., *Leucas cephalotes* Spreng., *Leucas mollissima* Wall., *Gloriosa superba* Linn., *Phyllanthus fraternus* Webster, *P. urinaria* Linn., *P. virgatus* J.G. Forst., *Digera muricata* (Linn.) Mart., *Bulbostylis barbata* Clarke, *Dactyloctenium aegypticum* Beauv., *Celosia argentea* Linn., *Gomphrena celosioides* Mart., *Lobelia alsinoides* Lamk., *Triumfetta pentandra* A. Rich., *T. rhomboidea* Jacq., *Zornia gibbosa* Span., *Indigofera hirsuta* Linn., *I. trifoliata* Linn., *Crotalaria medicaginea* Lamk., *Alysicarpus vaginalis* DC., *Desmodium gangeticum* DC., *Cyperus iria* Linn., *C. kyllingia* Endl., *Imperata cylindrica* Beauv., *Lindernia ciliata* (Colsm.) Pennell, *L. crustacea* (Linn.) F. Muell., *Chrysopogon aciculatus* (Retz.) Trin. etc. The twiners and climbers are represented by *Cayratia trifolia* (Linn.) Domin, *Trichosanthes cucumerina* Linn., *Coccinia grandis* Voigt, *Gymnopetalum cochinchinensis* Kurz etc.

The winter season plants are represented by species like *Anagallis arvensis* Linn., *Primula umbellata* (Lour.) Bentv., *Fumaria indica* Pugsley, *Asphodelus tenuifolius* Cav., *Coronopus didymus* (Linn.) Sm., *Salvia plebeia* R.Br., *Stellaria media* Vill., *Veronica anagallis-aquatica* Linn., *Medicago polymorpha* Linn., *Melilotus alba* Desr., *Melilotus indica* All., *Oxalis corniculata* Linn., *Biophytum sensitivum* (Linn.) DC., *Knoxia sumatrensis* (Retz.) DC., *Mazus pumilus* (Burm.f.) van Steenis, *M. delavayi* Bonati, *Peristrophe bicalyculata* (Retz.) Nees, *Rungia pectinata* (Linn.) Nees, *Potentilla supina* Linn., *Ranunculus sceleratus* Linn., *R. cantoniensis* DC., *Vicoa indica* (Linn.) DC., *V. vestita* (Wall. ex DC.) Benth. & Hook.f., *Ageratum conyzoides* Linn., *Sigesbeckia orientalis* Linn., *Vernonia cinerea* Less., *Cirsium arvensis* Scop., *Phalaris minor* Retz., *Polypogon monspeliensis* Desf., *Rumex dentatus* Linn., *Hygrophila auriculata* (Sch.) Heine, *Heliotropium strigosum* Willd., *Hemiadelphis polyspermus* Nees, *Sonchus oleraceus* Linn., *Sphaeranthus indicus* Linn. etc. Some of the climbers are *Cocculus hirsutus* (Linn.) Diels, *Cuscuta reflexa* Roxb., *Porana paniculata* Roxb. etc.

During the summer season, there are species like *Chenopodium ambrosioides* Linn., *Euphorbia dracunculoides* Lamk.,

E. acaulis Roxb., *Bothriospermum tenellum* Fisch. & Mey., *Glinus lotoides* Linn., *Cnidium monnieri* (Linn.) Cusson., *Emilia sonchifolia* (Linn.) DC., *Youngia japonica* (Linn.) DC., *Gnaphalium luteo-album* Linn. subsp. *affine* (D.Don) Koster, *Grangea maderaspatana* (Linn.) Poir., *Amberboa ramosa* (Roxb.) Jafri, *Argemone mexicana* Linn., *A. ochroleuca* Sweet, *Centaurium centaurioides* (Roxb.) Rolla Rao & Hemadri, *Polygonum plebeium* R.Br., *Hemigraphis hirta* T. Anders., *Phyla nodiflora* (Linn.) Greene, *Alternanthera pungens* H.B. & K., *Dichanthium annulatum* Stapf etc.

The common species in moist shady places are *Piper longum* Linn., *Oldenlandia paniculata* Linn., *O. corymbosa* Linn., *Pouzolzia pentandra* Benn., *P. zeylanica* (Linn.) Benn., *Centella asiatica* (Linn.) Urban, *Canscora decussata* Roem. & Schult., *Ardisia solanecea* Roxb., *Adenostemma lavenia* (Linn.) O. Ktze., *Cnidium monnieri* (Linn.) Cusson., *Eranthemum purpurescens* Wt. ex Nees, *Hybanthus enneaspermus* (Linn.) F. Muell., *Acrocephalus indicus* (Burm.f.) O. Ktze., etc. In drier areas, *Abutilon indicum* (Linn.) Sweet, *Colebrookia oppositifolia* Sm., *Scoparia dulcis* Linn., *Rauvolfia serpentina* Benth. ex Kurz, *Desmodium pulchellum* (Linn.) Benth., *Aerva sanguinolenta* (Linn.) Blume etc., are met with. In dried up beds of low-lying areas, species like *Bergia ammannioides* Roxb. ex Roth, *Sphaeranthus indicus* Linn., *Glinus lotoides* Linn., *Salvia plebeia* R.Br., *Ipomoea aquatica* Forsk., *Rumex dentatus* Linn., *Polygonum plebeium* R.Br., *Polycarpon prostratum* (Forsk.) Asch. & Sch., *Grangea maderaspatana* (Linn.) Poir., *Eleocharis atropurpurea* Kunth, *Scirpus mucronatus* Linn., *Alternanthera paronychioides* St. Hil. etc., are commonly met with. The common species along the banks of rivers, water streams and in marshes are: *Barringtonia acutangula* (Linn.) Gaertn., *Calamus tenuis* Roxb., *Carex fedia* Nees, *Kirganelia reticulata* (Poir.) Baill., *Phyla nodiflora* (Linn.) Greene, *Hydrolea zeylanica* (Linn.) Vahl, *Syzygium heyneanum* Wall. ex Duthie, *Tamarix dioica* Roxb., *Trewia nudiflora* Linn., *Typha angustata* Borry & Chaub., *Vetiveria zizanioides* (Linn.) Nash, *Phragmites karka* Trin., *Salix tetrasperma* Roxb., *oxystelma secamone* (Linn.) K. Shum., *Bischofia javanica* Blume, etc. are seen.

WEEDS AND ALIENS

The weed flora is found along roadsides, railway lines, in cultivated fields, gardens, orchards etc. The common weeds of wastelands and railway tracts are *Xanthium strumarium* Linn., *Cassia tora* Linn., *C. occidentalis* Linn., *Croton bonplandianum* Baill., *Cleome viscosa* Linn., *C. gynandra* Linn., *Solanum surattense* Burm.f., *Cannabis sativa* Linn., *Amaranthus spinosus* Linn., *A. virides* Linn., *Adhatoda zeylanica* Medic., *Scoparia dulcis* Linn., *Argemone mexicana* Linn., *Tephrosia purpurea* (Linn.) Pers., *Tribulus terrestris* Linn., *Urena lobata* Linn., etc. The weeds associated with winter seasons crops (Rabi) are as follows: *Nicotiana plumbaginifolia* Viv., *Fumaria indica* Pugsley, *Coronopus didymus* Sm., *Alternanthera paronychioides* St. Hil., *A. sessilis* DC., *Stellaria media* Vill., *Anagallis arvensis* Linn., *Asphodelus tenuifolius* Cav., *Solanum nigrum* Linn., *Phalaris minor* Retz., *Polypogon monspeliensis* Desf., *Euphorbia hirta* Linn., *E. dracunculoides* Lamk., *Medicago lupulina* Linn., *Melilotus alba* Desv., *M. indica* All. etc.

The common weeds associated with rainy season crops (Kharif) are as follows: *Ammannia baccifera* Linn., *A. multiflora* Roxb., *Bacopa monnieri* (Linn.) Wettst., *Caesulia axillaris* Roxb., *Cyperus iria* Linn., *C. alulatus* Kern, *Echinochloa colonum* Link., *E. crusgalli* Beauv., *Euphorbia hypericifolia* Linn., *Melochia corchorifolia* Linn., *Veronica anagallis-aquatica* Linn., *Hygroryza aristata* (Retz.) Nees, *Rumex dentatus* Linn., *Sphenoclea zeylanica* Gaertn., *Monochoria vaginalis* (Burm.f.) Presl, *Hydrolea zeylanica* Vahl etc.

There are a good number of exotic weeds that have been introduced in the district. The study reveals that the exotic weeds were introduced in the district, both adventitiously and intentionally. The most important amongst these are *Ageratum conyzoides* Linn., *Alternanthera paronychioides* St. Hil., *A. pungens* H.B. & K., *Anagallis arvensis* Linn., *Argemone mexicana* Linn., *A. ochroleuca* Sweet, *Cassia tora* Linn., *Chenopodium ambrosioides* Linn., *Coronopus didymus* (Linn.) Sm., *Croton bonplandianum* Baill., *Datura metel* Linn.,

Deeringia amaranthoides (Lamk.) Merr., *Eichhornia crassipes* (Mart.) Solms, *Elephantopus scaber* Linn., *Conyza bonariensis* (Linn.) *Cronquist, Conyza canadensis* (Linn.) Cronquist, *Gomphrena celosioides* Mart., *Hyptis suaveolens* Linn.) Poit., *Ipomoea carnea* Jacq. subsp. *fistulosa* (Mart. ex Choisy) Austin, *Jatropha curcas* Linn., *J. gossypifolia* Linn., *Lantana camara* Linn. var. *aculeata* (Linn.) Moldenke, *Malvastrum coromandelianum* (Linn.) Garcke, *Martynia annua* Linn., *Medicago lupulina* Linn., *M. polymorpha* Linn., *Melilotus alba* Desr., *M. indica* All., *Nicotiana plumbaginifolia* Viv., *Oxalis corniculata* Linn., *Polypogon monspeliensis* (Linn.) Desf., *Scoparia dulcis* Linn., *Volvulopsis nummularia* (Linn.) Roberty, etc.

9. FOREST PLANTATION AND INTRODUCTION

Among ornamental and economic trees and shrubs introduced by the Forest Department, mention is made of *Ailanthus excelsa* Roxb., *Morus alba* Linn., *Broussonetia papyrifera* Vent., *Populus* sp., *Eucalyptus citriodora* Hook.f., *Eucalyptus tereticornis* Smith, *Acacia auriculaeformis* A. Cunn., *Averrhoa carambola* Linn., *Cassia siamea* Lamk., *Delonix regia* (Boj. ex Hook.f.) Rafin, *Poinsettia pulcherrima* Willd., *Ficus pumila* Linn., *Grevillea robusta* A. Cunn., *Kigelia pinnata* DC., *Litchi chinensis* Sonner, *Pinus roxburghii* Sarg., *Prosopis juliflora* DC., *Prunus domestica* Linn. subsp. *institia* (Linn.) Schn., *Prunus persica* (Linn.) Stokes, *Spathodea campanulata* Beauv., *Terminalia arjuna* W.&A. etc.

In addition there are some plantation of *Tectona grandis* Linn.f. at Dudhwa, Mailani, Bankati etc. Experimental cultivation of Poplar (*Populus* sp.) has been taken up at Dudhwa for its utilization in match box industry.

10. FLORA OF SPECIAL HABITATS

The flora of old walls and ancient buildings is characterised by species like *Andrographis paniculata* (Burm.f.) Wall. ex Nees, *Boerhavia diffusa* Linn., *Cayratia trifolia* (Linn.) Domin., *Euphorbia hirta* Linn., *E. chamaesyce* Linn., *Phyllan-*

thus fraternus Webster, *Tridax procumbens* Linn., *Lindenbergia indica* (Linn.) Vatke etc. *Peperomia pellucida* H.B. & K. occurs in moist shady situation on the walls and trees. In addition, saplings of *Ficus benghalensis* Linn. and *F. religiosa* Linn. are also noticed on the walls.

Epiphytes and Parasites: The orchid species, *Vanda tessellata* (Roxb.) Hook. ex G.Don is usually found in the tree trunks of *Madhuca longifolia* var. *latifolia*. Among the parasites may be mentioned *Dendrophthoe falcata* (Linn.f.) Ettings., and *Cuscuta reflexa* Roxb.

Aquatic and Wet Vegetation: The common habitats of hydrophytes and marshland vegetation are the lakes, ponds and low-lying areas of the district, which remain submerged throughout the year. The following six categories are recognised, depending upon the nature of habitats and their contact with soil, water and air. **(Plate VI).**

(a) **Free-floating hydrophytes:** These are represented by *Pistia stratiotes* Linn., *Spirodela polyrrhiza* (Linn.) Schleid., *Hygroryza aristata* Nees, *Trapa bispinosa* Roxb., *Wolffia arrhiza* (Linn.) Wimmn., *Lemna perpusilla* Torr., *Eichhornia crassipes* (Mart.) Solms etc.

(b) **Suspended hydrophytes:** These are rootless submerged hydrophytes, e.g., *Ceratophyllum demersum* Linn., *Utricularia aurea* Lour.

(c) **Submerged attached hydrophytes:** These are represented by *Ottelia alismoides* (Linn.) Pers., *Potamogeton crispus* Linn., *P. nodosus* Poir., *P. pectinatus* Linn., *Vallisneria spiralis* Linn., *Hydrilla verticillata* (Linn. f.) Royle, etc.

(d) **Floating attached hydrophytes:** These include species like *Aponogeton crispus* Thunb., *Nelumbo nucifera* Gaertn., *Nymphaea nouchali* Burm.f., *N. stellata* Willd., *Nymphoides indicum* (Linn.) O. Ktze., *Ludwigia adscendens* (Linn.) Hara, *Ipomoea aquatica* Forsk. etc.

(e) **Emergent, amphibious hydrophytes:** The common species are *Amiscophacelus axillaris* (Linn.) Rolla Rao et Kam., *Ammannia baccifera* Linn., *A. multiflora* Roxb., *Aeschynomene aspera* Linn., *A. indica* Linn., *Hydrolea zeylanica* Vahl, *Sphenoclea zeylanica* Gaertn., *Monochoria vaginalis* (Burm.) Presl, *Oryza rufipogon* Griff., *Hemarthria compressa* R.Br., *Ischaemum rugosum* Salisb., *Polygonum limbatum* Meissn., *P. hydropiper* subsp. *microcarpum* Danser var. *lenticularis* Danser, *P. barbatum* Linn. subsp. *gracile* Danser, *P. lanigerum* R.Br., *Tenagocharis latifolia* (D.Don) Buchen., *Cyperus compactus* Retz., *C. exaltatus* Retz., *Eleocharis dulcis* (Burm.f.) Henschel, *Bergia ammannioides* Roxb. ex Roth, etc.

(f) **Wetland hydrophytes:** The common examples are *Mella hamiltoniana* (Benth.) Pennell, *Bacopa monnieri* (Linn.) Wettest., *Alternanthera paronychioides* St. Hil., *A. sessilis* (Linn.) DC., *Hygrophila auriculata* (Schum.) Heine, *Limnophila indica* (Linn.) Druce, *Ranunculus sceleratus* Linn., *Caesulia axillaris* Roxb., *Lindernia ciliata* (Colsm.) Pennell, *Justicia quinqueangularis* Koen. ex Roxb., *Typha angustata* Bory & Chaub., *Veronica anagal+lis-aquatica* Linn., *Phyla nodiflora* (Linn.) Greene, *Polygonum plebeium* R.Br., *Phragmites karka* Trin., *Fimbristylis bisumbellata* (Forsk.) Bub., *Rumex dentatus* Linn. etc.

11. CULTIVATED PLANTS AROUND THE PARK

The plant species grown by the Tharu tribe of the area for economic purpose around tribal villages of the Park are enumerated under following heads:

A. Cereals, pulses and pseudocereals

1. *Oryza sativa* Linn. (Paddy)
2. *Cajanus cajan* Millsp. (Pigeonpea)
3. *Pennisetum typhoides* (Burm.f.) Stapf. (Bulrush millet)
4. *Sorghum bicolor* (L.) Moench (Sorghum)
5. *Zea mays* Linn. (Maize),
6. *Triticum aestivum* Linn. (Wheat),

7. *Hordeum vulgare* Linn. (Barley),
8. *Cicer arietinum* Linn. (Gram)
9. *Pisum sativum* Linn. (Pea)
10. *P. arvense* Linn. (Pea)
11. *Lens esculenta* Moench (Lentil)
12. *Vigna radiata* (L.) Wilczck (Green gram)
13. *V. mungo* (L.) Hepper (Black gram)
14. *Eleusine corocana* (L.) Gaertn. (Finger millet)
15. *Saccharum officinarum* Linn. (Sugarcane)
16. *Amaranthus caudatus*. Linn (Amaranth)
17. *Lathyrus sativus* Linn.

B. **Vegetables**

1. *Solanum tuberosum* Linn. (Potato)
2. *S. melongena* Linn. (Egg plant or Brinjal)
3. *Trapa natans* subsp. *bispinosa* Makino (Water chestnut)
4. *Abelmoschus esculentus* (L.) Moench. (Lady's finger)
5. *Amorphophallus campanulatus* (Roxb.) Bl. ex Decne (Elephant yam)
6. *Artocarpus heterophyllus* Lamk. (Jack fruit)
7. *Benincasa hispida* (Thunb.) Cogn. (Wax gourd)
8. *Colocasia esculenta* (L.) Schott. (Arum)
9. *Cucurbita maxima* Duch. (Pumpkin)
10. *Hibiscus sabdariffa* Linn.
11. *H. cannabinus* Linn.
12. *Lablab purpureus* (L.) Sweet (Bean)
13. *Lagenaria siceraria* (Mol.) Standl. (Bottle gourd)
14. *Luffa acutangula* (L.) Roxb. (Angled loofaf)
15. *L. cylindrica* (L.) Roem. (Smooth loofaf)
16. *Lycopersicon lycopersicum* (L.) Kars. (Tomato)
17. *Momordica charantia* Linn. (Bitter gourd)
18. *M. dioca* Roxb. ex Willd.
19. *Spinacia oleracea* Linn (Spinach)
20. *Trichosanthes anguina* Linn. (Snake gourd)
21. *T. dioica* Roxb.
22. *Vicia faba* Linn. (Broad bean)
23. *Vigna unguiculata* (L.) Walp. (Cowpea)
24. *Raphanus sativus* Linn. (Radish)

C. **Fruits:**

1. *Anona squamosa* Linn. (Custard apple)
2. *Artocarpus lakoocha* Roxb.
3. *Carica papaya* Linn. (Papaya)
4. *Citrus medica* Linn. (Lemon)
5. *Cucumis sativus* Linn. (Cucumber)
6. *Mangifera indica* Linn. (Mango)
7. *Morus alba* Linn. (Mulberay)
8. *Musa paradisiaca* Linn. (Banana)
9. *Prunus persica* (L.) Stokes (Peach)
10. *Psidium guajava* Linn. (Guava)
11. *Punica granatum* Linn. (Pomergranate)

D. **Condiments**

1. *Allium cepa* Linn. (Onion)
2. *A. sativum* Linn. (Garlic)
3. *Anethum graveolens* Linn. (Dill)
4. *Capsicum frutescens* Linn. (Chilli)
5. *Coriandrum sativum* Linn. (Coriander)
6. *Curcuma longa* Linn. (Turmeric)

E. **Oilseeds**

1. *Linum usitatissimum* Linn. (Linseed)
2. *Brassica campestris* var. *toria* Duthie & Fuller (Indian rape)
3. *Sesamum indicum* Linn. (Sesame)
4. *Ricinus communis* Linn. (Castor oil)

F. **Narcotic**

1. *Nicotiana tabacum* L. (Tobacco).

12. FACTORS AFFECTING THE VEGETATION

A number of biotic and abiotic factors have been found to influence the vegetation and floristic composition of the forests. These are indicated below:

1. **Fire:** Fire is by far the most serious factor affecting the forests of the district. It causes severe damage to the

herbaceous flora and young saplings of various trees. In many cases, fires that occur are incendiary, caused by graziers interested in getting fresh clumps of grasses for the cattle. Fire also spreads in North Kheri Forest division from the Nepal border. It has, however, been commonly observed that species like *Pygmaeopremna herbacea* (Roxb.) Moldenke, *Grewia sapida* Boxb., *Olax nana* Wall. ex Benth., *Erythrina resupinata* Roxb., *Combretum nanum* Buch.-Ham. ex G.Don and *Curcuma angustifolia* Roxb. begin to appear on the burnt ground.

2. **Flood:** Floods cause severe damage to the vegetation of the Park. The Suheli river, Sarda river and other *nalas* and *tals* are overflooded during and sometimes after the rains, which inundate a major portion of the forest areas and cause considerable damage to the forests.

3. **Browsing and grazing:** Grazing by domestic animals causes much injury to the ground flora. The wild animals like elephants, pigs, spotted deer, blue buck and others cause damage to the forests by grazing and browsing the young twigs, branches and tubers of many species. During the summer season, the intense heat spell is generally followed by windstorm for few days with occasional thunder showers. The velocity of windstorm is sometimes so high that it uproots the trees and shrubs. The worst damage is, however, caused by the storm in destroying the flowers and immature fruits. The forests are also being continuously lopped for fodder and fuel-wood by the villagers. **(Plate VII).**

4. **Weeds and pests:** The forests of both the divisions are infested by a large number of climbers that cause great damage to the valuable timber trees. These climbers reach almost to the top of the forest canopy and hamper proper growth of young trees and shrubs. The parasitic plant like *Dendropthoe falcata* (Linn.f.) Ettings., and *Cuscuta reflexa* Roxb., are found on a wide range of host trees like *Adina cordifolia* (Roxb.) Hook.f. ex Brandis, *Bombax ceiba* Linn., *Xeromphis spinosa* (Thunb.) Keay, *Ziziphus xylopyrus* (Retz.) Willd., *Z. mauritiana* Lamk.

etc., and cause considerable damage to the forests. Besides, there are a large number of parasitic fungi and insects causing severe damage to Sal and other economic plants of the forest.

13. ETHNOBOTANY OF THE THARU TRIBE

While conducting floristic studies, an attempt was made to gather folklore data on plants, used either in indigenous system of medicine or by the Tharu tribe and local people. The data were gathered locally from the Tharu tribe, local vaidyas, experienced foresters, etc. The biological screening of 21 species collected from the forest areas was carried out at the Central Drug Research Institute, Lucknow. Table I gives the details of the biological activity observed in different species. The vegetable resources of the district are quite rich and can be utilized in setting up forest-based cottage industries, especially of *Katha,* paper, timber and minor forest products, in the region for human welfare. The survey of useful and medicinal plants was carried out extensively and regularly in different seasons. (See also Maheshwari *et al.,* 1981).

Plant Wealth of the Park

The economic and medicinal plant species are categorised as follows on the basis of different types of products yielded by them:

(A) Edible products;
(B) Fibres;
(C) Oils;
(D) Timber and wood work;
(E) Tannin and Dye materials;
(F) Gums and resins;
(G) Medicinal plants;

A. EDIBLE PRODUCTS:

1. *Emblica officinalis* Gaertn.
Local name : Amla

The fruit is rich in Vitamin C. The dried fruits have good market, as it is edible and is used in indigenous system of medicine.

2. *Dioscorea pentaphylla* Linn. and *D. bulbifera* Linn.

The tubers of these plants are used as vegetable by the local people.

3. The leaves of *Amaranthus spinosus* Linn., *Chenopodium album* Linn., *Ipomoea aquatica* Forsk. are frequently cooked. The fruits of *Xeropmhis uliginosa* (Retz.) Mahesh. are used as vegetable by the Tharus.

4. *Madhuca longifolia* (Koen.) Macb. var. *latifolia* Chev.
Local name : Mahua
The flowers and fruits are eaten by local people.

5. *Tamarindus indica* Linn.
Local name : Imli
The fruits are edible and are collected by the local people.

6. *Buchanania lanzan* Spreng.
Local name : Piyal
The kernel and fruits are edible.

7. *Curcuma angustifolia* Roxb.
Local name : Tikhur.
The tuber of this plant is a rich source of starch.

B. **FIBRES:**

Plants like *Eulaliopsis binata* (Retz.) Hubb., *Bauhinia vahlii* W. & A., *Helicteres isora* Linn., *Corchorus capsularis* Linn., *C. olitorius* Linn., *Sida cordata* (Burm.f.) Borss., *Sterculia villosa* Roxb. *Cannabis sativa* Linn., *Calotropis gigantea* R.Br., *Desmostachya bipinnata* (Linn.) Stapf etc. yield valuable fibres which are used by the Tharus for preparing ropes, animal straps, fishing nets, etc.

C. **OILS:**

The seeds of species like *Celastrus paniculatus* Willd., *Madhuca longifolia* (Koen.) Macb. var. *latifolia* Chev. *Acacia* rugata (Lour.) Merr., *Ricinus communis* Linn., *Mallotus philippensis* (Lamk.) Muell.-Arg., *Schleichera oleosa* Oken, *Semecarpus anacardium* Linn.f. *Ventilago denticulata* Willd. etc., are good source of oils which are prepared and used by the local people in making soaps, hair oil and for burning and edible purposes.

D. TIMBER AND WOOD WORK:

The species, *Tectona grandis* Linn.f., *Acacia nilotica* ssp *indica* Brenan, *Shorea robusta* Gaertn.f., *Dalbergia sissoo* Roxb., *Albizia lebbeck* (Linn.) Benth., *Terminalia alata* Heyne ex Roth, *Dendrocalamus strictus* Nees, *Adina cordifolia* (Roxb.) Hook.f. ex Brandis, *Lagerstroemia parviflora* Roxb., *Toona ciliata* Roem., *Gmelina arborea* Roxb., *Schleichera oleosa* (Lour.) Oken etc. form good quality timber and are used in making carts, wheels, and other agricultural implements by the local people.

E. TANNINS AND DYE MATERIALS:

The bark of species like *Bridelia squamosa* Gehrm., *Cassia fistula* Linn., *Shorea robusta* Gaertn.f. and fruits of *Terminalia chebula* Retz. and *T. bellirica* (Gaertn.) Roxb. yield tannin which is used in the leather industry.

Among the dye-yielding species may be mentioned *Acacia catechu* (Linn.f.) Willd., *Lawsonia inermis* Linn., *Butea monosperma* (Lamk.) Taub., and *Mallotus philippensis* (Lamk.) Muell.-Arg. *Woodfordia fruticosa* Kurz etc.

F. GUMS AND RESINS:

These products are obtained from a number of species inhabiting the forests, e.g., *Acacia nilotica* subsp. *indica* (Benth.) Brenan, *A. catechu* (Linn.f.) Willd., *Azadirachta indica* A. Juss., *Butea monosperma* (Lamk.) Taub., *Ficus religiosa* Linn., *Lannea coromandelica* (Houtt.) Merr., *Sterculia villosa* Roxb., *Shorea robusta* Gaertn., *Buchanania lanzan* Spreng. etc.

G. MEDICINAL PLANTS:

1. *Abrus precatorius* Linn. (Fabaceae)
   Local name : Ratti, Gumchi

The seed paste is used externally against skin diseases. The poultice of seeds placed in cotton pad and inserted into the vagina, brings about abortion.

2. *Achyranthes aspera* Linn. (Amaranthaceae)
   Local name : Latjeera

The decoction of root is used in stomach pain, fever and cough.

3. *Acorus calamus* Linn. (Araceae)
   Local name : Buch

The decoction of rhizome is used in the prevention of fever and dysentery in children. The rhizome is chewed to cure bronchitis and asthma.

4. *Adhatoda zeylanica* Medic. (Acanthaceae) = *Justicia adhatoda* L.
   Local name : Arusa

The decoction of leaves is used to cure bronchitis and cough. The leaves are also used as an insecticide.

5. *Aegle marmelos* (Linn.) Corr. (Rutaceae)
   Local name : Bel

The ripe fruits are used in summer season for cooling and stomach disorder. The root bark is used in fever. The alcoholic extract of fruits and roots showed hypoglycaemic activity in albino rats.

6. *Aeschynomene indica* Linn. (Fabaceae)
   Local name : Jhilmili

The alcoholic extract of the whole plant showed spermicidal activity.

7. *Andrographis paniculata* (Burm.f.) Nees (Acanthaceae)
   Local name : Kalmegh

The species is useful in fever, as anthelmintic and is a stomach tonic. It has shown hepatoprotective activity.

8. *Argemone mexicana* Linn. (Papaveraceae)
Local name : Katili, Bharbhar

The latex is used in the treatment of eye and skin troubles. The poultice of the seeds is applied externally to cure various skin diseases. The oil is used in skin diseases.

9. *Azadirachta indica* A. Juss. (Meliaceae)
Local name : Neem

The juice of the leaves is used in the treatment of eye troubles. The poultice of leaves and bark is applied to boils. The oil is used in diabetes. The decoction of bark, gum, leaves and seeds is used in snake bite and scorpion sting.

10. *Bacopa monnieri* (Linn.) Wettst. (Scrophulariaceae)

The decoction of plant is used in the treatment of asthma and snake bite. It has shown anti cancer activity.

11. *Barringtonia acutangula* (Linn.) Gaertn. (Lecythidaceae).
Local name : Paniyala, Neora

The alcoholic extract of root showed hypoglycaemic activity on albino rats and some pharmacological effect on central nervous system.

12. *Blumea lacera* (Burm.f.) DC. (Asteraceae)
Local name: Kukrondha

The juice of the leaves is used to cure wounds and to kill the worms in children.

13. *Boerhavia diffusa* Linn. (Nyctaginaceae)
Local name : Gadapurna

The decoction of root is used in the treatment of jaundice and eye diseases.

14. *Bridelia squamosa* Gehrm. (Euphorbiaceae)
Local name : Khaja

The bark mixed in gingelly oil is used as liniment in rheumatism. The alcoholic extract of stem bark showed antiviral and anticancer activity.

15. *Buchanania lanzan* Spreng. (Anacardiaceae)
Local name : Piyal

The paste of the kernel is used as an ointment in the treatment of skin diseases. The gum is used in diarrhoea.

16. *Butea monosperma* (Lamk.) Taub. (Fabaceae)
Local name : Dhak

The young leaves coated with mustard oil and warmed, are bandaged on rheumatic pains. The decoction of seeds is used as an anthelmintic.

17. *Calotropis gigantea* (Linn.) R.Br. (Asclepiadaceae)
Local name : Akahua

The flower buds mixed with salt are used in the treatment of stomach constipation.

18. *Calotropis procera* R.Br. (Asclepiadaceae)
Local name : Madar

The ash of the roots is used to remove pus from the gums. The ash of mature leaves mixed with sugar is used to cure asthma and bronchitis.

19. *Cassia fistula* Linn. (Caesalpiniaceae)
Local name : Amaltas

The pods are used as purgative. The leaf paste is used to cure skin diseases. The alcoholic extract of the stem bark and pods showed antiviral activity.

20. *C. surattensis* Burm.f. var. *suffruticosa* Chatt. (Caesalpiniaceae).

The alcoholic extract of the plant showed diuretic and some pharmacological activity on central nervous system.

21. *C. tora* Linn. (Caesalpiniaceae)
Local name : Chakwad

The decoction of the plant is used as a laxative. The seed paste is externally used for healing wounds and cuts.

22. *Cardiospermum halicacabum* Linn. (Sapindaceae)
Local name : Phutaki, Kanphol

The leaf paste is used in the treatment of rheumatism. The root juice is used as a laxative.

23. *Centella asiatica* (Linn.) Urban (Apiaceae)
Local name : Brahmibuti

The decoction of plant is used as a tonic, in diseases of skin, leprosy, nervous disorders etc. The leaves chewed with leaves of *Ocimum sanctum* Linn. early in the morning are said to improve the memory.

24. *Centipeda minima* (Linn.) A.Br. & Aschers. (Asteraceae)
Local name : Nakchikni

The powdered leaves and seeds induced sneezing and a snuff made from them is used in the treatment of cold and headache. The paste of the plant is used to cure toothache.

25. *Chrysopogon aciculatus* Trin. (Poaceae)

The alcoholic extract of the whole plant showed diuretic and gross effect on central nervous system.

26. *Clausena pentaphylla* (Roxb.) DC. (Rutaceae)
Local name : Ratanjot, Bannibu

The decoction of the leaves used in cough and cold.

The alcoholic extract of the plant showed spasmolytic activity.

27. *Clerodendrum serratum* (Linn.) Moon (Verbenaceae)

The leaves boiled with oil and butter and made into an ointment is useful inophthalmia,and in snake bite.

28. *Cocculus hirsutus* (Linn.) Diels (Menispermaceae)
Local name : Chareta

The leaf juice when mixed with water forms a jelly which is taken as a cooling medicine for gonorrhoea and is used externally for eczema.

29. *Colebrookia oppositifolia* Sm. (Lamiaceae)
Local name : Daya

The leaves are applied externally for healing wounds and cuts.

30. *Cordia dichotoma* Forst.f. (Boraginaceae)
Local name : Lasoora

The fruits are used in infection of urinary passage, diseases of lungs and spleen. The decoction of leaves mixed with wheat husk and salt is used internally to relieve chest congestion and influenza.

31. *Cryptolepis buchanani* Roem. & Schult. (Asclepiadaceae)
Local name : Dudhi bel

The plant juice is given to children to cure rickets.

32. *Curculigo orchioides* Gaertn. (Amaryllidaceae)
Local name : Kali musli

The tuberous rootstock mixed with the roots of *Sterculia villosa* Roxb. is given for inducing strength and vigour. The alcoholic extract of the whole plant showed hypoglycaemic and anticancer activity.

33. *Diospyros exsculpta* Buch.-Ham. (Ebenaceae). The bark is used in the treatment of dysentery.
Local Name : Tendu

The alcoholic extract of the bark showed antiprotozoal, antiviral and hypoglycaemic activity.

34. *Eclipta prostrata* (Linn.) Linn. (Asteraceae) syn. *E. alba* Hassk.
Local name : Bhangra

The poultice of the whole plant is used for healing wounds and cuts. Leaf juice given in liver disorder.

35. *Conyza canadensis* (Linn.) Cronquist (Asteraceae).
Local name : Jarupriya

The alcoholic extract of the whole plant showed gross effect on central nervous system.

36. *Euphorbia acaulis* Roxb. (Euphorbiaceae) syn. *E. fusiformis* Buch.-Ham. ex D.Don.
Local name : Banmuli

The leaf juice is applied on the burns. The poultice of the rootstock is applied on joints in relieving pain. It has shown antiarthritic and anticancer activity.

37. *Glycosmis mauritiana* (Lamk.) Tanaka (Rutaceae)
Local name : Makranda

The roots pounded and mixed with sugar is given in fever.

38. *Glinus lotoides* Linn. (Aizoaceae)

The alcoholic extract of the whole plant showed gross effect on central nervous system and diuretic activity.

39. *Gmelina arborea* Roxb. (Verbenaceae)
Local name : Gamhar

The leaf decoction is given in fever.

The alcoholic extract of the stem bark showed antiviral and hypoglycaemic activity.

40. *Gomphrena celosioides* Mart. (Amaranthaceae)

The alcoholic extract of whole plant showed diuretic activity.

41. *Helicteres isora* Linn. (Sterculiaceae)
Local name : Murra, Bendu

The stem bark is diuretic and is used to cure various skin diseases.

42. *Holarrhena antidysenterica* (Roth) Wall. ex DC. (Apocynaceae) = H. pubescens Wall. ex G. Don.
Local name : Kurchi

The decoction of bark is used in the treatment of dysentery.

43. *Hygrophila auriculata* (Schumach.) Heine (Acanthaceae)
Local name : Talmakhana

The decoction of the plant is used to cure skin allergy. The seeds are used in gonorrhoea.

44. *Hyptis suaveolens* Poit. (Lamiaceae)
Local name : Bantulsi

The plant is pounded and applied in the treatment of parasitical cutaneous diseases. The seed oil is used in skin diseases and rheumatism.

45. *Flemingia paniculata* Wall. ex Benth. (Fabaceae)
Local name : Banrahari

The alcoholic extract of the plant showed gross effect on central nervous system and diuretic activity. The leaves are used in diabetes.

46. *Moringa oleifera* Lamk. (Moringaceae)
Local name : Sahjan

The fresh leaf juice when taken three times a day reduces high blood pressure.

47. *Piper longum* Linn. (Piperaceae)
Local name : Peeper, Pipermul

The fruits are mixed with honey and taken orally to cure cough and cold. The decoction of root is used by women after delivery period. The alcoholic extract of the whole plant showed hypoglycaemic and spasmolytic activity.

48. *Plumbago zeylanica* Linn. (Plumbaginaceae)
Local name : Chitrak, Chit.

The paste of the plant is used to heal wounds and cuts. The root is used as abortifacient and in fever.

49. *Rauvolfia serpentina* Benth. ex Kurz (Apocynaceae)
Local name : Dhawarbarua

The root is used in the treatment of blood pressure and insanity. The powdered root along with seeds of black pepper when sniffed into the nose remove fits.

50. *Ricinus communis* Linn. (Euphorbiaceae)
Local name : Rendi

The kernel of seed, milk and sugar mixed together is given to cure waist pain. The oil is used in the treatment of stomach constipation.

51. *Scindapsus officinalis* (Roxb.) Schott (Araceae)
Local name : Gajpipal

The paste of the plant is used in bone fracture.

The alcocholic extract of the fruits showed hypoglycaemic activity on albino rats.

52. *Semecarpus anacardium* Linn.f. (Anacardiaceae)
Local name : Bhelawa

The kernel of the nuts brings abortion and is used as a vermifuge. The oil is used externally in the treatment of rheumatism. The alcoholic extract of the fruits showed anti-cancer activity.

53. *Terminalia bellirica* Gaertn. (Combretaceae)
Local name : Bahera

The powder of the fruits mixed with honey is used as a laxative.

54. *Terminalia chebula* Retz. (Combretaceae)
Local name : Harra

The powder of the fruits mixed with honey is used as a purgative.

55. *Tribulus terestris* Linn. (Zygophyllaceae)
Local name : Gokhru

The powdered fruits taken with water internally release much urine.

56. *Tylophora indica* (Burm.f.) Merr. (Asclepiadaceae)

One leave chewed daily for six days is said to cure asthma.

57. *Uraria picta* Desv. (Fabaceae)

The alcoholic extract of the whole plant showed gross effect on central nervous system and antistrychnine activity.

58. *Vetiveria zizanioides* (Linn.) Nash (Poaceae)
Local name : Khas

The decoction of the roots mixed with old jaggery is given in menstrual disorders. The alcoholic extract of root showed gross effect on central nervous system.

59. *Zingiber capitatum* Roxb. (Zingiberaceae)

The alcoholic extract of whole plant showed diuretic and anticancer activity.

Table-1: Biological screening of plants of Kheri district

Sl. No.	Name of the species	Plant parts used	Date of collection	Voucher	Biological activity observed specimens number
1.	*Aegle marmelos* (Linn.) Corr.	Fr,RT	19.6.73	KKS 4112	Hypoglycaemic
2.	*Aeschynomene indica* Linn.	PL	28.8.72	KKS 2968	Spermicidal
3.	*Barringtonia acutangula* (Linn.)	SB	20.11.73	KKS 4342	Hypoglycaemic, gross effect on central nervous system
4.	*Bridelia squamosa* Gehrm.	SB	27.8.72	KKS 2950	Antiviral, anticancer
5.	*Cassia fistula* Linn.	SB. Pod	22.11.73	KKS 4380	Antiviral
6.	*C. surattensis* var. *suffruticosa* Chatt.	PL	4.4.72	KKS 2926	Diuretic, gross effect on central nervous system
7.	*Chrysopogon aciculatus* Trin.	PL	1.9.72	KKS 2984	Diuretic, gross effect on central nervous system
8.	*Clausena pentaphylla* (Roxb.) DC.	PL	2.9.72	KKS 3000	Spasmolytic
9.	*Conyza canadensis* (Linn.) Cronquist	PL	5.9.72	KKS 3030	Gross effect on central nervous system
10.	*Curculigo orchioides* Gaertn.	PL	28.8.72	KKS 2966	Hypoglycaemic, anticancer
11.	*Diospyros exsculpta* Buch.-Ham.	SB	28.8.72	KKS 2858	Antiprotozoal, antiviral, hypoglycaemic
12.	*Glinus lotoides* Linn.	PL	5.4.72	KKS 2927	Diuretic, gross effect on central nervous system
13.	*Gmelina arborea* Roxb.	SB	5.4.72	KKS 2897	Antiviral, hypoglycaemic
14.	*Gomphrena celosioides* Mart.	PL	1.9.72	KKS 2985	Diuretic

Sl. No.	Name of the species	Plant parts used	Date of collection	Voucher	Biological activity observed specimens number
15.	*Flemingia paniculata* ex Benth (Wall.)	PX	5.4.72	KKS 2987	Diuretic, gross effect on central nervous system
16.	*Piper longum* Linn.	PL	2.9.72	KKS 3042	Spasmolytic, hypoglycaemic
17.	*Scindapsus officinalis* (Roxb.)	Fr.	6.1.76	KKS 7957	Hypoglycaemic
18.	*Semecarpus anacardium* Linn.f.	Fr.	27.8.72	KKS 2958	Anticancer
19.	*Uraria picta* Desv.	PL	49.9.72	KKS 3001	Gross effect on central nervous system, antistrychnine
20.	*Vetiveria zizanioides* (Linn. Nash)	RT	1.9.72	KKS 2983	Gross effect on central nervous system
21.	*Zingiber capitatum* Roxb.	PL	5.9.72	KKS 2993	Diuretic, anticancer

Explanation of symbols:

PX ... Plants without root   PL ... Plants with root   RT ... Roots

SB ... Stem bark   Fr ... Fruits

14. STATISTICAL SYNOPSIS OF THE FLORA

The present study covers 821 species of angiosperms belonging to 489 genera and 120 families. The first ten dominant families occurring in the district as well as in the adjacent regions and in the country as a whole have been compared. The floristic diversity and analysis of families, genera and species of Dicotyledons and Monocotyledons is presented herein.

Table I: Statistical analysis of Families, Genera and species of Dicotyledons and Monocotyledons

	Group	Families		Genera		Species	
		No.	%	No.	%	No.	%
1.	Dicotyledons	99	82.50	386	78.93	635	77.46
2.	Monocotyledons	21	17.50	103	21.07	186	22.54
Total :		120		489		821	

The ratio of families belonging to Monocotyledons and Dicotyledons is 1:4.76, of genera 1:3.75 and of species 1:3.41 in the flora of Dudhwa National Park and adjacent areas of Kheri District.

Except for the families Poaceae, Cyperaceae, Orchidaceae and Commelinaceae, the Monocotyledons are rather poorly represented. Out of 186 species of Monocotyledons, 129 species belong to the above mentioned 4 families, while the remaining 57 species belong to 17 different families of Monocotyledons.

The species, *Cymbopogon osmastonii* Parker is endemic to the Park. A comparison of ten dominant families of Kheri forests in terms of the species richness is given in Table II.

Table II: Ten Dominant Families of Kheri forests

Family	Genera	Species
Leguminosae (Fabaceae) including Caesalpiniaceae and Mimosaceae	41	106
Poaceae (Gramineae)	54	76
Asteraceae (Compositae)	35	56
Cyperaceae	7	46
Euphorbiaceae	19	35
Convolvulaceae	8	23
Lamiaceae (Labiatae)	7	22
Acanthaceae	16	20
Scrophulariaceae	11	19
Rubiaceae	12	17
Malvaceae	9	17

The dominant plant genera of the district with more than 6 species are listed in Table III.

Table III: Dominant Plant Genera of Kheri forests

Genus	No. of species
1. *Cyperus*	24
2. *Ipomoea*	14
3. *Ficus*	11
4. *Desmodium*	11
5. *Cassia*	9
6. *Fimbristylis*	9
7. *Grewia*	8
8. *Crotalaria*	8
9. *Euphorbia*	7
10. *Blumea*	7
11. *Leucas*	7

The Dicotyledonous families represented by single genus or species are: Dilleniaceae, Nelumbonaceae, Elatinaceae, Dipterocarpaceae, Bombacaceae, Zỳgophyllaceae, Simaroubaceae, Burseraceae, Oleaceae, Moringaceae, Passifloraceae, Cactaceae, Alangiaceae, Sphenocleaceae, Plumbaginaceae, Salvadoraceae, Hydrophyllaceae, Cuscutaceae, Lentibulariaceae, Proteaceae, Loranthaceae, Ulmaceae, Cannabaceae, Salicaceae, and Ceratophyllaceae. The Monocotyledonous families comprising single genus or species are: Amaryllidaceae, Typhaceae, Aponogetonaceae and Eriocaulaceae. In order to get an insight into the relations of the flora of Kheri district with its neighbouring regions, a comparative study of first ten dominant families of Kheri district and its neighbouring regions was made (Table IV).

Table IV: Ten dominant Families of Kheri district and its neighbouring States

	Kheri district (Present study)	Gangetic Plain (Hooker, 1904) 1976)	Gorakhpur (Srivastav 1963)	Delhi (Maheshwari
1.	Leguminosae	Gramineae	Leguminosae	Gramineae
2.	Poaceae (Gramineae)	Leguminosae	Gramineae	Leguminosae
3.	Asteraceae (Compositae)	Cyperaceae	Cyperaceae	Compositae
4.	Cyperaceae	Compositae	Compositae	Cyperaceae
5.	Euphorbiaceae	Scrophulariaceae	Euphorbiaceae	Acanthaceae
6.	Convolvulaceae	Malvaceae	Acanthaceae	Euphorbiaceae
7.	Lamiaceae (Labiatae)	Acanthaceae	Rubiaceae	Convolvulaceae
8.	Acanthaceae	Euphorbiaceae	Labiatae	Malvaceae
9.	Scrophulariaceae	Convolvulaceae	Verbenaceae	Amaranthaceae
10.	(a) Rubiaceae (b) Malvaceae	Labiatae	Scrophulariaceae	Scrophulariaceae

The family Leguminosae occupies the first position, while Poaceae and Asteraceae occupy second and third position in the Flora of Dudhwa National Park Kheri district respectively. The flora of Kheri district shows greater resemblance with that of Gorakhpur, in the dominance of certain plant families. The families Leguminosae (including Caesalpiniaceae and Mimosaceae) and Poaceae are dominant in this region.

ABBREVIATIONS

For the sake of economy of space, the conventional abbreviations of the titles of journals, books, floras, periodicals, names of the authors, herbaria etc., have been condensed as follows:

CAL. Central National Herbarium, Botanical Survey of India, Sibpur, Howrah.

CDRI. Herbarium, Botany Division, Central Drug Research Institute, Lucknow.

DD. Herbarium, Systematic Botany Branch, Forest Research Institute, Dehradun.

LWG. Herbarium, National Botanical Research Institute. Lucknow.

BSD. Herbarium, Botanical Survey of India, Northern Circle, Dehradun.

FBI. The Flora of British India by Sir J.D. Hooker *et al.,* (1872-1897).

FUGP. Flora of Upper Gangetic Plain and of the adjacent Siwalik and Sub-Himalayan Tracts by J.F. Duthie *et al.,* (rep. ed. 1960).

Hk.f. or Hook.f. Hooker, *filius,* i.e. Sir J.D. Hooker.

JAA. Journal of the Arnold Arboretum

JASB. Journal of the Asiatic Society of Bengal

JBNHS Journal of the Bombay Natural History Society

JIBS Journal of Indian Botanical Society

RBSI Records of the Botanical Survey of India

TLS Transactions of the Linnean Society of London

Fl. Del. The Flora of Delhi by J.K. Maheshwari (1963)

Fl. Hass. Distt. Flora of Hassan District edited by C.J. Saldanha & D.N. Nicolson.

FG. Flora Gorakhpurensis by T.N. Srivastava (1976)

For.Fl. UP. A Forest Flora for the Plains of Uttar Pradesh by P.C. Kanjilal (1966 Part II & III).

Nom. Chang. Fl. Plants of India & Adjac. reg. (Bennet 1987).

Herb. Flora Dehradun (CR. Babu 1977).

15. PHYTOGEOGRAPHY

Wulff (1943, 1950), Cain (1944), Exell (1944), Good (1953); Schnell (1952) and others have emphasized the importance of phytogeographical studies of an area in the solution of problems of origin and affinities of the flora.

Legris and Meher-Homji (1971) recognized twenty-five floristic elements in the vegetation of India. The forests of Dudhwa National Park, Kheri district which fall within the Indo-Gangetic Plain of Uttar Pradesh are characterized by several floristic elements in the vegetation. Based on the distribution of 821 species occurring in the forests, the following floristic elements are recognized in the vegetation of the Park and adjacent areas of the district.

1. Indian element,
2. Indo-Malayan element,
3. Tropical American (or Neotropical) element,
4. Tropical Indo-African element,
5. Pantropical,
6. Element of Warm countries,
7. Ṣaharo-Sindian element,
8. Sudano-Rajasthanian element,
9. Mediterranean element,
10. Eurasian element,
11. Temperate element,
12. Pluri-regional or cosmopolitan element, and
13. Endemic element.

Some of the typical examples of the species representingdifferent floristic elements in the vegetation of the adjacent forest areas of Kheri district are given below.

1. **Indian element:** This is represented by species like: *Miliusa velutina* Hk.f. Thoms., *M. tomentosa* (Roxb.) Finet & Gagnep., *Tinospora cordifolia* Miers, *Capparis zeylanica* Linn., *Xylosma longifolia* Clos., *Polygala erioptera* DC., *Tamarix dioica* Roxb., *Shorea robusta* Gaertn.f., *Aegle marmelos* (Linn.) Corr., *Feronia limonia* (Linn.) Swingle, *Murraya koenigii* (Linn.) Spreng; *Ziziphus rugosa* Lamk., *Z. xylopyrus* Willd.,

Ochna pumila Buch.-Ham. ex D.Don, *Kydia calycina* Roxb., *Azadirachta indica* A. Juss., *Sterculia villosa* Roxb., *Corchorus capsularis* Linn., *Grewia hirsuta* Vahl, *Ampelocissus latifolia* (Roxb.) Planch., *Cissus repanda* Vahl, *Lannea coromandelica* (Houtt.) Merr., *Buchanania lanzan* Spreng., *Crotalaria mysorensis* Roth, *Indigofera cassioides* Rottl. ex DC., *Atylosia volubilis* Prain, *Erythrina resupinnata* Roxb., *Pueraria tuberosa* DC., *Moghania chapper* (Buch.-Ham.) O. Ktze., *Butea monosperma* (Lamk.) Taub., *B. parviflora* Roxb., *Ougeinia oojeinensis* Hochr., *Dalbergia sissoo* Roxb., *D. volubilis* Roxb., *Pterocarpus marsupium* Roxb., *Mimosa himalayana* Gamble, *Millettia extensa* (Benth.) Baker, *Bauhinia malabarica* Roxb., *B. vahlii* W.&A., *Acacia catechu* (Linn.f.) Willd., *Terminalia arjuna* W.&A., *T. alata* Heyne ex Roth, *Syzygium heyneanum* Wall. ex Duthie, *Careya arborea* Roxb., *Lagerstroemia parviflora* Roxb., *Casearia graveolens* Dalz., *Peucedanum dhana* Buch.-Ham. ex Clarke, *Alangium salvifolium* (Linn.f.) Wang., *Adina cordifolia* Hk.f. ex Brandis, *Xeromphis uliginosa* (Retz.) Mahesh., *Mitragyna parvifolia* Korth., *Wendlandia heynei* (R.&S.) Sant. & Merch. *Erigeron asteroides* Roxb., *Caesulia axillaris* Roxb., *Blumea hieracifolia* var. *hamiltonii* Hook.f., *Launea procumbens* (Roxb.) Rammayya & Rajgopal, *Campanula benthamii* Wall. ex Kitamura, *Madhuca longifolia* (Koen.) Macb., *Diospyros exsculpta* Buch.-Ham., *Linociera intermedia* Wt. var. *roxburghii* Clarke, *Holarrhena antidysenterica* (Roth) A.DC., *Vallaris solanacea* (Roth) O. Ktze., *Hemidesmus indicus* R.&S., *Cryptolepis buchanani* R.&S., *Centaurium centaurioides* (Roxb.) R. Rao & Hemadri, *Hygrophila auriculata* (K. Schum.) Heine, *H. salicifolia* (Vahl) Nees, *Andrographis paniculata* Wall. ex Nees, *Clerodendrum serratum* (Linn.) Moon, *Callicarpa macrophylla* Vahl, *Colebrookia oppositifolia* Sm., *Pogostemon benghalensis* O. Ktze, *Nepeta hindostana* (Roth) Haines, *Leucas cephalotes* (Roth) Spreng., *L. biflora* var. *procumbens* (Desf.) Gamble, *Piper longum* Linn., *Antidesma diandrum* (Roxb.) Roth, *Bridelia squamosa* Gehrm., *Glochidion assamicum* Muell.-Arg., *Securinega leucopyros* (Willd.) Muell.-Arg., *Ficus benghalensis* Linn., *F. racemosa* Linn., *Commelina hasskarlii* Clarke, *Phoenix acaulis* Buch.-Ham., *Dioscorea belophylla* Voigt, *Smilax ocreata* A.DC., *S. perfoliata* Lour., *Saccharum benghalense* Retz. etc. which are indigenous species of India.

2. **Indo-Malayan element:** The typical examples are *Naravelia zeylanica* (Linn.) DC., *Clematis gouriana* Roxb. ex DC., *Dillenia pentagyna* Roxb., *Tinospora sinensis* (Lour.) Merr., *Tiliacora acuminata* Miers, *Rorippa indica* (Linn.) Hiern., *R. montana* Small, *Flacourtia indica* (Burm.f.) Merr., *F. jangomas* (Lour.) Reaush., *Polygala arvensis* Willd., *Abelmoschus crinitus* Wall., *Bombax ceiba* Linn., *Helicteres isora* Linn., *Grewia elastica* Royle, *G. hainseana* Hole, *Glycosmis arborea* (Roxb.) DC., *Garuga pinnata* Roxb., *Toona ciliata* Roem., *Trichilia connaroides* (W.&A.) Benth., *Celastrus paniculatus* Willd., *Cassine glauca* (Rottl.) O. Kuntze, *Ventilago denticulata* Willd., *Ziziphus oenoplia* (Linn.) Mill., *Gouania leptostachya* DC., *Cayratia trifolia* (Linn.) Domin, *Leea asiatica* Ridsdale, *Schleichera oleosa* (Lour.) Oken, *Crotalaria bialata* Schrank, *C. prostrata* Roxb., *C. albida* Heyne ex Roth, *Indigofera glabra* Linn., *Tephrosia candida* DC., *Alysicarpus bupleurifolius* (Linn.) DC., *A. longifolius* W.&A., *Desmodium laxiflorum* DC., *D. heterocarpon* (Linn.) DC., *D. triangulare* (Retz.) Merr., *D. microphyllum* (Thunb.) DC., *D. motorium* (Houtt.) Merr., *Flemingia stobilifera* (Linn.) St. Hil. *F. macrophylla* (Willd.) O. Ktze., *Cassia fistula* Linn., *Bauhinia racemosa* Lamk., *Acacia rugata* (Lamk.) Merr., *Albizia procera* (Roxb.) Benth., *Rubus niveus* Thunb., *Potentilla indica* (Andr.) Wolf, *Terminalia chebula* Retz., *T. bellirica* (Gaertn.) Roxb., *Syzygium cumini* (Linn.) Skeels, *Barringtonia acutangula* (Linn.) Gaertn., *Rotala indica* (Willd.) Koehne, *Trichosanthes bracteata* (Lamk.) Voigt, *T. cordata* Roxb., *Momordica dioica* Roxb. ex Willd., *Actinostemma tenerum* Griff., *Hymenodictyon excelsum* Wall., *Borreria articularis* (Linn.f.) F.N. Wils, *Vernonia anthelmintica* (Linn.) Willd., *Blumea laciniata* (Roxb.) DC., *B. membranacea* DC., *Laggera aurita* (Willd.) Sch.-Bip., *L. falcata* (D.Don) O. Ktze., *Lobelia alsinoides* Lamk., *Lysimachia candida* Lindl. var. *obovata* R. Kunth, *Ardisia solanacea* Roxb., *Wrightia tomentosa* R.&S., *Ichnocarpus frutescens* (Linn.) R.Br., *Buddleja asiatica* Lour., *Nymphoides indicum* (Linn.) O. Ktze., *Hydrolea zeylanica* Vahl, *Cyanoglossum zeylanicum* Thunb. ex Lehm., *Heliotropium strigosum* Wimm., *Dregea volubilis* Benth. ex Hook.f., *Tylophora indica* (Burm.f.) Merr., *T. rotundifolia* Buch.-Ham., *Leptadenia reticulata* W.&A., *Cuscuta reflexa* Roxb., *Solanum surattense* Burm.f.,

Limnophila indica (Linn.) Druce, *Lindernia anagallis* Pennell, *Oroxylum indicum* (Linn.) Vent., *Stereospermum suaveolens* DC., *Adhatoda zeylanica* Medic., *Barleria cristata* Linn., *Gmelina arborea* Roxb., *Clerodendrum indicum* (Linn.) O. Ktze., *C. viscosum* Vent., *Tectona grandis* Linn.f., *Acrocephalus indicus* (Burm.f.) O. Ktze, *Anisomeles indica* (Linn.) O. Ktze., *Psilotrichum ferrugineum* (Roxb.) Moq., *Deeringia amaranthoides* Merr., *Aerva sanguinolenta* (Linn.) Blume, *Litsea monopetala* (Roxb.) Pers., *Dendropthoe falcata* (Linn.f.) Ettings, *Bischofia javanica* Blume, *Emblica officinalis* Gaertn., *Trewia nudiflora* Linn., *Antidesma diandrum* (Roxb.) Roth, *Baliospermum montanum* Muell.-Arg., *Debregesia longifolia* (Burm.f.) Wedd., *Ficus hispida* Linn.f., *F. rumphii* Blume, *Streblus asper* Lour., *Salix tetrasperma* Roxb., *Eulophia graminea* Lindl., *E. nuda* Lindl., *Vanda tessellata* (Roxb.) Hook. ex G.Don, *Dioscorea bulbifera* Linn., *Globba sessiliflora* Sims, *Costus speciosus* (Koen.) Sm., *Smilax ovalifolia* Roxb., *Amischophacelus axillaris* (Linn.) R. Rao & Kammathy, *Cyanotis cristata* (Linn.) D.Don, *Commelina paludosa* Blume, *Murdannia nudiflora* (Linn.) Brenan, *Plesmonium margaritiferum* Schott, *Aponogeton crispus* Thunb., *Cyperus pangorei* Rottl., *Fimbristylis tenera* var. *oxylepis* Clarke, *F. falcata* Kunth, *F. podocarpa* Nees, *Chionachne koenigii* (Spreng.) Thw., *Dendrocalamus strictus* (Roxb.) Nees, *Narenga porphyrocoma* Bor, *Eulaliopsis binnata* (Retz.) Hubb., *Ischaemum rugosum* Salisb., *Apluda mutica* Linn., *Hemarthria compressa* (Linn.) O. Ktze., *Sorghum nitidum* Pers., *Oxyza rufipogon* Griff., *Phragmites karka* Trin., *Arundinella benghalense* Druce, *Bothriochloa kuntzeana* (Hack.) Henr., *Brachiaria kurzii* A. Camus, *Eragrostis unioloides* Retz.) Nees ex Steud. etc., are of Indo-Malayan origin.

3. **Tropical American or Neotropical element:** The neotropical element is represented by species like *Argemone mexicana* Linn., *A. ochroleuca* Sweet. subsp. *ochroleuca* Ownbey, *Malvastrum coromandelianum* (Linn.) Garcke, *Cassia occidentalis* Linn., *Mimosa pudica* Linn., *Corchorus aestuans* Linn., *Passiflora foetida* Linn., *Opuntia dillenii* Haw., *Conyza bonariensis* (Linn.) Cronquist, *Tridax procumbens* Linn., *Parthenium hysterophorus* Linn., *Ipomoea hederifolia* Linn., *I. carnea*

Jacq. subsp. *fistulosa* (Mart. ex Choisy) Austin, *Volvulopsis nummularia* (Linn.) Roberty, *Physalis minima* Linn., *Nicotiana plumbaginifolia* Viv., *Datura metel* Linn., *Solanum viarum* Dunal, *Scoparia dulcis* Linn., *Martynia annua* Linn., *Alternanthera pungens* H.B.&K., *A. paronychioides* St. Hil., *Gomphrena celosioides* Mart., *Chenopodium ambrosioides* Linn., *Jatropha curcas* Linn., *J. gossypifolia* Linn., *Croton bonplandianum* Baill., *Hyptis suaveolens* Poit., *Lantana camara* Linn., var. *aculeata* Mold., *Peperomia pellucida* (Linn.) H.B.&K., *Eichhornia crassipes* (Mart.) Solms, *Chloris barbata* (Linn.) Sw., *Digitaria adscendens* Hern. etc.

4. **Tropical Indo-African element:** The common examples are *Cocculus hirsutus* (Linn.) Diels, *Hibiscus lobatus* (Murr.) O. Ktze., *Azanza lampas* (Cav.) Alef., *Grewia tiliaefolia* Vahl, *G. disperma* Rottl. ex Spreng., *Triumfetta pilosa* Roth, *T. pentandra* A. Rich, *Tephrosia pumila* Pers., *Uraria rufescens* (DC.) Schind., *Alysicarpus vaginalis* (Linn.) DC., *Indigofera trifoliata* Linn., *Albizia lebbeck* (Linn.) Benth., *Woodfordia fruticosa* (Linn.) Kurz, *Mukia maderaspatana* (Linn.) Cogn., *Coccinia grandis* Voigt, *Diplocyclos palmatus* (Linn.) Jeffrey, *Xeromphis spinosa* (Thunb.) Keay, *Oldenlandia paniculata* Linn., *Dichrocephala integrifolia* (Linn.f.) O. Ktze., *Sphaeranthus indicus* Linn., *Chrysanthellum indicum* DC., *Plumbago zeylanica* Linn., *Diospyros peregrina* (Gaertn.) Gurke, *Canscora decussata* (Roxb.) R.&S., *Arnebia hispidissima* (Lehm.) DC., *Ehretia acuminata* R.Br. var. *serrata Johns, Ipomoea rumicifolia* Choisy, *I. obscura* (Linn.) Ker-Gawl., *I. pestigridis* Linn., *Solanum violaceum* Ortega, *S. incanum* Linn., *Lindernia parviflora* (Roxb.) Haines, *Pedalium murex* Linn., *Nelsonia canescens* (Lamk.) Spreng., *Peristrophe bicalyculata* (Retz.) Nees, *Leonotis nepetifolia* (Linn.) R.Br., *Aerva lanata* Juss., *Digera muricata* (Linn.) Mart., *Basella alba* Linn., *Polygonum limbatum* Meissn., *Euphorbia thymifolia* Linn., *E. dracunculoides* Lamk., *Tragia involucrata* Linn., *Securinega virosa* (Roxb. ex Willd.) Baill., *Dioscorea pentaphylla* Linn., *Asparagus racemosus* Willd., *Scirpus michelianus* Linn., *Vetiveria zizanioides* (Linn.) Nash, *Bothriochloa pertusa* (Linn.) A. Camus, *Urochloa panicoides* Beauv., *Brachiaria ramosa* (Linn.) Stapf etc.

5. **Pantropical element:** The typical example are *Cleome gynandra* Linn., *C. viscosa* Linn., *Polycarpaea corymbosa* (Linn.) Lamk., *Portulaca quadrifida* Linn., *Bergia ammannioides* Roxb. ex Roth, *Sida spinosa* Linn., *S. cordifolia* Linn., *Melochia corchorifolia* Linn., *Corchorus olitorius* Linn., *Biophytum sensitivum* (Linn.) DC., *Tribulus terrestris* Linn., *Tephrosia purpurea* (Linn.) Pers., *Desmodium triflorum* (Linn.) DC., *D. gangeticum* (Linn.) DC., *Zornia gibbosa* Span., *Abrus precatorius* Linn., *Aeschynomene indica* Linn., *Mucuna prurita* Hook.f., *Cassia tora* Linn., *C. pumila* Lamk., *Glinus lotoides* Linn., *Trianthema portulacastrum* Linn., *Oldenlandia corymbosa* Linn., *Ageratum conyzoides* Linn., *Vernonia cinerea* Less., *Elephantopus scaber* Linn., *Conyza canadensis* (Linn.) Cronquist, *Emilia sonchifolia* (Linn.) DC., *Grangeea maderaspatana* (Linn.) Poir., *Sigesbeckia orientalis* Linn., *Ipomoea aquatica* Forsk., *I. pentaphylla* Jacq., *I. muricata* (Linn.) Jacq., *Barleria prionotis* Linn., *Leucas aspera* (Willd.) Spreng., *Phyla nodiflora* (Linn.) Greene, *Boerhavia diffusa* Linn., *Achyranthus aspera* Linn., *Alternanthera sessilis* DC., *Celosia argentea* Linn., *Polygonum glabrum* Willd., *Euphorbia hirta* Linn., *Phyllanthus fraternus* Webster, *Pistia stratiotes* Linn., *Cyperus haspan* Linn., *C. compressus* Linn., *C. cuspidatus* Kunth, *Fimbristylis ovata* (Burm.f.) Kern, *Eleocharis atropurpurea* (Retz.) Kunth, *Saccharum spontaneum* Linn., *Hackelochloa granularis* (Linn.) O. Ktze., *Dichanthium annulatum* (Forsk.) Stapf, *Oplismenus burmanii* (Retz.) Beauv., *O. compositus* (Linn.) Beauv., *Eragrostis tenella* (Linn.) Beauv., *Rottboellia exaltata* Linn.f. etc.

6. **Element of warm countries:** The common examples are *Cissampelos pareira* Linn., *Nymphaea stellata* Willd., *Ziziphus mauritiana* Lamk., *Indigofera linifolia* (Linn.f.) Retz., *Alytosia scarabaeoides* (Linn.) Benth., *Rhynchosia minima* (Linn.) DC., *Glinus oppositifolius* (Linn.) DC., *Mollugo pentaphylla* Linn., *Blumea mollis* (D.Don) Merr., *Youngia japonica* (Linn.) DC., *Bidens biternata* (Lour.) Merr. & Sch., *Wahlenbergia marginata* (Thunb.) DC., *Calotropis procera* R.Br., *Mazus pumilus* (Burm.f.) van Steenis, *Chrozophora rottleri* Juss. ex Spreng., *Hydrilla verticillata* (Linn.f.) Royle, *Ehretia laevis* Roxb., *Cyperus kyllinga* Endl., *C. rotundus* Linn., *Scirpus affinis* Roth, *Fimbristylis dichotoma* (Linn.) Vahl, *Imperata cylindrica* (Linn.) Beauv., *Setaria tomentosa* (Roxb.) Kunth, *Sorghum halepense* (Linn.) Pers., *Bothriochloa pertusa* (Linn.) A.

Camus, *Heteropogon contortus* Beauv., *Echinochloa colonum* (Linn.) Link., *E. crusgalli* (Linn.) Beauv. etc.

7. **Saharo-Sindian element:** The examples are *Leptadenia pyrotechnica* (Forsk.) Decne., *Ziziphus nummularia* (Burm.f.) W.&A., and *Convolvulus microphyllus* Sieb. ex Spreng.

8. **Sudano-Rajasthanian element:** The typical examples are *Acacia nilotica* subsp. *indica* (Benth.) Brenan, *Salvadora oleoides* Decne., *Arthraxon prionodes* (Steud.) Dandy, *Cymbopogon jwarancusa* (Jones) Schult., and *Desmostachya bipinnata* (Linn.) Stapf.

9. **Mediterranean element:** The typical examples are *Vicia sativa* Linn., *Saussurea heteromalla* (D.Don) Hand.- Mazz., *Scirpus articulatus* Linn., *S. mucronatus* Linn., *Phalaris minor* Retz., etc.

10. **Eurasian element:** The examples are *Medicago polymorpha* Linn., *M. lupulina* Linn., *Melilotus alba* Desr., *M. indicus* All., *Xanthium strumarium* Linn., *Sonchus oleraceus* Linn., *Cannabis sativa* Linn., *Setaria glauca* (Linn.) Beauv., etc.

11. **Temperate element:** The common examples are *Ranunculus scleratus* Linn., *R. cantoniensis* DC., *Fumaria indica* (Haussk.) Pugsley, *Stellaria media* (Linn.) Vill., *Potentilla supina* (Andr.) Wolf, *Malva sylvestris* Linn., *Crotalaria sessiliflora* Linn., *Cirsium arvense* (Linn.) Scop., *Artemisia nilagirica* (Clarke) Pamp., *Venonica anagallis-aquatica* Linn., *Ajuga bracteosa* Wall. ex Benth., *Acorus calamus* Linn., *Eleocharis palustris* R.Br., *Potamogeton crispus* Linn., *Polypogon monspeliensis* (Linn.) Desf.

12. **Pluri-regional or Cosmopolitan element:** These species are world-wide in distribution and occur in both temperate and tropical parts of the world. Some of the typical examples are *Coronopus didymus* (Linn.) Sm., *Portulaca oleracea* Linn., *Malva verticillata* Linn., *Triumfetta rhomboidea* Jacq., *Oxalis corniculata* Linn., *Indigofera hirsuta* Linn., *Sesbania bispinosa* (Jacq.) Faw. & Rendl, *Cassia absus* Linn., *Rotala mexicana* Cham. & Schlect, *Ludwigia octovalvis* (Jacq.) Raven subsp. *sessiliflora* (Mich.) Raven, *Gnaphalium luleo-album* Linn.

subsp. *affine* (D.Don) Koster, *Adenostemma lavenia* (Linn.) O. Ktze, *Bacopa monnieri* Pennell, *Chenopodium album* Linn., *Chenopodium murale* Linn., *Amaranthus spinosus* Linn., *A. viridis* Linn., *Ceratophyllum demersum* Linn., *Vallisneria spiralis* Linn., *Juncus bufonius* Linn., *Spirodela polyrhiza* (Linn.) Schleid., *Wolffia arrhiza* (Linn.) Horkel ex Wimmer, *Potamogeton nodosus* Poir., *P. pectinatus* Linn., *Cyperus brevifolius* (Rottb.) Hassk., *C. cyperoides* (Linn.) O. Ktze., *C. pumilus* Linn., *C. rotundus* Linn., *Scirpus tuberosus* Desf., *Schizachyrium brevifolium* (Sw.) Nees, *Cynodon dactylon* Pers., *Setaria verticillata* (Linn.) Beauv. etc.

13. **Endemic element:** The species, *Cymbopogon osmastonii* Parker is endemic to the Park.

It would appear from the above examples of different floristic elements that many species are either of tropical origin or are Indo-Malayan. It can be concluded that the Indo-Malayan and tropical elements form dominant components in the vegetation of the district. However, the Indo-Malayan element is better represented in the moist deciduous tropical forests of the district. The Indian element is also fairly well represented in the vegetation. The occurrence of species like *Convolvulus microphyllus* Sieb. ex Spreng., *Leptadenia pyrotechnica* (Forsk.) Decne., and *Ziziphus nummularia* (Burm.f.) W.&A. is interesting, as these represent Saharo-Sindian element in the vegetation of the Park and adjacent areas of Kheri district.

COMPARISON AND AFFINITIES OF THE FLORA

The flora of the Park and adjacent areas of the district has been compared with the adjoining regions of Himalayan foothills of Nepal, plains of Uttar Pradesh and Delhi.

1. *Affinity with the flora of Nepal:* The flora of Central Himalaya or Nepal Himalaya has been thoroughly explored by Banerji (1965), Kihara (1952-53) and others. In recent years, the medicinal and economic plants of the forest areas of Nepal have been studied by the Department of Medicinal Plants, Thapathali, Kathmandu, Nepal (Anon. 1967, 1969, 1973, 1974).

These form the basis for the comparison and relationship of the flora. The flora of Kheri district shows similarities with the flora of Nepal in the occurrence of a number of species like *Ranunculus scleratus* Linn., *Cissampelos pariera* Linn., *Tiliacora acuminata* (Lamk.) Miers, *Rorippa indica* (Linn.) Hiern, *Polygala crotalarioides* Buch.-Ham. ex DC., *Shorea robusta* Gaertn.,f., *Sida cordifolia* Linn., *S. cordata* (Burm.f.) Borss., *S. rhombifolia* Linn., *Azanza lampas* Alef., *Urena lobata* Linn., *Bombax ceiba* Linn., *Grewia sapida* Roxb., *Triumfetta rhomboidea* Jacq., *Aegle marmelos* (Linn.) Corr., *Toona ciliata* Roem., *Trichilia connaroides* (W.&A.) Bent., *Gouania leptostachya* DC., *Leea asiatica* Rids., *Crotalaria sessiliflora* Linn., *C. albida* Heyne ex Roth, *Indigofera linifolia* Retz., *I. cassioides* Rottl. ex DC., *Millettia extensa* (Benth.) Baker, *Uraria rufescens* (DC.) Schindl., *Desmodium laxiflorum* DC., *D. heterocarpon* (Linn.) DC., *D. microphyllum* (Thunb.) DC., *D. motorium* (Houtt.) Merr., *Flemingia strobilifera* (Linn.) St. Hil. ex Jacks., *Butea monosperma* (Lamk.) Taub., *Bauhinia vahlii* W. & A., *Acacia catechu* Willd., *Pyrus pashia* Buch.-Ham. ex D.Don, *Potentilla indica* (Andr.) Wolf, *Terminalia chebula* Retz., *Syzygium cumini* (Linn.) Skeels, *Woodfordia fruticosa* (Linn.) Kurz, *Lagerstroemia parviflora* Roxb., *Diplocyclos palmatus* (Linn.) Jeffrey, *Centella asiatica* (Linn.) Urban, *Adina cordifolia* Hook.f. ex Brandis, *Wendlandia heynei* (R.&S.) Sant. & Merch., *Knoxia sumatrensis* (Retz.) DC., *Vernonia squarrosa* (D.Don) Less., *V. anthelmintica* (Linn.) Willd., *Artemisia nilagirica* (Clarke) Pamp., *Elephantopus scaber* Linn., *Dichrocephala integrifolia* (Linn.f.) O. Ktze., *Cyathocline purpurea* (D.Don) O. Ktze., *Eclipta prostrata* (Linn.) Linn., *Cotula hemisphaerica* (Roxb.) Clarke, *Youngia japonica* (Linn.) DC., *Ardisia solanacea* Roxb., *Holarrhena antidysenterica* (Roth) A.DC., *Rauvolfia serpentina* Benth. ex Kurz, *Vallaris solanacea* (Roth) O. Ktze., *Cryptolepis buchananii* R.&S., *Tylophora rotundifolia* Buch.-Ham., *Buddleja asiatica* Lour., *Ehretia laevis* Roxb., *E. acuminata* R.Br. var. *serrata* Johns, *Evolvulus alsinoides* Linn., *Solanum erianthum* D.Don, *Verbascum thapsus* Linn., *Oroxylum indicum* (Linn.) Vent., *Lepidagathis incurva* D.Don, *Verbena officinalis* Linn., *Clerodendrum serratum* (Linn.) Moon, *C. indicum* (Linn.) O. Ktze., *Plectranthus ternifolius* D.Don, *Colebrookia oppositifo-

lia Sm., *Perilla frutescens* (Linn.) Britt., *Ajuga bracteosa* Wall. ex Benth., *Deeringia amaranthoides* (Lamk.) Merr., *Chenopodium ambrosioides* Linn., *Litsea monopetala* (Roxb.) Pers., *Mallotus philippensis* (Lamk.) Muell.-Arg., *Emblica officinalis* Gaertn., *Cannabis sativa* Linn., *Debregesia longifolia* (Burm.f.) Wedd., *Pouzolzia hirta* (Bl.) Hassk., *Ficus benghalensis* Linn., *F. semicordata* Buch.-Ham., *Salix tetrasperma* Roxb., *Curcuma angustifolia* Roxb., *Dioscorea bulbifera* Linn., *Asparagus racemosus* Willd., *Smilax perfoliata* Lour., *Arisaema tortuosum* (Wall.) Schott, *Potamogeton pectinatus* Linn., *Cyperus kyllingia* Endl., *C. cyperoides* (Linn.) O. Ktze., *C. cuspidatus* Kunth, *Eleocharis palustris* R.Br., *Scirpus mucronatus* Linn., *Apluda mutica* Linn., *Chrysopogon aciculatus* (Retz.) Trin., *Capillipedium assimile* (Steud.) A. Camus, *Paspalum distichum* Linn., *Phragmites karka* Trin., *Polypogon monspeliensis* (Linn.) Desf. etc.

2. *Comparison and affinities with the plains of Uttar Pradesh:* The flora of the district has also been compared with that of Bahraich, Gorakhpur and Hastinapur in the plains of Uttar Pradesh.

The relationship of the flora with that of Bahraich district is based on the work of Panigrahi *et al.*, (1971). Nearly 393 species of terai forests of Bahraich district are common in the terai forests of Kheri district.

Some of the common species represented in both the forest areas are *Polyalthia suberosa* (Roxb.) Thw., *Naravelia zeylanica* (Linn.) DC., *Tiliacora acuminata* (Lamk.) Miers, *Capparis zeylanica* Linn., *Flacourtia indica* (Burm.f.) Merr., *Casearia elliptica* Willd., *Shorea robusta* Gaertn., *Kydia calycina* Roxb., *Bombax ceiba* Linn., *Sterculia villosa* Roxb., *Helicteres isora* Linn., *Grewia hirsuta* Vahl, *Aegle marmelos* (Linn.) Corr., *Cassine glauca* (Rottl.) O. Ktze., *Ventilago denticulata* Willd., *Ziziphus xylopyrus* Willd., *Z. oenoplia* (Linn.) Mill., *Buchanania lanzan* Spreng., *Lannea coromandelica* (Houtt.) Merr., *Abrus precatorius* Linn., *Crotalaria sericea* Retz., *C. sessiliflora* Linn., *Dalbergia sissoo* Roxb., *Desmodium pulchellum* (Linn.) Benth., *D. triangulare* (Retz.) Merr.,

Indigofera cassioides Rottl. ex DC., *Butea monosperma* (Lamk.) Taub., *Millettia extensa* (Benth.) Baker, *Flemingia chappar* (Buch.-Ham.) O. Ktze., *F. strobilifera* (Linn.) St. Hil. ex Jacks., *Ougeinia oojeinensis* (Roxb.) Hochr., *Bauhinia vahlii* W. & A., *B. malabarica* Roxb., *Caesalpinia bonduc* (Linn.) Roxb., *Cassia fistula* Linn., *Acacia catechu* Willd., *Albizia lebbeck* Benth., *A. procera* Benth., *Mimosa himalayana* Gamble, *Potentilla indica* (Andr.) Wolf, *Terminalia alata* Heyne ex Roth, *T. bellirica* (Gaertn.) Roxb., *Barringtonia acutangula* (Linn.) Gaertn., *Syzygium cumini* (Linn.) Skeels, *Actinostemma tenerum* Griff., *Diplocyclos palmatus* (Linn.) Jeffrey, *Trichosanthes cucumerina* Linn., *Centella asiatica* (Linn.) Urban, *Alangium salvifolium* (Linn.f.) Wang., *Adina cordifolia* Hook.f. ex Brandis, *Hymenodictyon excelsum* Wall., *Mitragyna parvifolia* Korth., *Elephantopus scaber* Linn., *Lactuca polycephala* (Cass.) Clarke, *Sigesbeckia orientalis* Linn., *Gnaphalium luteo-album* Linn. subsp. *affine* (D.Don) Koster, *Sphaeranthus indicus* Linn., *Vicoa indica* (Linn.) DC., *Primula umbellata* (Lour.) Bentv., *Ardisia solanacea* Roxb., *Madhuca longifolia* (Koen.) Macbr., *Holarrhena antidysenterica* (Roth) A.DC., *Rauvolfia serpentina* Benth. ex Kurz, *Vallaris solanacea* (Roth) O. Ktze., *Hemidesmus indicus* (Linn.) R.Br., *Buddleja asiatica* Lour., *Ehretia laevis* Roxb., *Arnebia hispidissima* DC., *Bothriospermum tenellum* Fish. & Mey., *Operculina turpethum* (Choisy) Van Oostr., *Merremia hederacea* (Burm.f.) Hallier f., *Solanum erianthum* D.Don, *Stereospermum suaveolens* DC., *Eranthemum purpurascens* Wt. ex Nees, *Lepidagathis incurva* D.Don, *Tectona grandis* Linn.f., *Callicarpa macrophylla* Vahl, *Pogostemon benghalensis* (Burm.f.) O. Ktze., *Litsea monopetala* (Roxb.) Pers., *Bridelia squamosa* Gehrm., *Drypetes roxburghii* (Wall.) Hurus., *Emblica officinalis* Linn., *Trewia nudiflora* Linn., *Mallotus philippensis* (Lamk.) Muell.-Arg., *Deeringia amaranthoides* (Lamk.) Merr., *Piper longum* Linn., *Ficus benghalensis* Linn., *F. hispida* Linn.f., *F. rumphii* Blume, *Streblus asper* Lour., *Salix tetrasperma* Roxb., *Ceratophyllum demersum* Linn., *Vanda tessellata* (Roxb.) Hook. ex G.Don, *Zeuxine strateumatica* (Linn.) Schlect., *Asparagus adscendens* Roxb., *Gloriosa superba* Linn., *Smilax perfoliata* Lour., *Calamus tenuis* Roxb., *Ottelia alismoides* (Linn.) Pers., *Sagittaria sagittifolia* Linn., *Aponogeton crispus*

Thunb., *Carex fedia* Nees, *Cyperus compactus* Retz., *C. brevifolius* (Rottl.) Hassk., *Scirpus mucronatus* Linn., *S. articulatus* Linn., *Apluda mutica* Linn., *Bothriochloa intermedia* (R.Br.) A. Camus, *Chionachne koenigii* (Spreng.) Thw., *Chrysopogon aciculatus* (Retz.) Trin., *Coix lachryma-jobi* Linn., *Heteropogon contortus* (Linn.) Beauv., *Vetiveria zizanioides* (Linn.) Nash etc.

The flora of Gorakhpur has been studied by Dixit *et al.*, (1966), Singh (1961), Panigrahi *et al.* (1967), Srivastava (1976) and others. A comparison with the flora of Gorakhpur revealed that about 570 species are common to the forest of Kheri and Gorakhpur districts. The closeness of the floristic composition in these two districts is well established by the common species like *Clematis gouriana* Roxb., ex DC., *Dillenia pentagyna* Roxb., *Miliusa tomentosa* (Roxb.) Finet & Gagnep., *Polyalthia suberosa* (Roxb.) Thw., *Cissampelos pareira* Linn., *Cocculus hirsutus* (Linn.) Diels, *Tiliacora acuminata* (Lamk.) Miers, *Flacourtia indica* (Burm.f.) Merr., *Xylosma longifolia* Clos., *Tamarix dioica* Roxb., *Kydia calycina* Roxb., *Helicteres isora* Linn., *Shorea robusta* Gaertn., *Grewia polygama* Roxb., *G. tiliaefolia* Vahl, *G. hirsuta* Vahl, *Bombax ceiba* Linn., *Sida acuta* Burm.f., *S. cordata* (Burm.f.) Borss., *S. cordifolia* Linn., *S. rhombifolia* Linn., *Urena lobata* Linn., *U. palmata* Roxb., *Abutilon indicum* (Linn.) Sw., *Tribulus terrestris* Linn., *Biophytum sensitivum* (Linn.) DC., *Aegle marmelos* (Linn.) Corr., *Clausena pentaphylla* DC., *Feronia limonia* (Linn.) Swingle, *Murraya koenigii* (Linn.) Spreng., *Glycosmis mauritiana* (Lam.) Tanaka, *Toona ciliata* Roem., *Celastrus paniculatus* Willd., *Cassine glauca* (Rottl.) O. Ktze., *Ziziphus nummularia* (Burm.f.) W.&A., *Z. oenoplia* (Linn.) Mill., *Ampelocissus latifolia* (Roxb.) Planch., *Leea alata* Edgew., *Schleichera oleosa* (Lour.) Oken., *Buchanania lanzan* Spreng., *Lannea coromandelica* (Houtt.) Merr., *Spondias pinnata* (Linn.f.) Kurz, *Semecarpus anacardium* Linn.f., *Abrus precatorius* Linn., *Butea monosperma* (Lamk.) Taub., *B. parviflora* Roxb., *Crotalaria albida* Heyne ex Roth, *C. medicaginea* Lamk., *Dalbergia sissoo* Roxb., *D. volubilis* Roxb., *Desmodium gangeticum* DC., *D. motorium* (Houtt.) Merr., *D. pulchellum* (Linn.) Benth, *D. triangulare* (Retz.) Merr., *Medicago polymorpha* Linn., *Millettia extensa* (Benth.) Baker, *Indigofera cassioides* Rottle. ex DC., *Flemingia strobilifera* (Linn.) St. Hil. *Ougeinia oojeinensis* (Roxb.) Hochr., *Bauhinia racemosa* Lamk., *B. vahlii* W. & A., *Albizia lebbeck* Benth., *Acacia catechu* Willd., *Mimosa himalayana* Gamble, *Terminalia alata* Heyne ex Roth, *T. bellirica* (Gaertn.) Roxb., *T. chebula* Retz., *Syzygium cumini* (Linn.) Skeels, *S. heyneanum* Wall. ex Duthie, *Barringtonia acutangula* Gaertn., *Lagerstroemia parviflora* Roxb., *Woodfordia fruticosa* (Linn.) Kurz, *Casearia elliptica* Willd., *Centella asiatica* (Linn.) Urban, *Cnidium monierii* (Linn.) Cusson., *Alangium salvifolium* (Linn.) Wang., *Adina cordifolia* Hook.f. ex Brandis, *Borreria articularis* (Linn.f.) F.N. Wils., *B. stricta* (Linn.f.) K. Schum., *Wendlandia heynei* (R.&S.) Sant. & Merch., *Hymenodictyon excelsum* Wall., *Mitragyna parvifolia* Korth., *Xeromphis spinosa* (Thunb.) Keay, *Adenostemma lavenia* (Linn.) O. Ktze., *Chrysanthellum indicum* DC., *Cotula anthelmintica* Linn., *Elephantopus scaber* Linn., *Gnaphalium luteo-album* Linn. subsp. *affine* (D.Don) Koster, *Grangea maderaspatana* Poir., *Laggera falcata* (D.Don) O. Ktze., *Saussurea heteromalla* (D.Don) Hand.-Mazz., *Sphaeranthus indicus* Linn., *Cirsium arvensis* (Linn.) Scop., *Youngia japonica* (Linn.) DC., *Campanula benthamii* Wall. ex Kitamura, *Lobelia alsinoides* Lamk., *Plumbago zeylanica* Linn., *Primula umbellata* (Lour.) Bentv., *Lysimachia candida* Lindl. subsp. *obovata, Ardisia solanacea* Roxb., *Madhuca longifolia* (Koen.) Macbr., *Holarrhena antidysenterica* (Roth) A.DC, *Ichnocarpus frutescens* (Linn.) R.Br., *Rauvolfia serpentina* Benth. ex Kurz, *Calotropis gigantea* (Linn.) R.Br., *Oxystelma secamone* (Linn.) K. Shum., *Canscora decussata* R.&S., *Heliotropium strigosum* Willd., *Trichodesma sedgwickianum* Banerjee, *Porana paniculata* Roxb., *Mella hamiltoniana* (Benth.) Pennell, *Buddleja asiatica* Lour., *Nymphoides indicum* (Linn.) O. Ktze., *Cordia dichotoma* Forst.f., *Ajuga bracteosa* Wall. ex Benth., *Adhatoda zeylanica* Medic., *Andrographis paniculata* (Burm.f.) Wall. ex Nees, *Lepidagathis incurva* D.Don, *Callicarpa macrophylla* Vahl, *Clerodendrum indicum* (Linn.) O. Ktze., *Premna latifolia* Roxb. var. *mucronata, Tectona grandis* Linn.f., *Colebrookia oppositifolia* Sm., *Leucas aspera* (Willd.) Link., *L. cephalotes* (Roth) Spreng., *Achyranthes aspera* Linn., *Alternanthera paronychioides* St. Hil.,

Polygonum limbatum Meissn., *P. barbatum* Linn. subsp. *gracile, Litsea glutinosa* (Lour.) Robins., *Baliospermum montanum* (Willd.) Muell.-Arg., *Bischofia javanica* Blume, *Bridelia squamosa* Gehrm., *B. stipularis* Blume, *Antidesma diandrum* Roth, *A. ghaesembilla* Gaertn., *Croton oblongifolius* Roxb., *Emblica officinalis* Gaertn., *Tragia involucrata* Linn., *Pouzolzia zeylanica* (Linn.) Benn., *Ficus racemosa* Linn., *F. rumphii* Blume, *F. semicordata* Buch.-Ham., *F. heterophylla* Linn.f., *Streblus asper* Lour., *Ceratophyllum demersum* Linn., *Ottelia alismoides* (Linn.) Pers., *Gloriosa superba* Linn., *Costus speciosus* Sm., *Curculigo orchioides* Gaertn., *Smilax prolifera* Lour., *Dioscorea bulbifera* Linn., *D. pentaphylla* Linn., *Calamus tenuis* Roxb., *Eriocaulon sieboldianum* Sieb., & Zucc. ex Steud., *Eleocharis palustris* R.Br., *E. dulcis* (Burm.f.) Henschel, *Brachiaria ramosa* (Linn.) Stapf, *Cyrtococum pratens* (Linn.) A. Camus, *Daclytoctenium aegypticum* (Linn.) Beauv., *Digitaria adscendens* (H.B.&K.) Henr., *Echinochloa colonum* (Linn.) Link., *E. crusgalli* (Linn.) Beauv., *Hackelochloa granularis* (Linn.) O. Ktze., *Themeda arundinacea* (Roxb.) Ridley, *Hemarthria compressa* (Linn.f.) R.Br., *Ischaemum rugosum* Salisb., *Phragmites karka* Trin., *Narenga porphyrocoma* (Hance) Bor, *Paspalum commersonii* Lamk., *Rottboellia exaltata* Linn.f., *Urochloa panicoides* Beauv., *Vetiveria zizanioides* (Linn.) Nash etc.

The flora of Hastinapur in Meerut district is worked out by Murty & Singh (1961). A comparison with the flora of Kheri is attempted here with species found in the flora of Hastinapur. Nearly 400 species are common to both the floras. Some of the commonly represented species in these areas are *Tinospora cordifolia* (Willd.) Miers, *Cocculus hirsutus* (Linn.) Diels, *Cissampelos pareira* Linn., *Rorippa indica* (Linn.) Hiern, *Capparis zeylanica* Linn., *Flacourtia indica* (Burm.f.) Merr., *Tamarix dioica* Roxb., *Sida spinosa* Linn., *S. acuta* Burm.f., *S. rhombifolia* Linn., *Urena palmata* Roxb., *Hibiscus vitifolius* Linn., *Bombax ceiba* Linn., *Helicteres isora* Linn., *Melochia corchorifolia* Linn., *Grewia sapida* Roxb., *G. hirsuta* Vahl, *Corchorus trilocularis* Linn., *Murraya koenigii* (Linn.) Spreng., *Citrus medica* Linn., *Feronia limonia* (Linn.) Swingle, *Aegle marmelos* (Linn.) Correa, *Toona ciliata* Roem.,

Ziziphus mauritiana Lamk., *Z. oenoplia* (Linn.) Mill., *Ampelocissus latifolia* (Roxb.) Planch., *Crotalaria calycina* Shrank., *C. mysorensis* Roth, *C. medicaginea* Lamk., *Medicago polymorpha* Linn., *M. lupulina* Linn., *Butea monosperma* (Lamk.) Taub., *Abrus precatorius* Linn., *Dalbergia sissoo* Roxb., *Zornia gibbosa* Span., *Desmodium gangeticum* DC., *D. triflorum* DC., *Cassia fistula* Linn., *C. sophera* Linn., *C. absus* Linn., *Bauhinia racemosa* Lamk., *Mimosa pudica* Linn., *Acacia catechu* Willd., *A. rugata* Merr., *Albizia lebbeck* Benth., *Terminalia bellirica* (Gaertn.) Roxb., *Syzygium cumini* (Linn.) Skeels, *Casearia elliptica* Willd., *Trichosanthes cucumerina* Linn., *Momordica dioica* Roxb. ex Willd., *Actinostemma tenerum* Griff., *Centella asiatica* (Linn.) Urban, *Borreria articularis* (Linn.f.) Wils., *B. stricta* (Linn.f.) K. Schum., *Pluchea tomentosa* DC., *Sigesbeckia orientalis* Linn., *Artemisia scoparia* Waldst. & Kit., *A. nilagirica* (Clarke) Pamp., *Saussurea heteromalla* (D.Don) Hand.-Mazz., *Amberboa ramosa* (Roxb.) Jafri, *Campanula benthamii* Wall. ex Kitamura, *Plumbago zeylanica* Linn., *Madhuca longifolia* (Koenig.) Macbr., *Wrightia tomentosa* (Roxb.) R.&S., *Ichnocarpus frutescens* (Linn.) R.Br., *Oxystelma secamone* (Linn.) K. Schum., *Leptadenia reticulata* W.&A., *L. pyrotechnica* (Forsk.) Decne., *Hydrolea zeylanica* Vahl, *Arnebia hispidissima* DC., *Cordia dichotoma* Forst.f., *Withania somnifera* (Linn.) Dunal, *Striga angustifolia* (D.Don) Saldanha, *Adhatoda zeylanica* Medic., *Gmelina arborea* Roxb., *Nepeta hindostana* (Roth) Haines, *Salvia plebeia* R.Br., *Chenopodium ambrosioides* Linn., *Polygonum barbatum* Linn., subsp. *gracile* Danser, *Breynia rhamnoides* (Retz.) Muell.-Arg., *Emblica officinalis* Gaertn., *Ficus palmata* Forsk., *Salix tetrasperma* Roxb., *Ceratophyllum demersum* Linn., *Zeuxine strateumatica* (Linn.) Schlect., *Dioscorea bulbifera* Linn., *Gloriosa superba* Linn., *Monochoria vaginalis* (Burm.f.) Presl, *Juncus bufonius* Linn., *Typha angustata* Bory & Chaub., *Desmostachya bipinnata* (Linn.) Stapf, *Polypogon monspeliensis* (Linn.) Desf., *Echinochloa colonum* (Linn.) Link., *E. crusgalli* (Linn.) Beauv., *Urochloa panicoides* Beauv., *Narenga porphyrocoma* (Hance) Bor, *Saccharum spontaneum* Linn., *S. bengalense* Retz., *Eulaliopsis binata* (Retz.) Hubb., *Hackelochloa granularis* (Linn.) O. Ktze., *Mnesithea laevis* (Retz.) Kunth, *Sorghum halepense* (Linn.) Pers., *Heteropogon contortus* (Linn.) Beauv., *Vetiveria zizanioides* (Linn.) Nash etc.

A comparison with the flora of Delhi published by Maheshwari (1963, 1966) revealed that about 430 species are common to both the regions. Some of the interesting common plants are *Cissampelos pareira* Linn., *Cocculus hirsutus* (Linn.) Diels, *Tinospora cordifolia* Miers., *Tiliacora acuminata* (Lamk.) Miers, *Nymphaea nouchali* Burm.f., *N. stellata* Willd., *Hybanthus ennespermus* (Linn.) F. Muell., *Flacourtia indica* (Burm.f.) Merr., *Polygala erioptera* DC., *Stellaria media* (Linn.) Vill., *Tamarix dioica* Roxt., *Sida rhombifolia* Linn., *S. acuta* Burm.f., *Abutilon indicum* (Linn.) Sweet, *Urena lobata* Linn., *Helicteres isora* Linn., *Corchorus trilocularis* Linn., *C. aestuans* Linn., *C. capsularis* Linn., *Triumfetta rhomboidea* Jacq., *Feronia limonia* (Linn.) Swingle, *Ziziphus nummularia* (Burm.f.) W.&A., *Z. oenoplia* Mill., *Cayratia trifolia* (Linn.) Domin, *Cardiospermum halicacabum* Linn., *Crotalaria medicaginea* Lamk., *C. mysorensis* Roth, *Medicago polymorpha* Linn., *M. lupulina* Linn., *Indigofera linifolia* Retz., *Tephrosia pumila* Pers., *Desmodium triflorum* (Linn.) DC., *D. triquetrum* Baker, *D. gangeticum* (Linn.) DC., *Alysicarpus longifolius* W.&A., *Butea monsperma* (Lamk.) Taub., *Zornia gibbosa* Span., *Rhynchosia rothii* Benth. ex Ait., *Cassia pumila* Lamk., *C. fistula* Linn., *Potentilla supina* Linn., *Ludwigia perennis* Linn., *Trichosanthes cucumerina* Linn., *Glinus lotoides* Linn., *Borreria stricta* (Linn.) K. Schum., *B. hispida* K. Schum., *Centella asiatica* (Linn.) Urban, *Youngia japonica* (Linn.) DC., *Grangea maderaspatana* (Linn.) Poir., *Conyza canadensis* (Linn.) Cronquist, *C. bonariensis* (Linn.) Cronquist, *Pulicaria crispa* Gaertn., *Blumea mollis* (D.Don) Merr., *Bidens biternata* (Lour.) Merr., *Cotula hemisphaerica* (Roxb.) Wall. ex Benth., *Artemisia scoparia* Waldst. & Kit., *Amberboa ramosa* (Roxb.) Jafri, *Plumbago zeylanica* Linn., *Holarrhena antidysenterica* (Roth) A. DC., *Oxystelma secamone* (Linn.) K. Schum., *Dregea volubilis* Benth., ex Hook.f., *Leptadaenia reticulata* W.&A., *Hydrolea zeylanica* (Linn.) Vahl, *Heliotropium strigosum* Willd., *Arnebia hispidissima* (Lehm.) DC., *Ehretia laevis* Roxb., *Operculina turpethum* (Linn.) Silva-Manso, *Ipomoea muricata* (Linn.) Jacq., *I. carnea*

Jacq. var. *fistulosa* (Mart. ex Choisy) Austin, *I. eriocarpa* R.Br. *I. obscura* (Linn.) Ker-Gawl., *Physalis minima* Linn., *Verbascum thapsus* Linn., *Andrographis paniculata* (Burm.f.) Wall. ex Nees, *Hygrophila polysperma* (Roxb.) T. Anders., *Adhatoda zeylanica* Medic., *Phyla nodiflora* (Linn.) Greene, *Verbena officinalis* Linn., *Anisomeles indica* (Linn.) O. Kuntze, *Nepeta hindostana* (Roth) Haines, *Alternanthera pungens* H.B.&K., *Polygonum lanigerum* R.Br., *Polygonum barbatum* Linn. subsp. *gracile* Danser, *Jatropha gossypifolia* Linn., *Ficus palmata* Forsk., *Streblus asper* Lour., *Cannabis sativa* Linn., *Ceratophyllum demersum* Linn., *Zeuxine strateumatica* (Linn.) Schltz., *Asparagus racemosus* Willd., *Eichhornia crassipes* (Mart.) Solms, *Commelina benghalensis* Linn., *Lemna pauci-costata* Hegel, *Sagittaria guayanensis* H.B.&K., *Cyperus kyll-inga* Endl., *C. niveus* Retz., *C. difformis* Linn., *Elceocharis atropurpurea* (Retz.) Kunth, *E. palustris* R.Br., *Apluda mutica* Linn., *Phragmite karka* Trin., *Eragrostis tennella* (Linn.) R.&S., *Paspalum distichum* Linn., *Digitaria adscendens* (H.B.&K.) Henr., *Hemarthria compressa* (Linn.f.) R.Br., *Echinochloa crusgalli* (Linn.) Beauv., *Sorghum halepense* (Linn.) Pers., *Polypogon monspeliensis* (Linn.) Desf., *Panicum paludosum* Roxb., *Vetiveria zizanioides* (Linn.) Nash etc.

From the above comparison and affinities of the flora it is evident that the flora of Dudhwa National Park, Kheri district much resembles with the flora of Gorakhpur and Bahraich districts of Uttar Pradesh.

16. KEY TO THE FAMILIES OF ANGIOSPERMS

1. Leaves net-veined; flowers usually 4 to 5-merous; embryo with two cotyledons:

    2. Perianth biseriate i.e. calyx and corolla distinct:

        3. Petals free:

            4. Aquatic plants:

                5. Leaves notched, floating; ovary composed of many cells... ....5. **Nymphaeaceae**

                5. Leaves not notched raised above the surface of water; carpels distinct, sunk in a fleshy torus...6. **Nelumbonaceae**

4. Terrestrial plants:

6. Carpels many, apocarpous or reduced to single:

7. Herbs or climbing shrubs ....1. **Ranunculaceae**

7. Trees:

8. Perianth 2-seriate, 5-merous ....2. **Dilleniaceae**

8. Perianth 3-seriate, 3-merous ....3. **Annonaceae**

6. Carpels united; ovary syncarpous (except in Rosaceae and Menispermaceae):

9. Sepals distinct, free from ovary; ovary superior:

10. Disc inconspicuous; stamens borne below the ovary:

11. Flowers 3-merous; ovary apocarpous....4. **Menispermaceae**

11. Flowers 4 to 5-merous; ovary syncarpous:

12. Ovary unilocular:

13. Placentation parietal:

14. Leaves reduced to scales......16. **Tamaricaceae**

14. Leaves well developed:

15. Ovary borne on a gynophore.
.....................................10. **Capparaceae**

16. Flowers actinomorphic.7. **Papaveraceae**

16. Flowers zygomorphic:

17. Sepals 5; anterior petals spurred
..............................11. **Violaceae**

17. Sepals 2; lateral petals spurred.....
............................8. **Fumariaceae**

13. Placentation free-central:

18. Sepals 2; leaves fleshy................................
....15. **Portulacaceae**

18. Sepals 5; leaves not fleshy..14. **Caryophyllaceae**

12. Ovary bi - to multilocular:

19. Placentation parietal:

20. Stamens 6, tetradynamous; fruit a siliqua or silicula....................................9. **Brassicaceae**

20. Stamens numerous; fruit a berry or capsule.....12. **Flacourtiaceae**

19. Placentation axile:

21. Inner two sepals enlarged and often petaloid......13. **Polygalaceae**

21. Inner two sepals not as above:

22. Stamens monadelphous:

23. Anthers monothecous:

24. Leaves simple; carpels 5 to many ....19. **Malvaceae**

24. Leaves digitately compound; carpels 2-5...20. **Bombacaceae**

22. Stamens distinct, free:

25. Fruit a winged capsule ....18. **Dipterocarpaceae**

25. Fruit a capsule or drupe, not winged:

26. Stamens many; sepals valvate ....22. **Tiliaceae**

26. Stamens less than 10, in two bundles; sepals imbricate ....17. **Elatinaceae**

10. Disc a ring, cushion or gland; Stamens borne on the edge of the disc:

27. Placentation parietal; fruit an elongated capsule ....37. **Moringaceae**

27. Placentation axile; fruit not as above:

28. Styles 1-5, free ....24. **Oxalidaceae**

28. Styles more or less combined or solitary:

29. Leaves gland-dotted ....25. **Rutaceae**

29. Leaves not gland-dotted:

30. Stamens monadelphous ....29. **Meliaceae**

30. Stamens free, distinct:

31. Flowers unisexual or polygamous:

32. Fruit a drupe ....36. **Anacardiaceae**

32. Fruit a capsule, indehiscent, lobed or winged...35. **Sapindaceae**

31. Flowers bisexual or polygamous:

33. Leaves simple:

34. Herbs or undershrubs:

35. Flowers white, solitary axillary ....30. **Olacaceae**

35. Flowers yellow, in an umbel ....27. **Ochnaceae**

34. Shrubs or trees:

36. Stamens opposite the petals:

37. Plants spiny ....32. **Rhamnaceae**

37. Plants not spiny:.

38. Climbing shrubs with tendrils ....33. **Vitaceae**

38. Erect shrubs without tendrils. ....34. **Leeaceae**

36. Stamens alternate the petals ....31. **Celastraceae**

33. Leaves compound:

39. Trees; fruits not spiny:

40. Fruit a samara ....26. **Simaroubaceae**

40. Fruit a drupe ....28. **Burseraceae**

39. Herbs; fruits spiny ....23. **Zygophyllaceae**

9. Sepals united, often adnate to ovary, ovary half inferior or inferior:

41. Carpel one; fruit a pod or legume:

42. Flowers actinomorphic; petals valvate; stamens numerous.......... ....40. **Mimosaceae**

42. Flowers zygomorphic; petals imbricate; stamens definite:

43. Corolla descending imbricate; stamens diadelphous.............. ....38. **Fabaceae**

43. Corolla ascending imbricate; stamens free.. ....39. **Caesalpiniaceae**

41. Carpels more than one; fruit not a pod or legume:

44. Stamens usually indefinite:

45. Leaves reduced; plants succulent with jointed phylloclades, spiny. ....50. **Cactaceae**

45. Leaves well-developed; plants not as above:

46. Carpels distinct, apocarpous; leaves stipulate................. ....41. **Rosaceae**

46. Carpels united, syncarpous; leaves exstipulate:

47. Calyx free from the ovary:

48. Petals present, crumpled in the bud; ovary 2 to 6-locular........ ....45. **Lythraceae**

48. Petals absent; ovary 1 to 2-locular:

49. Leaves opposite, fleshy ....52. **Aizoaceae**

49. Leaves in false whorls ....51. **Molluginaceae**

47. Calyx more or less adnate to the ovary:

50. Leaves opposite, gland-dotted, with intra marginal vein ....43. **Myrtaceae**

50. Leaves alternate, except (Quisqualis-opposite) not gland-dotted:

51. Fruit a drupe, leathery ....42. **Combretaceae**

51. Fruit a berry ....44. **Lecythidaceae**

44. Stamens definite, not more than 10:

52. Plants aquatic; fruits spiny ....46. **Trapaceae**

52. Plants terrestrial; fruits not spiny:

53. Herbs, twiners or climbers:

54. Plants with tendril:

55. Flowers bisexual; gynophore present .....49. **Passifloraceae**

55. Flowers unisexual; gynophore absent ....48. **Cucurbitaceae**

54. Plants without tendrils:

56. Flowers in simple or compound umbels ....53. **Apiaceae**

56. Flowers solitary, axillary ....46. **Onagraceae**

53. Trees ....54. **Alangiaceae**

3. Petals united:

57. Ovary inferior:

58. Anthers united; ovary unilocular; inflorescence capitulum............ ....56. **Asteraceae**

58. Anthers free; ovary unilocular inflorescence not as above:

59. Leaves opposite with intra petiolar stipules ......55. **Rubiaceae**

59. Leaves alternate, exstiputate:

60. Amphibious herbs; flowers in dense cylindrical spikes ....58. **Sphenocleaceae**

60. Terrestrial herbs; flowers solitary or in panicled clusters..57. **Campanulaceae**

57. Ovary superior:

61. Leaves absent; stem-parasites ....74. **Cuscutaceae**

61. Leaves present; plant not a parasite:

62. Plants carnivorous aquatics; leaves pinnately divided into capillary segments ....77. **Lentibulariaceae**

62. Plant not carnivorous; terrestrials:

63. Leaves generally alternate:

64. Flowers unisexual; stamens inserted on the receptacle..63. **Ebenaceae**

64. Flowers bisexual; stamens inserted on the corolla:

65. Ovules numerous in each locule:

66. Styles usually deeply 2-cleft...70. **Hydrophyllaceae**

66. Styles linear, not deeply 2-cleft ....75. **Solanaceae**

65. Ovules definite in each locule:

67. Styles 5; placentation free-central ....59. **Plumbaginaceae**

67. Styles not 5; placentation otherwise:

68. Habit chiefly climbing or twining ....73. **Convolvulaceae**

68. Habit erect or prostrate:

69. Herbs and undershrubs:

70. Fruits of 4-seeded nutlets; style ....gynobasic 71. **Boraginaceae**

70. Fruit a globose drupe, style not gynobasic ....61. **Myrsinaceae**

69. Trees:

71. Fruit a berry; corolla 4-9 lobed...62. **Sapotaceae**

71. Fruit a drupe or of 4 nutlets ....72. **Ehretiaceae**

63. Leaves generally opposite:

72. Flowers actinomorphic; ovary of 2-carpels, sometimes 1 or 3:

75. Milky latex present in plants; fruit a pair of follicle:

76. Styles 2; corona present, gynostegium present..67. **Asclepiadaceae**

76. Style 1; corona absent; gynostegium absent..66. **Apocynaceae**

75. Milky latex absent in plants; fruits not as above:

77. Stamens 2...64. **Oleaceae**

77. Stamens 4-5:

78. Herbs:

79. Stamens alternate to petals; placentation parietal...69. **Gentianaceae**

79. Stamens opposite to petals; placentation free-central...60. **Primulaceae**

78. Trees and shrubs:

80. Ovary usually unilocular; ovules 1-2; placentation basal ....65. **Salvadoraceae**

80. Ovary usually bilocular; ovules many in each locule; placentation axile ....68. **Loganiaceae**

72. Flowers zygomorphic; ovary of 2-4 carpels:

81. Trees; fruit an elongated capsule; seeds winged..78. **Bignoniaceae**

81. Herbs; fruits not elongated; seeds not winged:

82. Fruits spiny and hooked:

83. Placentation axile ....79. **Pedaliaceae**

83. Placentation parietal ....80. **Martyniaceae**

82. Fruits not as above:

84. Styles gynobasic ....83. **Lamiaceae**

84. Styles terminal:

85. Bracts conspicuous; stem with swollen nodes...81. **Acanthaceae**

85. Bracts minute or absent; stem without swollen nodes:

86. Ovules 1 or 2 in each locule; placenta not swollen....82. **Verbenaceae**

86. Ovules many in each locule; placenta swollen.....76. **Scrophulariaceae**

2. Perianth uniseriate or absent:

87. Perianth absent; stamens in 2 whorls ....89. **Piperaceae**

87. Perianth 1-seriate; stamens usually in one whorl:

88. Flowers hermaphrodite (unisexual in *Litsea):*

89. Parasitic plants; ovary inferior ....92. **Loranthaceae**

89. Non-parasitic plants; ovary superior:

90. Stamens opening by valves ....90. **Lauraceae**

90. Stamens not opening by valves:

91. Leaves with ocreate stipules...........88. **Polygonaceae**

91. Leaves exstipulate:

92. Herbs or shrubs:

93. Perianth bract-like:

94. Leaves not succulent; filaments connate; bracts scarious ....85. **Amaranthaceae**

94. Leaves succulent; filaments not connate; bracts not scarious ....86. **Chenopodiaceae**

92. Climbers or trees:

93. Perianth petaloid:

95. Succulent plants; flowers in spikes, racemes or panicles...........87. **Basellaceae**

95. Non-succulent plants; flowers in heads, cymes or umbels..........84. **Nyctaginaceae**

88. Flowers unisexual or polygamous:

96. Male and female flowers in catkins or spikes:

97. Sap milky..97. **Moraceae**

97. Sap watery:

.............................96. **Urticaceae**

98. Trees ....98. **Salicaceae**

96. Flowers not in catkins or spikes:

99. Aquatics; leaves finely dissected ....99. **Ceratophyllaceae**

99. Terrestrials; leaves not as above;

100. Ovary trilocular, with 1 or 2 ovules in each locule.................. ....93. **Euphorbiaceae**

100. Ovary unilocular:

101. Plants aromatic; fruit an achene ....95. **Cannabaceae**

101. Plants not aromatic; fruit a samara ....94. **Ulmaceae**

1. Leaves parallel-veined (except in *Smilax* and *Dioscorea);* flowers 3-merous; embryo with one cotyledon:

102. Perianth absent or reduced to scales or bristles, not petaloid:

103. Florets subtended by glumes or scales:

104. Inflorescence a spikelet:

105. Culms solid, not noded, triquetrous or terete; leaf-sheath not split; fruit an achene or urticle ....119. **Cyperaceae**

105. Culms hollow, noded, cylindrical or compressed; leaf-sheath split; fruit caryopsis...................120. **Poaceae**

104. Inflorescence a head or panicled corymb:

106. Flowers in heads;style 1 ....118. **Eriocaulaceae**

106. Flowers in panicled cyme; style 1-3.....109. **Juncaceae**

103. Flowers not subtended by glumes:

107. Plants woody; leaves pinnately or palmately divided......... ....110. **Arecaceae**

107. Plant herbaceous; leaves various, not as above:

108. Plants minute, thalloid, free-floating ...113. **Lemnaceae**

108. Plants with differentiated leaves:

109. Flowers solitary ....100. **Hydrocharitaceae**

109. **Flowers in spikes:**

110. Flowers scattered on spikes: ....117. **Potamogetonaceae**

110. Flowers densely arranged on spikes:

111. Flowers subtended by an usually large bract-like spathe, fleshy........112. **Araceae**

111. Flowers subtendec by caducous bract-like spathe, dry ....111. **Typhaceae**

102. Perianth petaloid at least in part, not reduced to bristles or scales:

112. Carpels one-many, apocarpous:

113. Perianth biseriate, 3 sepaloid and 3 petaloid:

114. Carpels 1-seeded ....114. **Alismataceae**

114. Carpels more than 1-seeded ....115. **Butomaceae**

113. Perianth uniseriate,1-3, all petaloid......116. **Aponogetonaceae**

112. Carpel 1, united:

115. Ovary superior:

116. Perianth with 3 petaloid tepals..........108. **Commelinaceae**

116. Perianth of 6 petaloid tepals:

117. Plants aquatic ....107. **Pontederiaceae**

117. Plants terrestrial:

118. Plants woody climbers; leaves 3-5 nerved, reticulately veined; tendril present; flowers unisexual ....106. **Smilacaceae**

118. Plants herbaceous; leaves with parallel venation; tendril absent; flowers bisexual........105. **Liliaceae**

115. Ovary inferior:

119. Flowers actinomorphic:

120. Twining plants; flowers arranged in spikes or racemes ....104. **Dioscoreaceae**

120. Herbs; flowers arranged in scape..103. **Amaryllidaceae**

119. Flowers zygomorphic:

121. Androecium and gynoecium confluent in a column; ovary twisted.................................101. **Orchidaceae**

121. Andrecium and gynoecium not confluent; ovary not twisted.......................................102. **Zingiberaceae**

17. SYSTEMATIC TREATMENT OF DICOTYLEDONS

1. RANUNCULACEAE

1. Herbs with alternate leaves; flowers yellow ....**Ranunculus**

1. Cliimbing shrubs with opposite leaves; flowers greenish-white:

2. Leaves terminating in tendrils; fruit a head of tailed achenes .... **Naravelia**

2. Leaves not terminating in tendrils; fruit a head of achene with feathery style .... **Clematis**

Ranunculus Linn.

1. Plants glabrous; achenes turgid and not margined .... **R. scleratus**

1. Plants hairy; achenes flattened, provided with an intermarginal rib.. **R. cantoniensis**

R. scleratus Linn. Sp. Pl. 551. 1753; FBI 1:19; FUGP 1:21; Mukerjee in Bull. Bot. Surv. India 2: 105. 1960; Srivastava, FG. 30. 1976.

Erect glabrous, succulent annual herbs, 30 cm long, with ribbed branches. Radical leaves long-petiolate, 3-partite with obovate, cuneate segments. Cauline leaves shortly petiolate, trifid with linear segments. Flowers yellow with reflexed sepals. Achenes many in oblong heads, turgid, not margined, obliquely obovate.

Local name : Jaldhania
Flowers and Fruits: March-June

Occurs throughout the area in wet and marshy places and along river beds. Singh, 2810, 7120, 7226 (CDRI); Saran 26409 (LWG).

R. cantoniensis DC. Prodr. 1: 43. 1824; Hand. Mzt. in Act. Hort. Gotob. 13: 165. 1939; Hara in J. Jap. Bot. 18: 459. 1942 et 30: 322. 1955; Mukerjee in Bull. Bot. Surv. India 2: 102. 1960. *R. pensylvanicus* Hook.f. & Thoms. in FBI 1: 19. 1872 (non Linn.f. 1781); FUGP 1:22.

Erect hairy annual herbs, 30-40 cm tall. Radical leaves long-petiolate, three-partite; segments deeply divided, linear; uppermost leaves sessile with sheathing base; leaflets hairy, lobed and coarsely serrate. Flowers bright yellow with reflexed sepals. Achenial heads up to 1 cm long, globose-ovoid with glabrous, distinctly margined and flattened achenes.

Flowers and Fruits: April-May

Occurs rarely in Bankati range of the Park in wet and muddy soils along the Indo-Nepal Border. Singh 7164 (CDRI).

Naravelia DC.

N. zeylanica (Linn.) DC. Syst. 1:167. 1817; FBI 1:7; P.C. Kanjilal, For. Fl. UP. 2: 3. 1966; Raizada, Suppl. FUGP 2. 1976. *Atragene zeylanica* Linn. Sp. Pl. 542. 1753.

Woody climbers, with adpressed pubescent branches. Young parts silky tomentose. Leaves opposite with a single pair of leaflets; the rachis ending in a long three-fid tendril. Leaflets *ca* 10.5 x 5 cm, ovate, acuminate, entire or distinctly toothed, tomentosely villous beneath. Flowers greenish-white, in axillary and terminal panicles. Achenes *ca* 4.5 cm long, twisted, thin tailed with feathery fine spreading hairs.

Flowers and Fruits: September-November.

Found frequently on small trees forming a canopy along forest roads. Singh 4254, 7950 (CDRI)

Clematis Linn.

C. gouriana Roxb. ex DC. Syst. 1: 138. 1817; FBI 1:4; FUGP 1: 19; Santapau in RBSI 16(1):1. 1953; P.C. Kanjilal, For. Fl. UP. 2: 2. 1966; Srivastava, FG. 30. 1976.

Climbing shrubs; young parts adpressed pubescent. Leaves 2-pinate with entire or coarsely toothed, ovate, acuminate leaflets. Leaflets 5-6 x 5 cm, membranous, glabrous. Flowers cream-coloured or white, in axillary and terminal trichotomous panicles. Achenes *ca* 4 cm long, hairy with a slender feathery style.

Flowers and Fruits : September-October.

Occurs occasionally on shrubs and small trees along forest roads. Singh 4287, 7273, 7773, 7925 (CDRI).

2. DILLENIACEAE

Dillenia Linn.

D. pentagyna Roxb. Pl. Corom. 1:21, t. 20. 17. 1795; FBI 1: 38; FUGP 1: 23; Hoogl. in Fl. Males. 1, 4:172. 1951 et Blumea 7(1): 117. 1952; P.C. Kanjilal, For. Fl. UP. 2: 6. 1966.

Moderate-sized, deciduous trees. Leaves 30-60 x 20-22 cm, oblong-lanceolate or elliptic, acute or obtuse, denticulate, decurrent into a short sheathing petiole, lateral nerves many paired. Flowers yellow, in umbels along the leafless branches.

Local name : Aggai
Flowers and Fruits: March-May

Found occasionally in sal forest in damp situations. The leaves are used for thatching. Singh 7519 (CDRI).

3. ANNONACEAE

1. Carpels subconnate; tepals in two series or the third series very small; fruit a fleshy syncarp...**Annona**

1. Carpels free; tepals in three series; fruit a drupel:

    2. Outer petals longer than the sepals; connective concealing the anther cells...**Polyalthia**

    2. Outer petals small, similar to sepals; connective not concealing the anther cells ....**Miliusa**

Annona Linn.

A. squamosa Linn. Sp. Pl. 537. 1753; FBI 1: 78; FUGP 1:25; Benthall, Trees Calc. 12; P.C. Kanjilall For. Fl. UP. 2: 13. 1966; Srivastava, FG. 32. 1976.

Shrubs or small trees. Leaves oblong-lanceolate, 7-10 x 2-3 cm, glabrous. Flowers greenish-white, drooping, solitary or 2-4 together on short leaf opposed peduncles. Fruits fleshy, syncarp, globular, yellowish-green, pulp sweet. Seeds black.

Local name - Sharifa
Flowers and Fruits : April-November.

Occasionally cultivated in the district. The leaves possess insecticidal properties. The fruits are edible. Singh 7742 (CDRI)

Polyalthia Blume

P. suberosa (Roxb.) Thw. Enum. 398. 1814; Sinclair in Gdns. Bull. 14:298. 1955; Santapau & Wagh, in Bull. Bot. Surv. India 5:2:108. 1963; P.C. Kanjilal, For. Fl. UP. 2:9.1966. *Uvaria suberosa* Roxb. Pl. Corom. 1: t. 34,1795 et Fl. Ind. 2:567. 1824. *Polyalthia suberosa* (Benth. & Hook.) Hook.f. & Thoms. in FBI 1:65. 1872; FUGP 1:24.

Small trees or shrubs. Young parts rusty-villous. Leaves alternate, 5-7 x 2.5-3 cm, shortly petiolate, oblong or oblong-obovate, obtuse, membranous, glabrous, usually deep green above. Flowers drooping, greenish-yellow on slender pedicels. Fruits in clusters of rounded drupels, sparsely hairy, 1 to 2-seeded.

Flowers and Fruits: March-June.

Found frequently along the edge of the sal forest. Singh 4313, 7528, 8438 (CDRI).

Miliusa Leschen.

1. Ovules two in each carpel; inner series of petal base not saccate. .... **M. velutina**

1. Ovules six in each carpel; inner series of petal base saccate **M.tomentosa**

M. velutina (Dunal) Hook.f. & Thoms. Fl. Ind. 1: 151. 1835 et in FBI 1: 87; King, in Ann. R. Bot. Gdns. Calcutta 4:158. t. 206, 1893; Finet & Gagnep. in Bull. Soc. Bot.

Fr. 53: 4: 150. 1906; P.C. Kanjilal, For. Fl. UP. 2:11. 1966. *Uvaria velutina* Dunal, Mon. Fam. Anon. 91, 1817.

Medium-sized trees. Young branches, leaves and inflorescence densely fulvous tomentose. Leaves 7-10 x 5 cm, ovate-oblong or subrounded, shortly apiculate, softly tomentose on both surfaces. Flowers solitary or 2-3 from a leaf opposed peduncle, greenish-yellow on slender pedicel. Sepals 3, Petals 6 in two series. Stamens numerous slightly apiculate. Carpels many inserted towards the top of the torus, two ovuled.

Local name : Choti-Kari
Flowers and Fruits : March-August.

Occurs frequently within the sal forest. The wood is used for timber. Singh 7346 (CDRI)

M. tomentosa (Roxb.) Finet & Gagnep. in Bull. Soc. Bot. Fr. 53: 4: 1153, 1906; Sinclair, Gdns. Bull. 14:378. 1955; Raizada in Indian For. 84:8:497. 1958; P.C. Kanjilal, For. Fl. UP. 2:12.1966; Srivastava, FG. 33. 1976. *Uvaria tomentosa* Roxb. Pl. Corom. 1. t. 35, 1795 et Fl. Ind. 2:667.1824; W.&A. Prodr. 8:1834. *Saccopetalum tomentosum* Hook.f. & Thoms. Fl. Ind. 1: 152. 1855; FBI 1:88; FUGP 1:26; Santapau in RBSI 16(1):2. 1953.

Moderate-sized deciduous trees; young parts tomentose. Leaves 16-22 x 11-12.5 cm, thickly membranous, ovate or ovate-oblong, shortly acuminate, softly hairy when young. Flowers leaf-opposed, 1-2 together on slender pedicels. Sepals three, Petals 6 in two series, inner usually saccate at the base, dark purple. Stamens many. Fruits in umbels of rounded fleshy carpels. Ovules 6 in each carpel.

Local name : Barikari
Flowers & Fruits : April-August.

Commonly met within the sal forest. It is used as timber wood. Singh 4106, 4135 (CDRI)

4. MENISPERMACEAE

1. Leaves peltate; carpel one .... Cissampelos
1. Leaves not peltate; carpels three or more:
   2. Seeds ovoid or reniform .... Tinospora
   2. Seeds horse-shoe-shaped:
      3. Leaves ovate-oblong, villous .... Cocculus
      3. Leaves ovate-cordate, glabrous .... Tiliacora

Cissampelos Linn.

C. pareira Linn. Sp. Pl. 1031. 1753; FBI 1:104; FUGP 1:31; Santapau in RBSI 16(1): 5. 1953; P.C. Kanjilal, For. Fl. UP. 2:22.1966; Rhodes in Phytologia 30(6): 446. 1975; Srivastava, FG. 34. 1976.

Twining shrubs with slender, softly grey hairs. Leaves 7-10 x 4-6 cm, triangularly broad cordate, suborbicular or peltate, mucronate. Male flowers in axillary branched slender cymes, densely tomentose. Female flowers 1-2, in the axil of orbicular accrescent bracts arranged on axillary racemes. Drupes 4-5 mm subglobose, hispid, red when ripe. The leaves are very variable in size, shape and degree of hairiness.

Local name : Porainparhi

Flowers and Fruits : August-November

Found frequently twining on hedges and on small trees in open situation. The plant is used in stomach-pain, fever and cuts. Singh 2980, 4311, 7447, 8398 (CDRI) Saran 26208, 26639 (LWG).

Tinospora Miers

1. Leaves glabrous; drupels pisiform .... T. cordifolia
1. Leaves tomentose; drupels rounded .... T. sinensis

T. cordifolia (Willd.) Miers in Ann. & Mag. nat. Hist. ser. 2,7:38.1851; FBI 1:97; FUGP 1:27; Santapau in RBSI

16(1):3.1953; P.C. Kanjilal, For. Fl. UP. 2:16.1966; Srivastava, FG. 35.1976. *Menispermum cordifolium* Willd. Sp. Pl. 4:826.1806.

Glabrous, climbing succulent shrubs. Leaves 9.5 x 7.3 cm, long petiolate, cordate, shortly acuminate, glabrous. Male flowers in fascicles. Female flower solitary. Drupels pisiform.

Local name : Giloe, Guruch
Flowers and Fruits : August-September

Occurs occasionally on trees near the edge of the forests. The plant is used in fever. Singh 7509, 7649, 8416 (CDRI) Saran & Party 2619 (LWG).

T. sinensis (Lour.) Merr. in Sunyatsenia 1:193. 1934 et in Trans Am. Philosop. Soc. Philad. n.s. 24: 158. 1935; Raizada, Suppl. FUGP 4. 1976. *Campylus sinensis* Lour. Fl. Cochinch. 113. 1790. *Tinospora malabarica* Miers in Ann. Nat. Hist. II 7:38.1851, Hook.f., & Thoms. in FBI 1:96.1872; P.C. Kanjilal, For. Fl. UP. 2:17.1966.

Large, soft wooded climbers. Stems greyish-green, covered with large lenticels. Branches and young parts hairy. Leaves *ca* 19 cm long, rounded cordate, apex acuminate, grey-tomentose beneath. Flowers greenish-yellow. Male flowers 1-4 in fascicles from the axil of bracts. Female flower solitary. Carpels 3. Drupels round or broadly ovoid, scarlet when ripe.

Local name : Guruch
Flowers and Fruits : March-June

Found frequent on bushes and small trees in open areas of the Park. Singh 7272, 7569 (CDRI).

Cocculus DC. *nom. cons.*

C. hirsutus (Linn.) Diels in Pfreich. 46:236.1910; Santapau in RBSI 16(1): 5.1953; P.C. Kanjilal, For. Fl. UP.

2:20.1966; Srivastava, FG. 35. 1976. *Menispermum hirsutum* Linn. Sp. Pl. 341. 1753. *Cocculus villosus* DC. Syst. 1:525. 1818; FBI 1:101; FUGP 1:29.

Slender, densely villous climbers. Leaves triangular to ovate-oblong, 3-5 x 2-3 cm, dark green above, densely villous beneath, apex mucronate. Male flowers in short axillary panicles. Female flowers 1-3 on short peduncles. Drupe dark purple, compressed, subreniform.

Flowers and Fruits : February-April

Occurs occasionally on hedges and small trees in open situation. The plant is used in ayurvedic medicine. Singh 4367, 8302 (CDRI)

Tiliacora Colebr.

T. acuminata (Lamk.) Miers in Ann. & Mag. nat. Hist. ser. 2,7:39.1851; Diels in Pfreich. 46:61; P.C. Kanjilal, For. Fl. UP. 2:18.1966; Srivastava, FG. 35. 1976. *Menispermum acuminatum* Lamk. Encycl. 4:101. 1797. *Tiliacora racemosa* Colebr. in TLS 13:67.1822; FBI 1:99; FUGP 1:28.

Evergreen woody climbers with glabrous branches. Leaves 9 x 6-9 cm, ovate, acuminate, dark green, glabrous. Flowers yellow, unisexual in axillary panicles. Carpels 3, surrounded by a minute staminode. Drupels subglobose, usually 2-3 together, yellow or balck when ripe.

Local name : Karwanth
Flowers and Fruits : April-November

Found commonly in the forest climbing on sal trees and also spreading on the ground. The twigs are used for making basket. Singh 3047, 4228, 7367, 7780, 7994D (CDRI) Saran 26515, 6849 (LWG).

5. NYMPHAEACEAE

Nymphaea Linn. *nom. cons.*

1. Sepals ribbed; anthers without appendages ....**N. nouchali**

1. Sepals not ribbed; anthers with appendages..............................**N. stellata**

N. nouchali Burm.f. Fl. Ind. 120. 1768; Merr. in Phillip. J. Sci. (Bot.) 19:350. 1921 et Enum. 2:140.1923; Srivastava, FG. 37. 1976. *N. pubescens* Willd. Sp. Pl. 2:1154.1799; Santapau in RBSI 16(10)1953; P.C. Kanjilal, For. Fl. UP. 2:24. 1966. *N. lotus* Hook.f. & Thoms. in FBI 1:114. 1872 (non Linn.)

Aquatic herbs. Leaves deeply cordate, suborbicular, sharply toothed, pubescent beneath. Flowers white with cream centre. Sepals ribbed. Petals oblong. Anthers without appendages. Stigmatic rays with clubbed appendages. Seeds broadly ellipsoid.

Local name : Kokaveli, Kumodini
Flowers and Fruits : August-October

Found commonly in ponds and Jhils during the rainy season. The flowers remain open up to 1 P.M. during the day. Singh 7507 (CDRI)

N. stellata Willd. Sp. Pl. 2:1153.1799; Biswas & Calder 21, t. 1.f.3. 1955; FBI 1:114; FUGP 1:34; P.C. Kanjilal, For. Fl. UP. 2:24.1966; Srivastava, FG. 37. 1976.

Aquatic herbs. Leaves orbicular or peltate, entire or wavy, often blotched with purple beneath. Flowers white or blue. Sepals oblong-lanceolate, not ribbed. Petals narrow. Anthers tipped with long appendages. Fruits globular.

Local name : Chota Kamal
Flowers and Fruits : August-October

Occurs occasionally in ponds and lakes, often mixed with *N. nouchali* Burm.f. The tubers are cooked as vegetable Singh 7508, 7696 (CDRI)

6. NELUMBONACEAE

Nelumbo Adans.

N. nucifera Gaertn. Fruct. 1:73.t.19, f.2. 1788; Santapau in RBSI 16(1):7. 1953; P.C. Kanjilal, For. Fl. UP. 2:25.1966; Srivastava, FG. 36. 1976. *Nymphaea nelumbo* Linn. Sp. Pl. 511. 1753.
Nelumbium speciosum Willd. Sp. Pl. 2:1258.1799; FBI 1:116; FUGP 1:35.

Aquatic herbs with milky juice and horizontal creeping root stocks. Leaves 60-90 cm about a meter high above the water surface, orbicular or peltate. Flowers rose-coloured, 10-16 cm across, fragrant and solitary. Carpels sunk in the torus, oblong. Seeds with spongy seed coat.

Local name : Kamal
Flowers and Fruits : April-October

Found frequently in Jhils and ponds. The seeds are used as tonic. The rootstocks are eaten as vegetable Singh 3007 (CDRI).

7. PAPAVERACEAE

Argemone Linn.

1. Flower yellow ....**A. mexicana**

1. Flower white or cream coloured ....**A. ochroleuca** subsp. **ochroleuca**

A. mexicana Linn. Sp. Pl. 508. 1753; FBI 1:117; FUGP 1:37; van Steenis in Fl. Males. 5(1):114. 1953; Srivastava, FG. 38. 1976.

Prickly robust herbs or undershrubs. Leaves 8-9 cm long, sessile, amplexicaul, spiny, sinuate-pinnatifid, glaucous. Flowers yellow. Ovary covered with soft spines; stigmas red. Capsules erect, 3.5 cm long, dehiscing by valves. Seeds black.

Local name : Bharbhand
Flowers and Fruits : February-October

Occurs commonly in waste places along roads and in fields. The oil from the seeds is applied medicinally on skin diseases and for burning purpose. The latex is used in eye troubles and jaundice. Singh 7148 (CDRI); Saran & Party 26446, 26730 (LWG).

A. ochroleuca subsp. **ochroleuca** Jafri & Qaiser in Fl. W. Pak. 61:22.1974. A ochroleuca Sweet, Brit. Fl. Gard. 3:t, 242. 1828; Ownbey in Mem. Torr. Bot. Cl. 21. 1958; Srivastava, FG. 38. 1976; Raizada, Suppl. FUGP 5. 1976.

Similar to the preceding species except that it has white or lemon-yellow flowers. It has been reported cytologically as a distinct species, being octoploid (Malhotra, 1960).

Flowers and Fruits : February-October

Found occasionally on wastelands, often associated with *A. mexicana* Linn. Singh 76100 (CDRI).

8. FUMARIACEAE

Fumaria Linn.

F. indica (Haussk.) Pugsley in JLS 44:313.1919; Srivastava, FG. 38. 1976. *F. vaillantii* Loisel. var. *indica* Haussk. in Flora 56:443. 1873.
F. parviflora W.&A. Prodr. 18.1834 (non Lamk.)
F. parviflora Lamk. subsp. *vaillantii* Hook.f. in FBI 1:128. 1872; FUGP 1:37. (non F. vaillantii Loisel 1809).

Diffuse, much-branched, glaucous annuals. Latex watery. Leaves finely divided with flat segments. Flowers rosy-purple, about 6 x 2 mm. Sepals 2, minute. Petals 4, outer 2, the anterior of which flat, posterior spurred at the base, 2 inner narrow, cohering by their tips, keeled at the back. Fruits globose, indehiscent, 1-seeded.

Local name: Papara
Flowers and Fruits : November-February

Occurs commonly in cultivated fields and in forest plantations on moist sandy soil. Singh 7646 (CDRI).

9. BRASSICACEAE (*nom. alt.* CRUCIFERAE)

1. Pods dehiscent:

2. Flowers yellow; pod narrow cylindrical ....**Rorippa**

2. Flowers greenish-yellow with purple vein; pod oblong-ovoid ...**Eruca**

1. Pods indehiscent, 2-lobed ....**Coronopus**

Rorippa Scop.

1. Leaves toothed ....**R. indica**

1. Leaves entire ....**R. montana**

R. indica (Linn.) Hiern, Cat. Afr. Pl. Welw. 26, Addit. & Corr. 26. 1898; Santapau & Wagh in Bull. Bot. Surv. India 5:108.1963.
Sisymbrium indicum Linn. Mant. 1:93.1767.
Nasturtium indicum (Linn.) DC. Prodr. 1:139.1824; FBI 1:134; FUGP 1:39.pp.

Erect, glabrous or hairy annuals. Radical leaves pinnatifid, lobes usually toothed. Upper leaves lyrate. Flowers minute, yellow in racemes. Pods 1.2-2 cm long, narrow cylindrical.

Flowers and Fruits : February-August

Found occasionally in moist places in waste ground. Singh 2933, 4306, 7330 (CDRI).

R. montana Small, Fl. S.E. U.S. 1336, 1913. *Nasturtium montanum* Wall. ex Hook.f. & Thoms. in JLS 5:139. 1861; FBI 1:134; FUGP 1:39.

Erect, branched herbs. Stems slightly hairy. Radical leaves sinuate, pinnatifid, lobes entire. Flowers yellow minute, bracteate. Pods 2-4 cm long, narrow cylindrical.

Flowers and Fruits : February-August

Occurs occasionally in damp places in wastelands. Singh 7275 (CDRI).

Eruca Adans.

E. sativa Mill. Gard. Dict. ed. 8.n.1.1768; FBI 1:158; FUGP 1:45. Brassica eruca Linn. Sp. Pl. 667. 1753.

Erect branched herbs. Leaves lyrate-pinnatifid, margin entire or coarsely toothed. Petioles winged. Flowers greenish-yellow with purple veins. Petals clawed. Pods ovoid-oblong, 2.5 cm long, turgid provided with flat beak. Seeds reddish-brown.

Local name : Tara-mira
Flowers and Fruits : February-March

Found intermixed with cold season crops in the fields and as an escape. The seeds yield an oil which is used for burning. The plant is used as fodder. Singh 7616 (CDRI).

Coronopus Zinn *nom. cons.*

C. didymus (Linn.) Sm. Fl. Brit. 2:691. 1800; Exell, Cot. Vas. Pl. S. Tome. 105. 1944; Srivastava, FG. 41. 1976. *Lepidium didymum* Linn. Syst. 2:344.1754. *Senebiera pinnatifida* DC. in Mem. Soc. Hist. Nat. Paris 144.t.19.1799; FUGP 1:46.

Prostrate, smelling slightly hispid herbs. Leaves pinnatifid or pinnate-partite, lobes spreading. Flowers pale-green, minute, diandrous. Pods 1 x 2 mm, indehiscent, laterally compressed. Seeds reniform, brown.

Flowers and Fruits : January-April

A winter season weed, found occasionally in the moist places on sandy soil. Singh 7792 (CDRI).

10. CAPPARACEAE (nom. alt. CAPPARIDACEAE)

1. Erect herbs with capsular fruits..**Cleome**

1. Climbing shrubs with berry fruits ....**Capparis**

Cleome Linn.

1. Gynophore present ....**C. gynandra**

1. Gynophore absent ....**C. viscosa**

C. gynandra Linn. Sp. Pl. 671. 1753; Jacobs in Fl. Males. (ser.1)6: 101. 1960; Ernest in JAA 44:88.1963; Srivastava, FG. 42. 1976. *Gynandropsis gynandra* (Linn.) Briq. in Ann. Cons. Jard. Bot. Geneve 17:382. 1914; Santapau in RBSI 16(1); 10. 1953. *G. pentaphylla* DC. Prodr. 1:238. 1824; FBI 1:171; FUGP 1:49.

Erect glandular pubescent herbs upto 70 cm long. Leaves digitately 5-foliolate, long petiolate. Leaflets 5-7 x 3-3.5 cm, unequal, obovate. Flowers white or cream-coloured, in corymbose racemes, very viscid. Sepals erect. Gynophore present. Capsules 4-6 x 0.3-0.4 cm.

Flowers and Fruits : June-September

Found frequently in the wasteland and in cultivated fields. Singh 4382 (CDRI).

C. viscosa Linn. Sp. Pl. 672.1753; Iltis in brittonia 12:281. 1960; FBI 1:170; FUGP 1:48; Jacobs in Fl. Males. ser. 1; 6:103. 1960; Srivastava, FG. 43. 1976. *C. icosandra* Linn. Sp. Pl. 672. 1753. *Polanisia viscosa* DC. Prodr. 1:242. 1824.

Erect, branched, viscid glandular herbs, 30-60 cm long. Leaves 3 to 5-foliolate; leaflet sessile, ovate or obovate. Flowers long pedicelled, yellow, solitary axillary or in leaf bearing terminal racemes. Stamens 15-20. Capsules sticky pubescent, subarched, short beaked.

Local name : Hurhur
Flowers and Fruits : August-September

Occurs commonly on wastelands and in fields. An oil is extracted from the seeds. Singh 2986, 4215 (CDRI).

Note: De Candolle (Prodr. 1:242.1824) was the first to unite *C. viscosa* and *C. icosandra* under the genus *Polanisia* Raf. and adopted its former epithet `viscosa' for the combined species. Therefore, *C. viscosa* Linn. is the correct name for this plant, according to the Art. 57 of the ICBN 1966.

Capparis Linn.

C. zeylanica Linn. Sp. Pl. (ed.2) 720. 1762; Dunn in Kew Bull. 62. 1916; Jacobs in Fl. Males. 6:87. 1960; P.C. Kanjilal, For. Fl. UP. 2:29. 1966; Srivastava, FG. 41. 1976. (Plate VIII) *C. horrida* Linn.f. Suppl. 264. 1781; FBI 1:178; FUGP 1:52.

Climbing shrubs. Young parts densely ferruginous-pubescent. Thorns recurved. Leaves 8-9.7 x 4.5 cm, ovate or oblong, acute, shining above. Flowers purplish. Petals oblong, shorter than stamens. Stamens many with purple filament. Ovary provided with a long gynophore. Fruits subglobose, reddish-brown on the stout gynophore.

Local name : Karralura
Flowers and Fruits : February-May

Found commonly in the open areas, straggling on trees. The fruits are cooked as vegetable. Singh 2876, 7193, 8320 (CDRI) Saran & Party 26114 (LWG).

11. VIOLACEAE

Hybanthus Jacq. nom. cons.

H. enneaspermus (Linn.) F. Muell. Fragm. Phyt. Aust. 10: 81. 1877; Jacobs & Moore in Fl. Males. 7(1): 197. 1971. *Viola enneasperma* Linn. Sp. Pl. 937. 1753. *Ionidium enneaspermum* (Linn.) Vent. Jard. Malm. Sub. t. 27, 1803.

I. suffruticosum Ging. in DC. Prodr. 1:311. 1824; FBI 1:185.

I. heterophyllum Vent. Jard. Malm. 27, 1803; FUGP 1:54.

Diffuse, perennial pubescent herbs, often woody at the base. Leaves *ca* 2.7 cm long, subsessile, linear or lanceolate, serrate with subulate stipules. Flowers solitary, red. Petals 5, variable, one having a long claw and an oval or orbicular limb. Seeds striate.

Flowers and Fruits : August-September

Occurs occasionally in sal forest on moist soil. The roots are used in medicine. Singh 7431 (CDRI).

12. FLACOURTIACEAE

1. Ovaries 2 to 8-locular, styles 2 .... **Flacourtia**
1. Ovaries 1 locular, style 1:
   2. Flowers yellow in axillary compact panicles .... **Xylosma**
   2. Flowers greenish-yellow in axillary clusters .... **Casearia**

Flacourtia Commers. ex L. Herit.

1. Leaves ovate-lanceolate, acuminate, glabrous .... **F. jangomas**
1. Leaves ovate, obtuse, usually pubescent .... **F. indica**

F. jangomas (Lour.) Raeusch. Nomencl. Bot. ed. 3:290.1797; Sleumer in Fl. Males. 5(1):72. 1954. P.C. Kanjilal, For. Fl. UP. 2:35.1966; Srivastava, FG. 45. 1976.
Stigmarota jangomas Lour. Fl. Cochinch. 2:634.1790.
Flacourtia cataphracta Roxb. ex Willd. Sp. Pl. 4:830.1806; FBI 1:193; FUGP 1:56.

This taxon is included in the Working Plan of North Kheri Forest Division. I have not seen this plant during my collection tours. Hence, a brief description is reproduced below:

Small trees, often with compound thorn. Leaves ovate or oblong-lanceolate, acuminate, crenate, glabrous. Flowers small, unisexual in irregular racemes. Fruit a size of a small plum, purple.

F. indica (Burm.f.) Merr. Interpret. Herb. Amb. 377. 1917 et Enum.3:112; Sleumer in Fl. Males. 1,5(1):76, f.30 h-i, 1954; Srivastava, FG. 44. 1976.
Gmelina indica Burm.f. Fl. Ind. 132, t. 39, f. 5, 1768.
Flacourtia ramontchi L. Herit. Strip. Nov.59, t. 30. 1786; FBI 1:193; FUGP 1:57; P.C. Kanjilal, For. Fl. UP. 35.1976.

Much branched, spinous small trees. Branches pubescent or glabrous. Leaves very variable, broadly ovate, obovate, elliptic-oblong or suborbicular often tomentose beneath, crenate-serrate. Flowers unisexual, greenish-yellow in short branched pubescent racemes. Fruits drupe, globose with 4-5 short styles, purple.

Local name : Kantia
Flowers and Fruits : March-May

Found frequently near the edge of the mixed forest in open situation. The ripe fruits are eaten. Singh 4366, 7146, 7192, 7594 (CDRI) Saran 26151 (LWG).

Xylosma Forst.f.

X. longifolia Clos. in Ann. Sc. Nat. ser. 4,8:230.1857; P.C. Kanjilal, For. Fl. UP. 2:37.1966; FUGP 1:60.

Moderate-sized evergreen trees with dark grey or black bark. Leaves alternate, 14 x 4 cm, oblong-lanceolate, obtusely serrate, acuminate, coriaceous. Flowers dioecious, yellow in short dense axillary panicles. Sepals 4-5. Petals absent. Stamens numerous, inserted inside the red annular disk.

Local name : Kantawa
Flowers and Fruits : December-April

Found occasionally in Dudhwa range of the Park. Singh 7905 (CDRI).

Casearia Jacq.

1. Leaves oblong, glabrescent .... **C. graveolens**

1. Leaves elliptic-lanceolate, tomentose .... **C. elliptica**

C. graveolens Dalz. in Hook. Bot. 4:107.1852; FBI 2:592; FUGP 1:330; Santapau in RBSI 16(1):116.1953; P.C. Kanjilal, For. Fl. UP. 2:253.1966.

Small deciduous trees. Bark dark grey with white specks. Leaves 9 x 11 x 4-4.5 cm, oblong-elliptic, acute, glabrous, slightly crenate. Flowers greenish-yellow, clustered in the axil of leaves. Fruits 3-valved, 2.5 cm long, ellipsoid.

Local name : Chilla, beri
Flowers and Fruits : April-August

Found occasionally in the forest areas. Singh 4131 (CDRI) Sis Ram 1104 (DD).

C. elliptica Willd. Sp. Pl. 4:628.1806; DC. Prodr. 2:51.1825. *C. tomentosa* Roxb. Fl. Ind. 2:421.1824; FBI 2:593; FUGP 1:330; P.C. Kanjilal, For. Fl. UP. 2:353. 1966; Srivastava, FG. 142. 1976.

Medium-sized deciduous trees with dark brown rough bark. Leaves 14-20 x 6-8 cm elliptic-oblong, lanceolate, crenate, acute, tomentose, subcoriaceous. Flowers greenish-yellow, numerous in axillary fascicles. Fruits 3-valved, ellipsoid, with persistent calyx.

Local name : Chilla
Flowers and Fruits : March-July

Found commonly in the forest areas and in open places. The fruits are used for stupefying fish. Singh 2824, 7145, 7159, 7201 (CDRI) Sis Ram 1100 (DD).

13. POLYGALACEAE

Polygala Linn.

1. Stems woody below; wings petaloid, longer than capsules. .... **P. crotalarioides**

1. Stem herbaceous; wings herbaceous, not longer than capsules:

   2. Flowers yellow .... **P. erioptera**

   2. Flowers pink .... **P. arvensis**

P. crotalarioides Buch.-Ham. ex DC. Prodr. 1:327. 1824; FBI 1:202; FUGP 1:59. Mukherjee in Bull. Bot. Soc. Beng. 12 (1&2): 47.1958.

Small perennial hairy herbs. Leaves 4-6.5 cm long, oblong-obovate, apex rounded or notched. Flowers shortly pedicelled, purple on short axillary raceme. Wings petaloid, obovate with rounded tips, twice as long as capsule. Capsules broader than long, notched, ciliate.

Flowers and Fruits : August-October.

Found frequently in moist places in sal forest and in grasslands. Singh 2963, 7312 (CDRI) Saran & Party 26367 (LWG).

P. erioptera DC. Prodr. 1:326.1824; FBI 1:203; FUGP 1:60. Mukherjee in Bull. Bot. Soc. Beng. 1 (1 & 2): 46.1958

Erect of decumbent annual herbs. Leaves 2-5 cm long, linear or linear-oblong. Flowers yellow, arranged in axillary or extra axillary racemes. Wings petaloid, obliquely oblong with a strong green midrib. Capsules ovate, notched, pubescent.

Flowers and Fruits: August - October

Occurs occasionally in grasslands and in sal forest on moist places. Singh. 3019, 7380, 7776, 7862 (CDRI).

P. arvensis Willd. Sp. Pl. 3,2:376. 1802; Burtt in Notes R. Bot. Gdn. Edinb. 32:304.1973.
P. chinensis auct. pl. non Linn. 1753.

Stout, diffuse annual herbs. Stems usually procumbent, herbaceous, pubescent. Leaves 2-3.5 x 0.5-0.7 cm, linear-lanceolate, entire. Flowers pink, pendulous in short racemes. Capsules oblong, notched, ciliate on the margins. Seeds oblong, silky, strophiole 3-lobed.

Flowers and Fruits: August - October

Occur occasionally in the areas amongst grasses. *Singh* 8456, 8548 (CDRI) *Malhotra* 23515 (BSD)

Polycarpon Linn.

P. prostratum (Forsk.) Pax in Engl. & Prantl, Nat. Pfam. 3:(1b): 87.1889; Cufodontis in Bull. Jard. Bot. Brux. Suppl. 23:97.1953; Santapau in RBSI 16(1):12.1967.
Alsine prostrata Forsk. Fl. Aegypt.-Arab. 207.1775.
Polycarpon loeflingiae Benth. & Hook.f., FBI 1:245.1874; FUGP 1:65. *P. indicum* Merr. in Philip. Jour. Sci. (Bot. 10:30.1905 et Enum. 2:139.

Prostrate, dichotomously branched glabrous herbs. Leaves flat, opposite, appearing in whorls, spathulate. Flowers small, in many flowered cymes. Sepals persistent. Petals hyaline. Capsules subglobose.

Flowers and Fruits : April-September.

Occurs occasionally on dry sandy soil in puddles. Singh 8342 (CDRI) Saran & Party 7090 (LWG).

Polycarpaea Lamk. nom. cons.

P. corymbosa (Linn.) Lamk. Encycl. 2:129. 1791; FBI 1:245; FUGP 1:65.
Achyranthes corymbosa Linn. Sp. Pl. 205. 1753.

Much branched, slender herbs. Leaves 1-2 cm long, clustered on nodes, narrowly linear, mucronate. Flowers white with pinkish tinge, crowded on much branched cymes forming heads. Sepals chaffy, silvery. Petals short. Capsules small, 3-valved, brown.

Flowers and Fruits : August-November

Occurs frequently on sandy soil in cultivated fields. Singh 7449 (CDRI) Saran 25777 (LWG).

15. PORTULACACEAE

Portulaca Linn.

1. Flowers solitary, terminal .... **P. quadrifida**
1. Flowers in terminal clusters .... **P. oleracea**

P. quadrifida Linn. Mant. 1:73.1767; FBI 1:247; FUGP 1:66; Geesink in Fl. Males. 7(1):127.1971; Srivastava, FG. 48.1976.

Prostrate, filiform, glabrous herbs, rooting at the nodes. Leaves subsessile, flat, succulent, ovate to ovate-lanceolate; stipules with a ring of white hairs. Flowers solitary, terminal, yellow. Style filiform, 4-fid.

Flowers and Fruits : July-September

A common weed found on moist soil in plantations and fields. Singh 7743 (CDRI).

P. oleracea Linn. Sp. Pl. 445. 1753; FBI 1:246; FUGP 1:66; Geesink in Fl. Males. 7(1): 129.1971; Srivastava, FG. 48. 1976.

Glabrous, more or less succulent annuals. Leaves cuneate, oblong-obovate or spathulate, truncate or retuse at the apex. Flowers sessile, in terminal clusters, yellow with fleshy acute, persistent sepals. Seeds dark brown, muriculate.

Local name : Kulfa
Flowers and Fruits : During the greater part of the year.

Occurs commonly in sandy and moist places in forest plantations and along railway lines. The leaves are cooked as vegetable. Singh 7315 (CDRI).

16. TAMARICACEAE

Tamarix Linn.

T. dioica Roxb. Hort. Beng. 22.1814, nom. nud. et Fl. Ind. 2:101. 1824; FBI 1:249; FUGP 1:68; P.C. Kanjilal, For. Fl. UP. 2:40.1966; Srivastava, FG. 49.1976.

Branched, drooping glaucous shrubs with dark brown bark. Leaves minute, appressed, sheathing with acuminate tips, margin white. Flowers pink, in dense loosely panicled spike, dioecious. Male flowers have 5 stamens. Female flowers have staminode. Capsules oblong, elongate.

Local name : Jhau
Flowers and Fruits : July-October

Commonly found in low-lying areas and along sandy river banks. The twigs are used for making basket. Singh 2900, 7121, 8405, 8562 (CDRI).

17. ELATINACEAE

Bergia Linn.

B. ammannioides Roxb. ex Roth, Nov. Pl. Sp. 219. 1821; FBI 1:251; FUGP 1:69; Santapau in RBSI 16(1):16.1953; Srivastava, FG. 49. 1976.

Erect or decumbent, hairy annuals, 10-30 cm long. Leaves sessile, linear-lanceolate or oblanceolate, acute, sharply serrate. Flowers small in dense axillary fascicles, red. Sepals 5, lanceolate, denticulate. Petals 5. Stamens 5. Capsules 5-loculed.

Flowers and Fruits : November-May

Occurs occasionally in wet places along with *Ammannia* spp. Singh 7791 (CDRI) Saran 26307 (LWG).

18. DIPTEROCARPACEAE

Shorea Roxb. ex Gaertn.f.

S. robusta Gaertn. f. Fruct. 3:48.t.186.1805; FBI 1:306; FUGP 1:71; P.C. Kanjilal, For. Fl. UP. 2:43.1966; Srivastava, FG. 49. 1976.

Large deciduous trees. Leaves 8-18 cm long, broadly ovate, shortly acuminate, entire, glabrous and shining. Flowers cream-coloured, unilateral on the branches of lax axillary and terminal, grey-tomentose panicles. Fruits ovoid, acuminate, 1.5 cm long, indehiscent and winged.

Local name : Shakhu, Sal
Flowers and Fruits : March-June

Found commonly throughout the area forming Sal forest along the Sub-Himalayan tract. Timber wood of high quality is obtained. Seed oil and gum is used. Singh 2330, 7314 (CDRI) Saran & Party 26453 (LWG).

19. MALVACEAE

1. Staminal tube divided into 5-bundles of stamens; trees with polygamous flowers .... **Kydia**
1. Staminal tube entire or shortly divided at the top; herbs or shrubs:
    2. Fruits schizocarpic; cocci separating from the axis:
        3. Bracteoles absent:
            4. Carpels 1-seeded .... **Sida**
            4. Carpels 2 or more seeded .... **Abutilon**
        3. Bracteoles present:
            5. Number of bracteoles three:

6. Flowers violet; stigmas linear .... **Malva**

6. Flowers orange; stigmas capitate .... **Malvastrum**

5. Number of bracteoles five or more.. **Urena**

2. Fruits capsular:

7. Stigmas spreading:

8. Calyx 5-deft, persistent .... **Hibiscus**

8. Calyx spathaceous, deciduous .... **Abelmoschus**

7. Stigmas cohering in club-shape .... **Thespesia**

Kydia Roxb.

K. calycina Roxb. Hort. Beng. 51:1814, nom.-nud. et Fl. Ind. 3:188.1832; FBI 1:348; FUGP 1:92; Santapau in RBSI 16(1):20.1953; P.C. Kanjilal, For. Fl. UP. 2:62.1966; Srivastava, FG. 53.1976.

Moderate-sized deciduous trees with pale-grey bark. Leaves 9-10.5 cm long, suborbicular, cordate, slightly lobed, 5-nerved, tomentose beneath. Flowers white in axillary or terminal panicles, polygamous often deciduous. Staminal tube divided into 5 branches. Capsules subglobose, covered with dense stellate hairs.

Local name : Pula
Flowers and Fruits : August-December

Occurs frequently in the forest mixed with sal. The wood is used for making plywood. Singh 4262, 7419, 7899 (CDRI), Saran 26412 (LWG).

Sida Linn.

1. Procumbent trailing herbs .... **S. cordata**

1. Erect herbs or undershrubs:

2. Spiny tubercles present below the petiole .... **S. alba**

2. Spines absent:

3. Carpellary awn shorter than the calyx:

4. Leaves lanceolate to oblong, glabrous, serrate ....**S. acuta**

4. Leaves rhomboid, lanceolate or obovate, hoary beneath.......... ....**S. rhombifolia**

3. Carpellary awn longer than calyx ....**S. cordifolia**

S. cordata (Burm.f.) Borss. in Blumea 14:182.1966. *Melochia cordata* Burm.f. Fl. Ind. 143.1768. *S. veronicifolia* Lamk. Encycl. 1:5.1783 (*veronicaefolia)* FUGP 1:75; P.C. Kanjilal, For. Fl. UP. 2:47.1966.
S. humilis Cav. Diss. 5:277. t. 134, f.2, 1768; FBI 1:322.

Procumbent trailing, hairy herbs. Leaves 2.5-4 x 1.7-3.5 cm, ovate-cordate, shortly acuminate, crenate-serrate, hairy. Flowers axillary, solitary or in pairs, pale yellow; peduncles jointed above the middle. Carpels 5, shortly cuspidate or mucronate.

Local name : Bhumi-Bariara, Fardbuti.
Flowers and Fruits : June-October

Found commonly in wastelands and in forest clearings. Plant is used in venereal diseases. Singh 3079, 4231, 7265 (CDRI).

S. alba Linn. Sp. Pl. ed. 2, 2: 960. 1763; Bennet. Nomen. Chang. Fl. Plants 522. 1987. *S. spinosa* Linn. Sp. Pl. 683.1753; FBI 1:323; (excl. syn. *alnifolia* Linn.); FUGP 1:76; Borss. in Blumea 14:191. 1966; P.C. Kanjilal, For. Fl. UP. 2:48. 1966.
S. alba Linn. Sp. Pl. ed. 2, 960.1763. Exell in Fl. Zamb. 1:477.1961.

Much branched herbs. Stem green, rough with minute stellate hairs having spiny tubercles at the nodes below the petiole. Leaves ovate to ovate-lanceolate, crenate-serrate, glabrous above, tomentose beneath. Flowers cream coloured. Ripe carpels 5, with 2 long, erect beaks.

Local name : Bariara
Flowers and Fruits : September-October

Scarce in wastelands and forest clearings. Saran 26691 (LWG).

S. acuta Burm.f. Fl. Ind. 147. 1768; Hara in Enum. Sperm. Jap. 3:150.1954; Exell in Fl. Zamb. 1:421.1960; FUGP 1:76. Santapau in RBSI 16(1):19.1953; P.C. Kanjilal, For. Fl. UP. 2:49.1966; Srivastava, FG. 54. 1976.
S. carpinifolia sensu Mast. in FBI 1:323.1875, pro parte (non Linn.f. 1781).

Erect undershrubs. Stems smooth or slightly rough with minute stellate hairs. Leaves 3.2-10 cm long, lanceolate or oblong, serrate, glabrous, acute. Stipules linear-subulate, longer than petiole. Flowers yellow; peduncles jointed near the middle. Carpels 5-10; rugose, 2-awned.

Local name : Kharenti
Flowers and Fruits : July-October

Found commonly throughout the area in wastelands and near forest in open area. The plant yields a good fibre. Singh 3068, 4242, 7152 (CDRI).

S. rhombifolia Linn. Sp. Pl. 684. 1753; FBI 1:323; FUGP 1:76; Borss. in Blumea 14:195.1966; P.C. Kanjilal, For. Fl. UP. 2.49.1966; Srivastava, FG. 55.1976.
S. orientalis Cav. Diss. 1:t.12.f.1:1785; Hu in Fl. Chin. Fam. 153.18.1955.
S. rhombifolia Linn. var. *rhomboidea* Mast. in Hook.f. FBI 1:324.1874.

Erect much branched undershrubs. Stems brown-red in colour. Leaves variable broadly obovate or rhomboid-lanceolate or nearly orbicular, coarsely serrate; base cuneate. Flowers pale-orange axillary or clustered at the end of the branches. Carpels usually 5, each shortly 2-awned.

Local name : Lal Bariara
Flowers and Fruits : August-October

Occurs commonly in waste grounds and forest clearings. The root and seeds are used in venereal diseases and strength. Fibre is also obtained. Singh 4189, 7228, 7322, 7373, 7710, 7816, 8389 (CDRI).

S. cordifolia Linn. Sp. Pl. 684.1753; FBI 1:324; FUGP 1:77; Borss. in Blumea 14:199.1966; P.C. Kanjilal, For. Fl. UP. 2:50.1966; Srivastava, FG. 55.1976.

Erect undershrubs, covered with long spreading hairs. Leaves 3-6 cm long, ovate, cordate, crenate, obtuse, tomentose on both surfaces. Flowers pale-orange. Carpels 10, each with 2 long awns which exceed the calyx.

Local name : Kharenti
Flowers and Fruits : August-November

Occurs commonly in waste lands on damp sandy soils. The seeds are used in medicine. Singh 7369 (CDRI).

Abutilon Gaertn.

A. indicum (Linn.) Sweet, Hort. Brit. 54.1827; FBI 1:326; FUGP 1:78; Santapau & Wagh in Bull. Bot. Surv. India 5:105.1963; Borss. in Blumea 14:171.1966; P.C. Kanjilal, For. Fl. UP. 2.51.1966; Srivastava, FG. 51.1976.
Sida indica Linn. Cent. Pl. 2:26.1756.

Branched shrubs. Leaves ovate, orbicular, cordate, dentate, shortly acuminate. Flowers orange-yellow, usually solitary on long peduncle and jointed near the top. Carpels 15-20, exceeding the calyx with short spreading awns, tomentose or subglabrous. Seeds black, minutely stellate hairy.

Local name : Kanghi
Flowers and Fruits : September-November

Occurs commonly in open areas and in wastelands. The roots are used in fever. Singh 2885, 3065, 4341 (CDRI).

MALVA Linn.

1. Petals twice the length of sepals .... M. verticillata
1. Petals 3-4 times the length of sepals .... M. sylvestris

M. verticillata Linn. Sp. Pl. 689.1753; FBI 1:320; FUGP 1:74.

Erect branched pubescent annuals. Leaves lobed, suborbicular, cordate, hairy. Flowers small sessile in dense axillary clusters. Bracteoles linear. Petals notched.

Specimens examined: Kheri district, *Inayat* 21601 (DD)

M. sylvestris Linn. Sp. Pl. 689. 1753; FBI 1:320; FUGP 1:74.

Erect glabrous herbs. Leaves cordate, orbicular lobed. Flowers axillary, dark purple or violet. Petals notched, 3-4 times as long as sepals.

Specimens examined: Kheri district. *Inayat* 21600 (DD).

Note: I have not seen these two plants in the fields but included here based on Inayat collection.

Malvastrum A. Gray *nom. cons.*

M. coromandelianum (Linn.) Garcke in Bonplandia 5:297. 1857; Merr. Enum. 3:33; Borss. in Blumea 14:152.1966; P.C. Kanjilal, loc. cit. 46. 1966; Srivastava, FG. 54. 1976.
Malva coromandeliana Linn. Sp. Pl. 687. 1753.
Malvastrum tricuspidatum A. Grey, Pl. Wright. 1:16.1852; FBI 1:321; FUGP 1:75.

Erect or ascending adpressed hairy herbs. Leaves ovate to ovate-lanceolate, crenate-serrate. Flowers yellow. Bracteoles three. Calyx lobes triangular, acute. Carpels 8-12, reniform with 3 projecting points.

Flowers and Fruits : Greater part of the year.

Found commonly in the waste places and in forest clearings. Singh 4337, 7749 (CDRI); Saran 26743 (LWG)

Urena Linn.

1. Carpels armed with hooked bristles .... **U. lobata**

1. Carpels unarmed and smooth .... **U. palmata**

U. lobata Linn. Sp. Pl. 692. 1753; FBI 1:329 (excl. Syn. *U. palmata); FUGP 1:85; Santapau in RBSI 16(1): 21.1953; Borss. in Blumea 14:138.1966; P.C. Kanjilal, For. Fl. UP. 2:53.1966; Srivastava, FG. 56. 1976.
U. sinuta Linn. subsp. *lobata* Borss.

Tall erect shrubs, covered with spreading hairs. Leaves variable, 5-11 x 4-9 cm, rounded, ovate, entire or slightly lobed, nerves prominent beneath. Flowers bright pink. Carpels armed with hooked bristles.

Local name : Lapetua
Flowers and Fruits : August-November

Occurs commonly on the edge of the forest near damp places and along railway lines. The stem yields good fibre. Singh 3035, 4162, 7439 (CDRI) Saran 26494 (LWG).

U. palmata Roxb. Fl. Ind. 3:182.1832; P.C. Kanjilal, For. Fl. UP. 2:53.1966.
U. repanda Roxb. Hort. Beng. 51.1814, nom. nud; FBI 1:330; FUGP 1:80.

Erect much branched, perennial, densely woolly herbs. Leaves 5-14 cm long, orbicular cordate, rough on the upper surface, repand, serrate, upper leaves lanceolate. Flowers pink in axillary racemes. Carpels smooth, unarmed.

Flowers and Fruits : July-November

Found commonly on the edge of the forests on moist places. The plant is used for making fibre. Singh 4317, 7389, 7729 (CDRI).

Hibiscus Linn. *nom. cons.*

1. Shrubs or small trees; staminal tube much exserted........**H. rosa-sinensis**
1. Herbs or undershrubs, staminal tube short:
    2. Flowers yellow ....**H. vitifolius**
    2. Flowers white or light pink ....**H. lobatus**

H. rosa-sinensis Linn. Sp. Pl. 694. 1753; FBI 1:344; Bailey, Man. Cult. Pl. 665; Srivastava, FG. 52. 1976; Raizada, Suppl. FUGP 30.1976.

Evergreen shrubs or small trees. Leaves ovate, acuminate, glabrous with linear stipules. Flowers solitary axillary, red or scarlet. Bracts 7-9, linear. Staminal tube exceeding the corolla.

Local name : Gurhal
Flowers and Fruits : February-October

Commonly planted in the district for its large ornamental flowers. The flowers are used in medicine. Singh 7744 (CDRI).

H. vitifolius Linn. Sp. Pl. 696. 1753; FBI 1:388; FUGP 1:85; P.C. Kanjilal, For. Fl. UP. 2:59. 1966.

Perennial, hispid herbs. Leaves 6 cm long, cordate, 3-lobed, serrate, acute, tomentose on both surfaces. Flowers axillary solitary, drooping, yellow with purple centre. Capsules short, subglabrous, 5-winged, hairy.

Local name : Kapasi
Flowers and Fruits : May-August

Occurs rarely in grasslands in Dudwa range of North Kheri Forest Division. Singh 7308 (CDRI); Saran 26258 (LWG).

H. lobatus (Murr.) O. Kuntze, Rev. Gen. 3:19. 1898; Blatt. in JBNHS 34:632.1930; Borss. in Blumea 14:77.1966; Rakshit & Kundu, Bull. Bot. Surv. India 12:169.1972.
Solandra lobata Murr. in Comm. Gotting. 6:24.t.1.1785.
Hibiscus solandra L'Herit. Stirp. 1:103,t.49, 1786; FBI 1:336; FUGP 1:84.

Erect, pubescent annuals up to 40 cm long. Leaves ovate-cordate, long petiolate below. Upper leaves deeply 3-lobed and linear-lanceolate in floral region. Flowers white. Calyx sticky. Capsules ovoid. Seeds black, tubercled.

Flowers and Fruits : September-October

Occurs occasionally on the edge of the forest. Singh 7733 (CDRI).

Abelmoschus Medic.

A. crinitus Wall. Pl. As. Rar. 1:39.f.44.1830; Backer & Bakh.f. Fl. Java 1:435.1963; Borss. loc. cit. 103. 1966.
Hibiscus cancellatus Roxb. Hort. Beng. 51.1814, nom. nud. et Fl. Ind. 3:201.1832; FBI 1:342; FUGP 1:87.

Erect bristly annual herbs. Leaves alternate, petiole, 3-12 cm long; blade rounded or obtusely 5-angled, crenate, hairy. Flowers yellow with a crimson centre. Bracts 2. Bracteoles 10-15, linear, hispid. Capsules ovoid, 3 cm long, hairy. Seeds reniform, black.

Flowers and Fruits : August-November

Found rarely in Kishenpur range of South Kheri and Sonaripur range of the Park. The roots and seeds are used for strength and vitality. Singh 7903, 8516 (CDRI).

Thespesia Sol. ex Corr.

Thespesia lampas (Cav.) Dalz. & Gibs. Bom. Fl. 19. 1861; FBI 1:345; Santapau in RBSI 16(1):23.1953. Borss. in

Blumea 14(1): 166. 1966; P.C. Kanjilal, For. Fl. UP. 2: 61. 1966. *Hibiscus lampas* Cav. Diss. 3: 154. t. 56. f. 2, 1787.
T. macrophylla non Blume; Bijdr 73. 1825; FUGP 1:86.
Azanza lampas (Cav.) Alef. in Bot. Zeit. 19:297.1861; Ecell in Fl. Zamb. 1:434.1961. (Plate IX).

Shrubs up to 3 m high. Leaves 8-12 cm long, slightly 3-lobed, cordate, lobes acuminate, stellately tomentose. Flowers 1-3 on axillary or terminal peduncles, yellow with crimson centre. Capsules 2-6 cm long, ovate, pointed, 5-valved. Seeds smooth.

Local name : Banbhindi
Flowers and Fruits : May-September

Found commonly along the edge of the sal forest. A strong fibre is obtained from stem bark. Singh 2812, 2971, 7394 (CDRI).

20. BOMBACACEAE

Bombax Linn.

B. ceiba Linn. Sp. Pl. 511. 1753; pro parte; A Robyns in Taxon 10:68. 1960 & 1961 et in Bull. Jard. Bot. Etat. Brux. 33:88.1963; Santapau in RBSI 16(1):20.1967; Srivastava, FG. 57.1976.
B. malabaricum DC. Prodr. 1:4779.1824; FBI 1:349; FUGP 1:92.
Salmalia malabarica (DC.) Schott & Endl. Melet. 35.1832; Benthall, Trees Calc. 43; P.C. Kanjilal, For. Fl. UP. 2:63.1966.

Tall, deciduous trees forming spreading crown and buttressed trunk. Leaflets 5-7, elliptic-lanceolate, acuminate, entire, glabrous. Flowers large, bright red, Capsules woody. Seeds smooth, surrounded by white silky cotton.

Local name : Semar

Flowers and Fruits : March-May

Found frequently in grasslands. The young roots are used as tonic and are edible. The flower buds are cooked as vegetable. Singh 7317 (CDRI).

21. STERCULIACEAE

1. Flowers polygamous; petals absent:
   2. Leaves cordate-ovate; follicles woody; seeds winged ....**Pterygota**
   2. Leaves simple, lobed; follicles coriaceous; seeds not winged............ ....**Sterculia**
1. Flowers bisexual; petals present:
   3. Fruit a spirally twisted follicle ....**Helicteres**
   3. Fruits a capsule:
      4. Petals deciduous; seeds winged ....**Pterospermum**
      4. Petals persistent; seeds not winged:
         5. Androecium tubular with 5 groups of 15 fertile stamens with staminodes..........**Pentapetes**
         5. Androecium only tubular at the base, stamens 5, without staminodes..........**Melochia**

Pterygota Endl.

P. alata (Roxb.) R.Br. in Benn. Pl. Jav. Rar. 234. 1844; Benthall, Trees Calc. 50; Santapau in RBSI 16(1):26.1953; P.C. Kanjilal, For. Fl. UP. 2:71.1966. *Sterculia alata* Roxb. Pl. Cor. 3:84, t. 287. 1820; FBI 1:360.

Tall trees. Leaves 2.5 x 17 cm, entire, cordate-ovate, 7-nerved. Flowers in short rusty-tomentose panicle. Follicles globose, depressed, woody. Seeds many, oblong, winged, compressed.

Local name : Burman teak
Flowers : March-April; Fruits : July-September

Planted as an ornamental tree in Dudhwa Forest Rest House campus. A good timber tree. Singh 7781 (CDRI).

Sterculia Linn.

S. villosa Roxb. Hort. Beng. 50. 1814, nom. nud. et Fl. Ind. 2:153. 1824; FBI 1:355; FUGP 1:94; P.C. Kanjilal, For. Fl. UP. 2:69.1966.

Large shrubs or small trees with smooth, grey bark. Young parts rusty tomentose. Leaves 30-40 cm, crowded at the end of the branches, usually 5 to 7-lobed; lobes ovate-oblong, abruptly acuminate. Flowers polygamous, yellow in drooping panicles. Follicles sessile, 3-5 cm long, rough with stellate pubescent hairs. Seeds oval, smooth.

Local name : Udar
Flowers and Fruits : March-May

Found commonly along the edge of the sal forests. The root is used for strength and vitality. Fibre obtained from the stem bark Singh 4292, 7313 (CDRI).

Helicteres Linn.

H. isora Linn. Sp. Pl. 693.1753; FBI 1:365; FUGP 1:94; Santapau in RBSI 16(1):27.1953; P.C. Kanjilal, For. Fl. UP. 2:72.1966 (Fig. 4).

Erect shrubs; young parts densely stellate, tomentose. Leaves 7.5-13.5 cm long, 5 to 7-nerved, obovate, suborbicular or obliquely cordate, shortly acuminate, irregularly serrate, pubescent beneath. Flowers red, zygomorphic, 3.5 cm long in few flowered, extra axillary cymes. Fruit 4-5 cm long, of 5 spirally twisted tomentose follicles on a long gynophore.

Local name : Murra, Bhendu
Flowers and Fruits : August-November

Occurs commonly in grasslands. The bark yields a strong fibre. The seeds, roots and bark are used in dysentery and dog bite. Singh 2981, 7960 (CDRI) Saran & Party 22165 (LWG).

Pterospermum Schreb. nom. cons.

P. acerifolium Willd. Sp. Pl. 3:729.1802; FBI 1:368; FUGP 1:97; Santapau in RBSI 16(1):28.1953; P.C. Kanjilal, For. Fl. UP. 2:73.1966; Srivastava, FG. 59. 1976.

Large evergreen trees with ash coloured bark. Young branches and calyx covered with ferruginous tomentum. Leaves large, deep cordate or peltate, often lobed, sinuate, glabrous above, grey-tomentose beneath. Flowers white, axillary, fragrant. Calyx fleshy deeply divided. Capsules 5-angled, woody. Seeds compressed, winged.

Local name : Kanakchampa
Flowers and Fruits : March-June

A handsome tree planted in Dudhwa Forest Rest House campus. Singh 7339 (CDRI); Saran & Party 26202 (LWG).

Pentapetes Linn.

P. phoenicea Linn. Sp. Pl. 698. 1753; Roxb. Fl. Ind. 3:157.1832; FBI 1:371; FUGP 1:99; P.C. Kanjilal, For. Fl. UP. 2:78.1966; Srivastava FG. 59.1976.

Erect shrubs. Leaves variable, 3-6 cm long, linear-lanceolate, strongly crenate-serrate. Flowers scarlet, solitary or in pairs. Calyx deeply divided, persistent in fruit, hairy. Capsules subglobose.

Local name : Dopahria
Flowers and Fruits : August-October

Ornamental plants. Occurs occasionally in moist situation near cultivated fields. Singh 4319, 8447 (CDRI).

Melochia Linn.

M. corchorifolia Linn. Sp. Pl. 675. 1753; FBI 1:374; FUGP 1:100; P.C. Kanjilal, For. Fl. UP. 2:78.1966; Srivastava, FG. 58.1976.

Erect herbs. Leaves 6-7 cm long, ovate or ovate-lanceolate, cordate at the base, crenate-serrate, glabrous above, thinly hairy beneath. Flowers pink or white in dense axillary and terminal clusters with many hispid bracteoles. Capsules small, depressed, globose, hispid.

Flowers and Fruits : July-September

Occurs commonly near wet and muddy places in open areas. Singh 7443, 7732, 8449 (CDRI).

22. TILIACEAE

1. Trees or shrubs .... **Grewia**
1. Herbs or undershrubs:
    2. Fruit dry echinate, indehiscent .... **Triumfetta**
    2. Fruit capsular, dehiscent .... **Corchorus**

Grewia Linn.

1. Drupes with crustaceous rind; leaves scabrous .... **G. sclerophylla**
1. Drupes fleshy with no distinct rind; leaves softly hairy beneath:
    2. Trees; peduncle shorter than the petiole:
        3. Leaves glabrescent .... **G. tiliaefolia**
        3. Leaves tomentose:
            4. Leaves obliquely ovate-cordate, finely serrulate..**G. elastica**
            4. Leaves distinctly cordate or sub-orbicular, serrulate.......... .... **G. hainesiana**
    2. Shrubs or undershrubs; peduncle much longer than the petiole.
        5. Dwarf undershrubs .... **G. sapida**

5. Tall shrubs:

   6. Flowers bisexual:

      7. Drupe rugose, 2-lobed .... **G. disperma**

      7. Drupe smooth 4-lobed .... **G. hirsuta**

   6. Flowers polygamous, mostly 1-sexual .... **G. polygama**

G. sclerophylla Roxb. ex G. Don, Gen. Syst. 4. 1831; P.C. Kanjilal, For. Fl. UP. 2:90.1966.
G. scabrophylla Roxb. Fl. Ind. 584.1832; FBI 1:387; FUGP 1:112.

Shrubs or undershrubs; young parts covered with densely brown, stellate tomentum. Leaves 8-11 cm long, ovate to suborbicular, irregularly serrate, scabrous on upper surface, densely stellate-tomentose beneath, base 3 to 5-nerved. Flowers white. Drupes globose, rough, stellate hairy. Pyrenes usually 4.

Local name : Dapher
Flowers and Fruits : April-October

Found commonly in grasslands. The stem yields a good fibre. The ripe fruits are edible. Singh 7293, 7468 (CDRI) Inayat 23573 (CAL).

G. tiliaefolia Vahl, Symb. Bot. 1:35.1790; FBI 1:386, pro parte; Santapau in RBSI 16(1):28.1953; P.C. Kanjilal, For. Fl. UP. 2:85.1966.
Microcos lateriflora Linn. Sp. Pl. 514.1753 (non *G. lateriflora* G.Don, 1831).

Moderate-sized trees; young portion hoary tomentose. Leaves 7.5-15 cm long, broadly ovate, obliquely cordate, acute or acuminate sometimes obscurely 3-lobed, irregularly serrate. Peduncles many in each axil, 3 to 4-flowered, shorter than petiole. Flowers yellow. Drupes globose.

Local name : Phalsa

Flowers and Fruits : April-September

Occurs rarely in the forest area. The ripe fruits are edible. Inayat 21650 (CAL).

G. elastica Royle, Ill. 104 t.22.1839; FUGP 1:106; P.C. Kanjilal, For. Fl. UP. 2:88.1966.

G. asiatica Mast. var. *vestita* Masters in FBI 1:387. 1874, pro parte.

Small trees. Young parts, inflorescence densely hoary or rufous with stellate pubescence. Leaves 5-12 x 4-9 cm, obliquely ovate, acuminate, crenate-serrate, fulvous tomentose when young, base rounded with 5-7 straight nerves. Flowers yellow, peduncles axillary or extra-axillary, longer than petiole, 3-flowered. Drupes small, obscurely lobed.

Flowers and Fruits : April-September

Found occasionally in the sal forest. Singh 7535 (CDRI).

G. hainesiana Hole in Indian For. 43:316.1917; Raizada, Suppl. FUGP 34.1976.

G. subinaequalis DC. Prodr. 1:511.1824; Narayanaswami & Seshagiri Rao in J. Indian Bot. Soc. 29:103.t.3c. 1950; Santapau in Fl. Saurashtra 68.1962; P.C. Kanjilal, For. Fl. UP. 2:87.1966.

G. asiatica Masters in FBI 1.386.1874 (non Linn.); FUGP 1:107.

Moderate-sized trees with smooth light grey or greenish bark. Young parts and inflorescence softly tomentose. Leaves 5-14 cm long, broad-ovate or suborbicular, apex rounded, acute or acuminate, crenate-serrate. Peduncles axillary, 3-flowered. Flowers yellow or reddish. Drupes globose, often 2-lobed, pilose, purple to almost black when ripe.

Local name : Phalsa
Flowers and Fruits : April-August

Found frequently in the forest area. The fruits are edible. Singh 7457 (CDRI) Saran 26414, 26209 (LWG) Inayat 23570 (CAL).

G. sapida Roxb. Hort. Beng. 42. 1814, nom. nud. et Fl. Ind. 2:590.1824; Royle, Ill. 103, FBI 1:387; FUGP 1:109; P.C. Kanjilal, For. Fl. UP. 2:90.1966.

Undershrubs with a woody underground rootstocks, from which a number of flowering shoots are given out annually. Young parts hoary-pubescent. Leaves 6-14 x 4-8 cm, ovate or obovate or suborbicular, somewhat obliquely rounded or cuneate at the base, slightly acuminate at the apex, irregularly serrate. Peduncles slender, 2 to 5-flowered, much longer than the petiole. Flowers yellow. Drupes of the size of a pea.

Local name : Bhumi Phalsa
Flowers : April-May; Fruits : October-November

Found commonly in the grasslands and along forest roads. Singh 2907, 7160, 7437, 8333 (CDRI).

G. disperma Rottl. ex Spreng. Syst. 2:579.1925; Santapau in RBSI 16(1):29.1953; Srivastava, FG. 61.1976. *G. laevigata* auct. plur. (non Vahl, 1790).

Small trees; young parts and inflorescence hairy. Leaves 7.5-12 cm long, lanceolate, acuminate, pilose on the nerves beneath, serrate, 3-nerved at the base, membranous, green on both surfaces. Flowers white, 2-6 together on axillary peduncle which is longer than petiole. Drupes 1 to 4-lobed, glabrescent, black when ripe.

Local name : Kakai, Haria
Flowers : August-September; Fruits : November-January

Found frequently in the forest areas in open places. Singh 4253 (CDRI).

G. hirsuta Vahl, Symb. Bot. 1:34.1790; FBI 1:391; FUGP 1:110; P.C. Kanjilal, For. Fl. UP. 2:85.1966; Srivastava, FG 61.1976.

Small spreading shrubs with slender stellately pubescent branches. Leaves 5-7 cm long, ovate or lanceolate, or oblong, acute or acuminate, irregularly serrate, hairy and scabrous above, densely tomentose beneath, 3-nerved at the base, stipules linear or subulate. Flowers white turning yellow, polygamous. Peduncles 1-3, axillary, 1 to 4-flowered. Drupes 4-lobed, brown, hirsute and shining.

Local name : Gursakari
Flowers and Fruits : April-November

Occurs commonly in the grasslands and in the forest areas.

The fruits are eaten by local people. The roots are used for strength and vitality. Singh 2954, 4238, 7534, 7565, 7551 (CDRI) Saran 7198 (LWG).

G. polygama Roxb. Hort. Beng. 42. 1814, nom. nud. et Fl. Ind. 2:588.1824; FBI 1:391; FUGP 1:110.

Shrubs with grey tomentose young branches. Leaves 7.5-9 cm long, subsessile, lanceolate, acuminate, rounded at the base, serrate, hairy on both surfaces, base 3-nerved. Flowers polygamous, mostly unisexual. Peduncles 1-5, axillary, 2-flowered. Drupes 4-lobed, brownish, shining.

Flowers and Fruits : July-November

Occurs frequently in grasslands. Singh 4230 (CDRI).

Triumfetta Linn.

1. Leaves lanceolate or ovate-lanceolate .... **T. pilosa**
1. Leaves orbicular and 3-lobed:
    2. Fruits oblong; spines hispid-ciliate .... **T. pentandra**
    2. Fruits globose; spines glabrous .... **T. rhomboidea**

T. pilosa Roth, Nov. Pl. Sp. 223. 1821; FBI 1:394; FUGP 1:111; Santapau in RBSI 16(1): 30.1953; P.C. Kanjilal, For. Fl. UP. 2:94.1966.

T. cana Blume, Bijdr. 113.1825; FBI 1:396.
T. tomentosa Noronha in Verh. Batav. Gen. 5:ed.1. Act. 4:27.1790.

I have included this taxon on the authority of P.C. Kanjilal loc.cit. who reports the occurrence of this plant from Kheri forest area. Hence, a brief description is reproduced below.

Undershrubs, covered with bulbose-base bristles. Leaves ovate-lanceolate, acuminate, base rounded, unequally serrate. stellately-tomentose. Flowers yellow in short peduncled lateral and axillary cymes. Fruits 4-locular, densely covered with hooked bristles.

T. pentandra A. Rich. in Guill & Perr. Fl. Seneg. Tent. 93.t.19.1830; Srivastava, FG. 62. 1976.
T. neglecta W.&A. Prodr. 1:75.1834; FBI 1:396; FUGP 1:112.

Annual herbs. Leaves roundish, 3 to 5-nerved at the base, lobed, irregularly toothed, tomentose, upper narrowed at both ends. Flowers small in extra axillary clusters. Fruits oblong, spiny, ciliate.

Flowers and Fruits : July-October

Found commonly on the edge of the forest and in waste places. Singh 7370 (CDRI) Saran 26346, 26582 (LWG).

T. rhomboidea Jacq. Enum. Pl. Carib. 22. 1760; FBI 1:395; FUGP 1:111; Shah in JBNHS 59(1):320.1952; Srivastava, FG. 62.1976.
T. bartramia Linn. Syst. ed. 10,1044.1759 (nom. ellegit.); P.C. Kanjilal, For. Fl. UP. 2:94.1966.
Bartamia indica Linn. Sp. Pl. 389.1753. (non *T. indica* Lamk. 1789).

Herbs or small shrubs. Leaves variable, often 3-lobed, orbicular-ovate, acuminate, irregularly serrate, hairy, cordate, rounded, 3 to 7-nerved, upper not lobed. Flowers yellow in dense lateral and terminal cymes. Fruits globular, clothed with hooked spines.

Local name : Lapetua

Flowers and Fruits : August-October

Found commonly in the area in open places and in forests. The stem yields a good fibre. Singh 4241, 7460, 7573 (CDRI) Saran 25791 (LWG).

Corchorus Linn.

1. Capsules cylindrical:

   2. Capsules ending in a beak, lobed at the top:

      3. Capsule 10-ribbed, glabrous, beak long .... **C. olitorius**

      3. Capsules 3 to 4-angled, scabrous; beak short ...... **C. trilocularis**

   2. Capsules ending in 3, radiating points .... **C. aestuans**

1. Capsules globose .... **C. capsularis**

C. olitorius Linn. Sp. Pl. 529.1753; FBI 1:397; FUGP 1:113; Santapau in RBSI 16(1):32.1953; P.C. Kanjilal, For. Fl. UP. 2:96.1966.

Erect annual herbs. Leaves ovate-lanceolate, 3 to 5-nerved, serrate, the basal two serratures prolonged into filiform appendages. Peduncles 1 to 3-flowered, very short. Flowers yellow. Capsules elongated, 10-ribbed, long beaked.

Local name : Jute
Flowers and Fruits : September-October

Occurs frequently in moist shady places and in cultivated fields. A good fibre jute is obtained from this plant. Singh 7442, 7737 (CDRI); Saran & Party 26640 (LWG).

C. trilocularis Linn. Mant. 77. 1767; FBI 1: 397; FUGP 1: 113; Maheshwari Fl. Del. 89. 1963.

Erect or diffuse, pilose undershrubs. Leaves ovate-oblong, 11 x 2 cm, crenate-serrate, with basal lobes. Flowers yellow, on leaf opposed peduncle. Capsules 2-3.5 x 0.2-0.4 cm, straight or curved, 3 to 4-angled; beak very short.

Flowers and Fruits : August-October

Occurs rarely in Sonaripur range of the Park. Singh 8518 (CDRI).

C. aestuans Linn. Syst. ed. 10.1079.1759 (non Forsk.); Santapau in RBSI 16(1):31.1953; P.C. Kanjilal, For. Fl. UP. 97.1966; Srivastava, FG. 61.1976.
C. acutangulus Lamk. Encycl. 2:104.1786; FBI 1:398; FUGP 1:114.

Ascending or erect hairy annuals. Leaves ovate-oblong, acuminate, serrate. Stipules linear-subulate, shorter than petiole. Flowers yellow on 2 to 3-flowered peduncles. Capsules short, 6-angled, 3 of the angles winged. Beak 3-fid.

Flowers and Fruits : August-October

Occurs commonly in wastelands and other shady places. Singh 3069, 7561, 7735, 8537 (CDRI).

C. capsularis Linn. Sp. Pl. 529.1753; FBI 1:397; FUGP 1:113; Santapau in RBSI 16(1):31.1953; P.C. Kanjilal, For. Fl. UP. 96.1966; Srivastava, FG. 60. 1976.

Erect, glabrous annual herbs. Leaves ovate-lanceolate, acuminate, serrate, the lowest serratures produced into filiform awns. Flowers yellow. Capsules subglobose, depressed at the apex, ridged, muricated.

Flowers and Fruits : September-October

Found frequently on moist places in sal forest and wastelands. Singh 3064 (CDRI); Saran & Party 26393 (LWG).

23. ZYGOPHYLLACEAE

Tribulus Linn.

T. terrestris Linn. Sp. Pl. 387. 1753; FBI 1:423; FUGP 1:119; P.C. Kanjilal, For. Fl. UP. 2:103.1966; Srivastava, FG. 63.1976.

Prostrate branched herbs, clothed with silky hairs. Leaves paripinnate; leaflet 12-14, hairy, mucronate. Flowers yellow on axillary short peduncles. Fruits shizocarpic, splitting into 4-6 cocci, each with 2 long and 2 short spines.

Local name : Gokharu
Flowers and Fruits : August-December

Found commonly in wastelands on sandy soils. The fruits are used in medicine. Singh 4121, 7368, (CDRI).

24. OXALIDACEAE

1. Herbs:
   2. Leaves 3-foliolate .... Oxalis
   2. Leaves pinnately compound .... Biophytum
1. Trees .... Averrhoa

Oxalis Linn.

O. corniculata Linn. Sp. Pl. 435. 1753; FBI 1:346; FUGP 1:122; Srivastava, FG. 65.1976.

Perennial pubescent herbs with long slender creeping or subterranean stems, rooting at the nodes. Leaves long petiolate, 3-foliolate; leaflets obcordate. Flowers yellow, subumbellate. Capsules sub-cylindrical, angular with deflexed fruiting pedicels. Seeds ribbed, brown.

Local name : Khatti-buti
Flowers and Fruits : November-March

Found commonly in wastelands and in forest plantations near moist places. The leaves are used in ophthalmic diseases. Singh 2837, 4218, 7119, 7601 (CDRI).

Biophytum DC.

B. sensitivum (Linn.) DC. Prodr. 1:690.1824; FBI 1:436; FUGP 1:123; Veldkamp in Fl. Males. ser. 1.7 (1): 162.

1971; Srivastava, FG. 64.1976.
Oxalis sensitiva Linn. Sp. Pl. 434. 1753.

Erect hispid herbs. Leaves 8 cm long; leaflet 5-15 pairs, oblong. Peduncles 3-8 cm long, hispid. Flowers shortly pedicelled, yellow. Calyx usually much longer than the capsule. Capsules elliptic. Seeds minute.

Flowers and Fruits : August-October

Occurs commonly in moist open places and along railway lines. Singh 4208, 7767, 8506 (CDRI).

Averrhoa Linn.

A. carambola Linn. Sp. Pl. 428.1753; FBI 1:439; FUGP 1:124; P.C. Kanjilal, For. Fl. UP. 2:104.1966; Srivastava, FG. 64.1976.

Medium-sized trees with dark-grey barks. Leaves alternate, imparipinnate; leaflets subopposite, 5-11, ovate-oblong, acuminate, base oblique. Flowers pink in short racemes. Fruits oblong, acutely angled.

Local name : Kamrak
Flowers and Fruits : September-October

Occasionally cultivated in the district for its edible fruits. Singh 7815 (CDRI).

25. RUTACEAE

1. Armed shrubs or trees:
   2. Pericarp leathery ....**Citrus**
   2. Pericarp woody:
      3. Leaves 3-foliolate; stamens many ....**Aegle**
      3. Leaves imparipinnate; stamens 10-12 ....**Limonia**
1. Unarmed shrubs or small trees:
   4. Leaves 1 to 5-foliolate, ovules solitary in each cell ....**Glycosmis**

4. Leaves imparipinnate, ovules 2 in each locule:

5. Style long; stamens not dialated below ....**Murraya**

5. Style short; stamens dialated below ....**Clausena**

Citrus Linn.

C. medica Linn. Sp. Pl. 782.1753; FBI 1:514; FUGP 1:131; P.C. Kanjilal, For. Fl. UP. 2:110.1966.

Small, armed trees or shrubs. Leaves 1-foliolate, 6-10 cm long, elliptic, oblong-lanceolate, margin crenate, serrate, glabrous. Flowers always tinged with pink. Rarely white flowered variety may also be seen. Fruits oblong or ovoid, mammilate at the apex.

Local name : Bijora
Flowers : April-May; Fruits : November-January

Found frequently in swampy places in the forest in South Kheri Forest division. The fruits are used medicinally. Singh 7165, 7361 (CDRI).

Aegle Correa nom. cons.

A. marmelos (Linn.) Correa in TLS 5:223. 1800; FBI 1:516; FUGP 1:134; P.C. Kanjilal, For. Fl. UP. 2:115.1966; Srivastava, FG. 66.1976.
Crataeva marmelos Linn. Sp. Pl. 444. 1753.

Medium-sized, deciduous trees, with grey-white bark. Spines 3 cm long, straight axillary, Leaves glabrous, 3-foliolate; leaflets usually three, ovate-lanceolate. Flowers greenish-white. Fruits large, globose, pericarp woody, pulp sweet.

Local name : Bel
Flowers : April-May; Fruits : April-August

Found frequently in the forest area. The ripe fruits are edible and used for preparing cooling drinks. Singh 4112, 7260 (CDRI) Saran & Party 26268 (LWG).

Limonia Linn.

Limonia acidissima Linn. Sp. Pl. ed.2, 554. 1762. Stone & Nicolson in Taxon 27.551, 1978; Bennet Fl. Howrah, 196. 1979.

Limonia elephantum (Corr.) Panigrahi, Taxon 26:577.1977.

Feronia elephantum Correa in TLS 5:225.1800; FBI 1:516; FUGP 1:134.

F. limonia (Linn.) Swingle in J. Wash. Acad. Sci. 4:328. 1914; P.C. Kanjilal, For. Fl. UP. 2:114.1966; Srivastava, FG. 68.1976.

Medium-sized armed, deciduous trees with axillary strong thorns. Leaves imparipinnate, cuneate or obovate, glabrous. Flowers polygamous, pale-green tinged with red in loose pubescent panicles. Fruits large, globose, rind woody.

Local name : Kaintha
Flowers : April-June; Fruits : November-January

Found occasionally in the mixed forests. The fruits are edible. Singh 7438, 8321 (CDRI).

Glycosmis Correa

1. Small trees; leaflets oblong-lanceolate, entire or wavy; fruits mammilate...**G. arborea**

1. Shrubs; leaflets elliptic or ovate-lanceolate, undulated or obscurely crenulate; fruits pitted, glandular ....**G. mauritiana**

G. arborea (Roxb.) DC. Prodr. 1:538.1824; Brizicky in JAA 43:90.1962.

Limonia arborea Roxb. Pl. Corom. 1:60 t.86.1798.

L. pentaphylla Retz. Obs. Bot. 5:24.1788.

Glycosmis pentaphylla auct. plur. (non DC. 1824).

Small evergreen trees upto 2 m high. Leaves imparipinnate, 5-foliolate; leaflets usually 7-12 x 4-5 cm, oblong-lanceolate, entire or slightly wavy, dark green. Flowers white in axillary and terminal 3-12 cm long panicles. Fruits globose, mammilate.

Local name : Makaranda
Flowers and Fruits : March-June

Found frequently in the sal forest. Singh 7151, 7197, 7621, 7662 (CDRI).

G. mauritiana (Lamk.) Tanaka in Bull. Soc. Bot. Fr. 75: 708. 1928; Brizicky in JAA 43:1962; Srivastava, FG. 68.1976.
Limonia mauritiana Lamk. Encycl. Bot. 3:517.1792.
Glycosmis pentaphylla DC. Prodr. 1:538.1824; FBI 1:499 FUGP 1:127; incorrectly given as *G. pentaphylla* (Retz.) Correa by Narayanaswami in RBSI 14(2):12.1941.

Shrubs upto 1 m high. Leaves 2-5 foliolate; leaflets alternate or subopposite, 8-10 x 4-4.5 cm, ovate-lanceolate or elliptic, acute, margin undulated or slightly crenulate. Flowers white, in erect and axillary pubescent panicles. Fruits globose, glandular and pitted.

Local name : Makaranda
Flowers and Fruits : February-August

Found commonly in the sal forest and along roadsides. The ripe fruits are edible. Singh 2815, 4119, 7202, 7586, 7814, 8334, 8483 (CDRI).

Murraya Koenig ex Linn. *nom. cons.*

M. koenigii (Linn.) Spreng. Syst. 2:315.1825; FBI 1:503; FUGP 1:120; P.C. Kanjilal, For. Fl. UP. 2:108.1966, Srivastava, FG. 69.1976.
Bergera koenigii Linn. Mant. 563.1771.

Large deciduous shrubs or small trees, with dark brown bark and pubescent young branches. Leaves 13-20 cm long; leaflets 9-25, ovate-lanceolate with an oblique base, entire or obscurely crenulate. Flowers white in terminal corymbose panicles. Berries ovoid, rugose, black when ripe.

Local name : Nimkathia
Flowers : February-May; Fruits : October-December

Occurs commonly near the edge of the forests. The leaves are used for flavouring curries. Singh 2877, 7153, 8387 (CDRI).

Clausena Burm.f.

C. pentaphylla (Roxb.) DC. Prodr. 1:538.1825; Brandis, For. Fl. 49; FBI 1:503; FUGP 1:129; P.C. Kanjilal, For. Fl. UP. 2:109.1966.
Amyris pentaphylla Roxb. Fl. Ind. 2:247.1824.

Deciduous shrubs; young parts silky tomentose. Leaves strongly aromatic, 3 to 7-foliolate; leaflets subopposite or alternate, 12-15 cm long, acuminate, repand, tomentose beneath. Flowers yellowish in erect terminal, downy panicles. Berries ovoid, papilose.

Local name : Ban Nimbu or Ratanjot
Flowers and Fruits : April-July

Found commonly in the sal forest as an undergrowth. Leaf decoction is given in cough and cold. Singh 3000, 4130, 7303, 7475, 8323 (CDRI); Saran & Party 26233 (LWG).

26. SIMAROUBACEAE

Ailanthus Desf. *nom. cons.*

A. excelsa Roxb. Pl. Cor. 1:t.23.1795; FBI 1:518; FUGP 1:136; Nooteboom in Fl. Males. 6(2):219.1962; P.C. Kanjilal, For. Fl. UP. 2:117.1966; Srivastava, FG.69.1976.

Large deciduous trees; young parts grey tomentose. Leaves imparipinnate; leaflets 6-14 pairs, obliquely-ovate or falcate, acuminate, coarsely toothed. Flowers yellowish in large lax often much branched panicles. Samaras 5 cm long, twisted at the centre, reddish-brown. Seed solitary at the centre.

Local name : Arua
Flowers : January-March; Fruits : April-May

Planted near the Rest House in Bankati range of the Park. The bark is used in medicine. Singh 7210 (CDRI).

27. OCHNACEAE

Ochna Linn.

O. pumila Buch.-Ham. ex D.Don, Prodr. 224.1825; FBI 1:524; FUGP 1:137; P.C. Kanjilal, For. Fl. UP. 2:120.1966.

Small undershrubs, giving number of annual shoots from a perennial root-stock. Leaves oblong-lanceolate, serrulate. Flowers bright yellow, 3-6 together in an umbel at the end of an axillary peduncle.

Local name : Bhui Champa
Flowers : March-June

Found frequently in the sal forest as undergrowth along forest roads and in grasslands. Singh 7345 (CDRI).

28. BURSERACEAE

Garuga Roxb.

G. pinnata Roxb. Hort. Beng. 33. 1814, nom. nud. et Pl. Corom. 3:5 t.208.1819; FBI 1:528; FUGP 1:139; Leenhouts in Fl. Males. 5:215 1958; P.C. Kanjilal, For. Fl. UP. 2:122.1966.

Moderate-sized deciduous trees; young parts and inflorescence pubescent. Leaves 30 cm long; leaflets 6 pairs, ovate, obtusely acuminate, oblique and crenate. Flowers greenish-yellow in long pubescent panicles. Drupes globose with 2-3 pyrenes.

Local name : Ghohar
Flowers and Fruits : April-October

Found frequently in the forest area associated with sal trees. Singh 2881, 7307 (CDRI).

29. MELIACEAE

1. Leaves toothed; fruit indehiscent drupe .... **Azadirachta**

1. Leaves entire, fruit dehiscent capsules:

   2. Seeds winged; stamens free .... **Toona**

   2. Seeds not winged; stamens forming tube .... **Trichilia**

Azadirachta A. Juss.

A. indica A. Juss. in Mem. Mus. Par. 19:221.1830; P.C. Kanjilal, For. Fl. UP. 2:124.1966; Srivastava, FG. 70. 1976.
Melia azadirachta Linn. Sp. Pl. 385.1753; FBI 1:544; FUGP 1:141.

Large trees with dark brown bark. Leaves long pinnate, crowded at the end of the branches; leaflets 9-15, subopposite, falcately lanceolate, acuminate, serrate, glabrous. Flowers white, fragrant, in numerous lax axillary panicles. Drupes ovoid-oblong, smooth.

Local name : Neem
Flowers and Fruits : March-July

Mostly planted near village side or self sown. All parts of the plants have medicinal value. Singh 7217 (CDRI); Saran & Party 26101 (LWG).

Toona Roem.

T. ciliata Roem. Syn. Hesp. 139.1846; Santapau in RBSI 16(1):45.1953; Srivastava, FG. 71.1976.
Cedrela toona Roxb. ex Rottl. & Willd. in Ges. Naturf. Fr. Neuc. Schr. 2:198.1903; FBI 1:568; FUGP 1:143; P.C. Kanjilal, For. Fl. UP. 2:130.1966.

Large deciduous trees with grey bark. Leaves paripinnate; leaflets entire or undulate, ovate or lanceolate, acuminate, glabrous and shining. Flowers white in terminal panicles. Calyx lobe ciliate. Petals oblong, ciliate. Capsules oblong, smooth.

Local name : Tun
Flowers and Fruits : March-June

Found frequently in Bankati range of the Park. Singh 2904, 7142, 7270, 8377 (CDRI).

Trichilia P. Browne *nom. cons.*

T. connaroides (W.&A.) Benth. Acta Bot. Neerl. 11:13.1962; Saldanha & Nicolson, Fl. Hass. Distt. 396.1976.
Zanthoxylum connaroides W.&A. Prodr. 148. 1834.
Walsura trijuga Kurz in J. Asia. Soc. Beng. 44(2):148.1875; Albert in Blumea 6:322.1948.
Heynea trijuga Roxb. ex Sims, Bot. Mag. 41:t.1738.1815; FBI 1:565; FUGP 1:142.

Small trees, with dark-brown bark. Leaves *ca* 30 cm long, rachis jointed; leaflets 5-11, ovate-oblong or lanceolate, acuminate, entire, glabrous. Flowers white in long panicled cymes. Staminal tube divided into 8-10 segments. Fruits capsular, 1-seeded, glabrous, globose.

Flowers and Fruits : April-August

Found occasionally in the sal forest of the Park. Singh 7141, 7267 (CDRI).

30. OLACACEA

Olax Linn.

O. nana Wall. ex Benth. in TLS 18:678.1838; FBI 1:576; FUGP 1:45; P.C. Kanjilal, For. Fl. 2:133.1966.

Small undershrubs, giving herbaceous shoots from a woody stock. Leaves nearly sessile, glabrous, elliptic-lanceolate, glaucos beneath, margin recurved. Flowers white, solitary on slender axillary peduncle. Fruits small, globose.

Flowers and Fruits : April-June

Found frequently in the grasslands. Due to periodical jungle fire the plant becomes dwarf. Singh 2920, 7336 (CDRI).

31. CELASTRACEAE

1. Climbing shrubs with capsular fruits ....**Celastrus**
1. Trees with drupaceous fruits ....**Cassine**

Celastrus Linn.

C. paniculatus Willd. Sp. Pl. 1125.1798; FBI 1:617; FUGP 1:148; Santapau in RBSI 16(1):48.1953; P.C. Kanjilal, For. Fl. UP. 2:136.1966; Srivastava, FG. 71.1976.

Large woody climbers with rough, pale-brown bark and numerous lenticels. Leaves ovate or obovate, shortly acuminate, crenate. Flowers yellowish-green, in terminal compound pubescent panicled cymes. Capsules globose, orange yellow when ripe, 3 to 6-seeded.

Local name : Kangani
Flowers and Fruits : April-August

Found commonly in open forest areas climbing on trees. The oil from the seed is used in medicine. Singh 4129, 7191, 7263, 7403 (CDRI); Saran & Party 26340 (LWG).

Cassine Linn.

C. glauca (Rottl.) O. Kuntze, Rev. Gen. Pl. 1:114.1891; Ding Hou in Fl. Males. ser 1.6:286.1962; Saldanha & Nicolson, Fl. Hass. Distt. 318.1976.
Mangifera glauca Rottl. Nye Saml. Kongel. Norske Vidensk. Selsk. Skr. 2:534.t.4,f.1.1783.
Elaeodendron roxburghii W.&A. Prodr. 157.1834; Wight, Ill. 178.t.71; P.C. Kanjilal, For. Fl. UP. 2:139.1966.
E. glaucum auct. (non Pers. 1805).

Moderate-sized trees. Leaves 10-13 cm long, elliptic, ovate or oblong, acuminate, crenate. Flowers in diffuse dichotomous cymes, greenish-white on long peduncles. Drupes 1.5 cm long, ovoid or obovoid, tipped with remains of style.

Local name : Chauri, Mamar.
Flowers : March-May; Fruits : October-January

Found occasionally in mixed forests in open situation. The bark is used in medicine. Singh 4350 (CDRI).

32. RHAMNACEAE

1. Small trees or woody climbers; ovary superior or half inferior:
    2. Armed shrubs or trees; fruits fleshy drupeaceous .... **Ziziphus**
    2. Unarmed climbing shrubs; fruits dry samaroid .... **Ventilago**
1. Climbers; ovary inferior .... **Gouania**

Ziziphus Mill.

1. Flowers in subsessile, fascicled, axillary cymes:
    2. Trees .... **Z. mauritiana**
    2. Shrubs:
        3. Leaves elliptic, tomentose beneath; fruit fleshy, oblong globose....................................... **Z. mauritiana var. fruticosa**
        3. Leaves orbicular; fruit dry, globose .... **Z. nummularia**

Flowers in axillary and terminal, panicled cymes, or in short cymes:

3. Small trees ....**Z. xylopyrus**

3. Straggling or climbing shrubs:

4. Leaves 2-3 cm long, silky pubescent; petals present........**Z. oenoplia**

4. Leaves 10-12 cm long, glabrous; petals absent ....**Z. rugosa**

Z. mauritiana Lamk. Encycl. 3:319.1789; P.C. Kanjilal, For. Fl. UP. 2:144.1966; Srivastava, FG. 72.1976.
Z. jujuba Lamk. Encycl. 3:318.1789; (non Mill. 1768); FBI 1:632; FUGP 1:152.

Medium-sized trees, young parts densely rusty or grey-tomentose. Leaves 2.5-3 x 8-10 cm, ovate or ovate-oblong, obtuse or acute, entire or serrulate. Flowers greenish-yellow in short axillary, subsessile cymes. Drupes oblong-ovoid.

Local name : Ber
Flowers and Fruits : September-December

Cultivated near village side for its edible fruits. **Singh** 7965, 7448 (CDRI).

Z. mauritiana Lamk. var. **fruticosa (Haines)** Seb. & Balak. in Indian For. 89:525.1963; **Raizada & Saxena** in Indian For. 92:326.1966.
Z. jujuba Lamk. var. *fruticosa* Haines, For. Fl. Chotanagpur 270.1910 et Bot. Bih. Orissa 195. 1927.

Densely branched thorny shrubs. Young branches and inflorescence densely tomentose. Leaves variable, 2-3.5 x 5-7 cm long, elliptic, serrulate, glabrous and dark green above, clothed with dense, pale tomentum beneath. Flowers greenish-yellow in cymes. Drupes fleshy, globose.

Local name: Jharberi
Flowers and Fruits : May - December

Common shrubs forming gregarious patches in grasslands. The fruits are edible and sold in the local market. **Singh** 2915, 3095, 7747, (CDRI); Saran & Party 22167, 25800 (LWG).

Z. nummularia (Burm.f.) W.&A. Prodr. 162.1834; FBI 1:633; P.C. Kanjilal, l.c. 145. 1966; Srivastava, FG. 73.1976.
Rhamnus nummularia Burm.f. Fl. Ind. 61.1768.
Ziziphus rotundifolia Lamk. Encycl. 3:319.1789; FUGP 1:153.

Thorny, tomentose shrubs with grey branches. Leaves shortly stalked, ovate to orbicular, serrate, dark green, velvety above, pale and tomentose beneath. Flowers in axillary compact cymes. Drupes globose and dry.

Local name : Jharberi
Flowers and Fruits : March-December

Found scarcely in the area on dry waste land. The ripe fruits are edible. Singh 3084 (CDRI); Saran & Party 46176 (LWG).

Z. xylopyrus Willd. Sp. Pl. 1:1104. 1789; Roxb. Fl. Ind. 1:611.1820. FBI 1:634; FUGP 1:154; P.C. Kanjilal, loc.cit. 146. 1966; Srivastava, FG. 73. 1976.

Small deciduous trees; young parts and inflorescence softly tomentose. Leaves ovate, broadly elliptic or orbicular, base oblique, serrulate, glabrous above, pubescent beneath. Flowers greenish-white in axillary peduncled cymes. Drupes globose, dry, grey-tomentose when young.

Local name : Banber, Kathber
Flowers : April-June; Fruits : November-February

Found frequently in scrubs and grasslands. The fruits and bark is used for tannin materials. Singh 4105, 4110, 7354, 7505 (CDRI).

Z. oenoplia (Linn.) Mill. Gard. Dict. n.3, 1768; FBI 1:634; FUGP 1:154; Santapau in RBSI 16(1):50.1953; P.C. Kanjilal, loc. cit. 147. 1966; Srivastava, FG. 73.1976.
Rhamnus oenoplia Linn. Sp. Pl. 194.1753.

Scandent shrubs, armed with prickles; young branches rusty-tomentose. Leaves 2-3 cm long, obliquely ovate-lanceolate, acute or shortly acuminate, entire or serrate, glabrous above, brown silky pubescent beneath. Flowers minute in short axillary subsessile cymes. Drupes black when ripe.

Local name : Makoi
Flowers and Fruits : April-July

Found commonly in scrub forest. The fruits are edible. Singh 2952, 3074, 3100 (CDRI).

Z. rugosa Lamk. Encycl. 3:319.1789; FBI 1:636; FUGP 1:154; Santapau in RBSI 16(1):51.1953; P.C. Kanjilal, loc.cit. 147.1966.

Large pubescent woody climbers. Leaves 13-15 cm long, broadly ovate or elliptic, rounded from an oblique often cordate base, serrulate, acute. Flowers yellowish-green in axillary or terminal panicle. Fruits obovoid, stony, 1-seeded.

Flowers and Fruits : March-June

Found occasionally in the forest areas, climbing on trees. Singh 4365, 7135 (CDRI).

Ventilago Gaertn.

V. denticulata Willd. Ges. Natur.f. Fr. Neue Schr. 3:417. 1801; Banerji & Mukherji in Indian For. 96(3):209.1970.
V. calyculata Tul. in Ann. Sc. Nat. ser. 4,8:124.1857; FBI 1:631; P.C. Kanjilal, For. Fl. UP. 2:142.1966.
V. maderaspatana Gaertn. var. *calyculata* King in J. Asiat. Soc. Beng. 65(2):378.1896; FUGP 1:152.

Large climbing shrubs; young parts softly pubescent. Leaves elliptic-oblong, base oblique, acute or acuminate, crenate or entire. Flowers green in long terminal pubescent panicle. Fruits roundish nut in cup-like calyx tube, which prolongs in a linear wing.

Local name : Hariabori, Keonti
Flowers and Fruits : March-June

Found frequently in mixed forest in open places. The oil from seed is edible. Singh 7125, 7680 (CDRI).

Gouania Linn.

G. leptostachya DC. Prodr. 2:40.1825; FBI 1:634; FUGP 1:157; Hara, Fl. E. Himal. 197.1966.

Climbing shrubs with tendrils at the end of the branches. Leaves 6-9 x 3-4 cm, ovate, acuminate, crenate, glabrous, rounded-cordate at the base. Flowers greenish, polygamous on long terminal and axillary panicles. Fruits 1 x 1.5 cm, 3-winged, separating into 3 coriaceous indehiscent carpels.

Flowers : August-October; Fruits : November-January

Occurs rarely in the Bankati range of the Park. Singh 4263 (CDRI).

33. VITACEAE

1. Flowers polygamous; petals 5, free not calyptrate ....**Ampelocissus**
1. Flowers hermaphrodite; petals 4:
    2. Leaves simple; berries 1-seeded ....**Cissus**
    2. Leaves 3-foliolate; berries 2 to 4-seeded ....**Cayratia**

Ampelocissus Planch. *nom. cons.*

A. latifolia (Roxb.) Planch. in DC. Monogr. Phan. 5:378.1887; Santapau in RBSI 16(1):52.1953; Hara, Fl. E. Himal. 198.1966; Srivastava, FG. 74.1976.
Vitis latifolia Roxb. Hort. Beng. 18.1814, nom. nud. et Fl. Ind. 1:661.1820; FBI 1:652; FUGP 1:160; P.C. Kanjilal, For. Fl. UP. 2:154.1966.

Large climbers with weak annual stems. Leaves long petiolate, 3 to 7-lobed, orbicular, cordate, dentate, base 5 to 7-nerved. Flowers reddish-brown in thyrsoid pubescent cymes on a stout peduncle, bearing a fork tendril, close to cymes. Berries 2-seeded, black.

Local name : Panibel, Jangli angur
Flowers : April-July; Fruits : August-October

Commonly found in the forest areas and along railway lines. The ripe fruits are edible. Singh 2973, 4127 (CDRI).

Cissus Linn.

C. repanda Vahl, Symb. 3:18.1794; Santapau in RBSI 16 (1): 53. 1953.
Vitis repanda W.&A. Prodr. 125.1834; FBI 1:648; FUGP 1: 162; P.C. Kanjilal, For. Fl. UP. 2:153.1966.

Large trailing herbs, clothed with rusty-villous tomentum. Leaves simple, broadly ovate-cordate, abruptly acuminate, crenate-serrate, rusty-tomentose beneath when young, afterwards glabrous. Flowers in umbel like clusters, pedicelled.

Flowers and Fruits : August-November

Found occasionally in the area. Singh 7459 (CDRI) Inayat 21355 (CAL & DD) 21356 (DD).

Cayratia Juss. *nom. cons.*

C. trifolia (Linn.) Domin. in Bibl. Bot. 89.371.1927; Suesseng. in Engl. & Prantl, Pfam. ed. 2.20d. 280.1953; Srivastava, FG. 75.1976; Raizada Suppl. 43.1976. (Plate X).
Vitis trifolia Linn. Sp. Pl. 203.1753; FBI 1:654; FUGP 1:163; P.C. Kanjilal, For. Fl. UP. 2:157.1966.
Cissus carnosa Lamk. Encycl. 1:31.1783.
Vitis carnosa (Lamk.) Wall. ex Law. in FBI 1:654.1875.
Cayratia carnosa (Lamk.) Gagnep. in Not. Syst. 1:347.1911.

Twining shrubs with slender stems and branched tendrils. Young parts pubescent. Leaves trifoliolate; leaflets ovate or ovate-lanceolate, crenate-serrate, obtuse or acute. Flowers greenish-white in axillary cymes. Berries globose, black, 2-4 seeded.

Local name : Jangli angur
Flowers : April-June; Fruits : Cold season

Occurs commonly in open areas and in hedges. The ripe fruits are edible. Singh 2974, 7506 (CDRI).

34. LEEACEAE

Leea D. Royen ex Linn. *nom. cons.*

1. Flowers brick red .... **L. alata**
1. Flowers greenish-white:
    2. Leaves simple, large .... **L. macrophylla**
    2. Leaves compound, 1-3 pinnate .... **L. asiatica**

L. alata Edgew. in TLS 20:36.1841; FBI 1:665; FUGP 1:164; P.C. Kanjilal, For. Fl. UP. 2:160.1966.

Stiff perennial herbs up to 1 m high, with a woody root stock. Leaves imparipinnate; leaflets subsessile, opposite, oblong-lanceolate, acute, sharply serrate up to 17 cm long. Flowers red in corymbose cymes. Fruits reddish-purple, succulent.

Flowers and Fruits : April-October

Found frequently near sal forest in moist places and in grasslands. Singh 3051, 7548, 8696 (CDRI).

L. macrophylla Roxb. Hort. Beng. 18.1814, nom. nud. et Fl. Ind. 1:653.1820; FBI 1:654; FUGP 1:164.

Annual herbs upto 1 m high. Leaves simple large, broadly ovate or suborbicular, coarsely serrate; stipules large persistent. Flowers small greenish-white in corymbose cymes. Berries 4 to 6-loculed.

Local Name : Hathikan
Flowers and Fruits : September-November

Found rarely in the grassland in Sonaripur range of the Park. The plant is used in bone fracture. Singh 8529 (CDRI).

L. asiatica (L.) Ridsdale in BHHM 189. 1980; Bennet Name. change in Fl. Plants 318. 1987.
Phytolacca asiatical Linn. Sp. Pl. 1: 474. 1753.
L. crispa Linn. Sp. Pl. 1753.
L. edgeworthii Santapau in RBSI 16(1):54.1953; P.C. Kanjilal, For. Fl. UP. 2:162.1966. (Plate XI).
L. aspera Edgew. in TLS 20:36.1841 (non Wall. 1824) FBI 1:665; FUGP 1:164; Gagnep. in Bull. Soc. Bot. Fr. 57:331-336.1910.

Shrubs about 1-2 m high. Leaves unipinnate or the lower pinnae 3-foliolate; leaflets 15-20 cm long, elliptic, acuminate, serrate, nerves many, close and parallel. Flowers greenish-white in cymes. Berries greenish-black, depressed, globose.

Flowers and Fruits : May-November

Occurs commonly in the sal forest along forest roads. The ripe fruits are edible. Singh 3005, 4104, 4279, 7897 (CDRI).

35. SAPINDACEAE

1. Herbs, climbing with tendrils ....**Cardiospermum**
1. Trees or shrubs:
    2. Flowers zygomorphic ....**Erioglossum**
    2. Flowers actinomorphic:
        3. Leaves alternate, simple; fruits winged capsule ....**Dodonaea**
        3. Leaves paripinnate-compound; fruit drupe, globose, indehiscent...**Schleichera**

Cardiospermum Linn.

C. halicacabum Linn. Sp. Pl. 366.1753; FBI 1:670; FUGP 1: 166; P.C. Kanjilal, For. Fl. UP. 2:169. 1966; Srivastava, FG. 76. 1976.

Annual climbing herbs with tendrils and furrowed stems. Leaves bipinnate; leaflets deeply incised, membranous, glabrous, acuminate. Flowers small, white on long slender peduncled cymes. Peduncles with a pair of opposite tendrils below the cymes, Capsules trigonous, bladder-like. Seeds black.

Local name : Kanphool, Phutaki
Flowers and Fruits : June-November

Occurs occasionally in the hedges in open areas. Singh 4354 (CDRI).

Erioglossum Blume

E. rubiginosum (Roxb.) Blume, Rumphia 3: 118.1847; Backer & Bakh.f. Fl. Java 2:134.1965; P.C. Kanjilal, For. Fl. UP. 2:164.1966.

Sapindus rubiginosum Roxb. Pl. Corom. 1:44,t.62.1795.

Erioglossum edule Bl. Rumphia 3: 118. 1847; FBI 1: 672. 1875.

Medium-sized trees; young parts and inflorescence rusty tomentose. Leaves *ca* 30 cm long; leaflets 8-12, subopposite, elliptic, base oblique, sparsely hairy beneath, acuminate. Flowers brownish-white in small clusters, arranged in terminal panicles. Fruits glabrous, ovoid-elliptic.

Local name : Anga-banga
Flowers and Fruits : April-June

Found frequently in the forest areas in shady places associated with sal of the Park. The ripe fruits are edible. Singh 2901, 7235 (CDRI); Inayat 21524 (DD).

Dodonaea Linn.

D. viscosa (Linn.) Jacq. Enum. Pl. Carib. 19.1760; FBI 1:697; FUGP 1:170; P.C. Kanjilal, For. Fl. UP. 2:168. 1966; Srivastava, FG. 76. 1976.

Ptelea viscosa Linn. Sp. Pl. 118.1753.

Evergreen shrubs or small trees. Leaves alternate, viscid, shining, oblanceolate with cuneate base. Flowers greenish-yellow in short terminal panicles. Capsules thin membranous, winged, yellowish-brown.

Flowers and Fruits : August-March

Commonly grown as hedged plant in Forest Rest House campus. Singh 7585 (CDRI); Saran & Party 26163 (LWG).

Schleichera Willd. *nom. cons.*

S. oleosa (Lour.) Oken. Allg. Naturgesch. (3)2:1341.1841; Santapau on RBSI 16(1):57.1953; P.C. Kanjilal, For. Fl. UP. 2:165.1966; Srivastava, FG. 78.1976.
Pistacia oleosa Lour. Fl. Cochinch. 2:615.1790.
Schleichera trijuga Willd. Sp. Pl. 4(2):1396.1806; FBI 1:681; FUGP 1:167.

Large deciduous trees. Leaves paripinnate; leaflets 2-4 pairs, opposite, oblong or elliptic-oblong, entire, acute. Flowers yellowish-green; male and bisexual flowers on different trees. Fruits indehiscent, ovoid, short pointed.

Local name : Kosum
Flowers and Fruits : April-June

Found frequently in the mixed forest. The ripe fruits are edible. Lac is obtained from the plant. The seed oil is used for burning. Singh 2864, 7262, 8386 (CDRI); Saran & Party 26534 (LWG).

36. ANACARDIACEAE

1. Leaves simple:

 2. Perfect stamens one .... Mangifera

 2. Perfect stamens 5-10:

3. Carpels three, united; drupes on fleshy receptacle.........**Semecarpus**

3. Carpels five, free; drupes not on fleshy receptacle ....**Buchanania**

1. Leaves compound:

4. Leaflets without intra-marginal veins; drupes small compressed...**Lannea**

4. Leaflets with intramarginal veins; drupes large ellipsoid ......**Spondias**

Mangifera Linn.

M. indica Linn. Sp. Pl. 200. 1753; FBI 2:13; FUGP 1:176; Merr. Enum. 2:468; P.C. Kanjilal, For. Fl. UP. 2:171. 1966; Srivastava, FG. 79.1976.

Large evergreen trees with dark coloured bark. Leaves crowded at the end of the branches, oblong-lanceolate, acuminate, margin undulate, glabrous, coriaceous. Flowers yellowish-green in terminal panicles, polygamous. Drupes ovoid.

Local name : Aam
Flowers and Fruits : March-June

Occasionally met in the forests and mostly cultivated in the district, for edible fruits. Singh 7175 (CDRI).

Semecarpus Linn.f.

S. anacardium Linn.f. Suppl. 182.1781; FBI 2:30; FUGP 1:176; Hara, Fl. E. Himal. 187.1966; P.C. Kanjilal, For. Fl. UP. 2:172.1966; Srivastava, FG. 79.1976.

Moderate-sized, deciduous trees with dark coloured bark. Young parts and inflorescence softly grey tomentose. Leaves oblong or obovate-oblong, rounded at the apex, cuneate at the base, entire, coriaceous, hispidly pubescent beneath. Flowers greenish-yellow, polygamous, subsessile, in fascicles arranged on terminal panicles. Fruit a drupe seated on a cupular fleshy receptacle.

Local name : Bhilawa
Flowers : April-May; Fruits : November-December

Found frequently in sal forest. The ripe receptacle is edible. The fruits are used in medicine. Singh 2958, 3087 (CDRI) Saran & Party 26545 (LWG).

Buchanania Roxb.

B. lanzan Spreng. in Schrad. Journ. 4:234.1801; P.C. Kanjilal, For. Fl. UP. 2:173.1966.
B. latifolia Roxb. Hort. Beng. 23.1814, nom. nud. et Fl. Ind. 2:385.1824; FBI 2:33; FUGP 1:173.

Moderate-sized, 13-20 m high trees with dark-grey bark. Leaves thick coriaceous, 13-17 x 4-6 cm, crowded at the end of branches, broadly oblong-ovate, obtuse, glabrous. Flowers greenish-white, sessile in terminal and axillary rusty-pilose panicles. Drupes subglobose, black when ripe.

Local name : Piyal, Chironjee
Flowers and Fruits : February-April

Found frequently in the sal forest associated with *Terminalia* spp. The ripe fruits and seeds are edible and bark has medicinal value. Singh 4111, 4152, 7348, 7641 (CDRI) Saran & Party 26276 (LWG).

Lannea A. Rich.

L. coromandelica (Houtt.) Merr. in JAA 19:353.1939; P.C. Kanjilal, loc.cit. 174. 1966; Srivastava, FG. 79.1976.
Dialium coromandelicum Houtt., Nat. Hist. 11,2: 39. t. 5. f2. 1774.
Odina wodier Roxb. Hort. Beng. 29.1814, nom. nud. et Fl. Ind. 2:293.1824; FBI 2:29; FUGP 1:175.

Moderate-sized, deciduous trees with smooth grey bark. Young parts stellate-tomentose. Leaves imparipinnate; leaflets opposite, 3-5 pairs, 6-13 cm long, oblong-ovate, acuminate. Flowers yellowish, crowded in corymbose fascicles. Male flowers in compound racemes. Female flowers in simple raceme. Drupes reniform, compressed, glabrous, red when ripe.

Local name : Jhingan
Flowers and Fruits : March-May

Found commonly in the forest areas. The bark exude is used in cuts. Singh 2860, 4375, 7208 (CDRI).

Spondias Linn.

S. pinnata (Linn.f.) Kurz, Pegu Rep. A.44.1875; Merr. Enum. 2:470; P.C. Kanjilal, For. Fl. UP. 2:176.1966; Srivastava, FG. 80.1976.
Mangifera pinnata Linn.f. Suppl. 156.1781.
Spondias mangifera Willd. Sp. Pl. 2:751.1799; FBI 2:42; FUGP 1:177.

Medium-sized, deciduous trees with grey bark. Leaves alternate, 20-30 cm long; leaflets subopposite, 3-6 pairs, elliptic-oblong, shortly acuminate, entire, secondary nerves 10-20 pairs, jointed with intramarginal veins. Flowers polygamous, greenish-white in panicles. Drupes ovoid, fleshy, yellow, 4-5 cm long.

Local name : Ambara
Flowers and Fruits : April-June

Occurs occasionally in the forest area of the Park. The fruits are edible and pickled. Singh 7567 (CDRI).

37. MORINGACEAE

Moringa Adans.

M. oleifera Lamk. Encycl. 1:398.11785; Benthall, Trees Calc. 137; P.C. Kanjilal, For. Fl. UP. 22:177.1966; Srivastava, FG. 80.1976.

Guilandina moringa Linn. Sp. Pl. 381.1753.
Moringa pterygosperma Gaertn. Fruct. 2:314.1791; FBI 2:45; FUGP 1:179.

Trees with grey-white trunk and corky bark. Leaves usually 3-pinnate on *ca* 30 cm long greenish rachis. Leaflets 1-1.7 cm long, opposite, elliptic-obovate, Flowers pale-white in panicles. Capsules long pendulous, 9-ribbed.

Local name : Sahjana, Saijan
Flowers : January-April; Fruits : May-June

Occasionally planted near village side. The leaves and pods are used medicinally in high blood pressure. The fruits are used as vegetable. Singh 7577, 7627 (CDRI).

Note: The pink flowered hybrid plant is also seen near villages.

38. FABACEAE (nom.alt. PAPILIONACEAE)

Key to the Tribes

A. Leaves simple, rarely digitately 3-foliolate; stamens monadelphous; pods not jointed..I. **Genisteae**

B. Leaves pinnately or digitately 3-foliolate, leaflets toothed; stamens diadelphous.. ....II. **Trifolieae**

C. Leaves imparipinnate; leaflets entire; stamens usually diadelphous; pods not jointed, usually dehiscent but when indehiscent 1 to 2-seeded ....III. **Galegeae**

D. Leaves imparipinnate or 1 to 3-foliolate; stamens dia or monadelphous; pods jointed or 1-seeded ....IV. **Hedysareae**

E. Leaves paripinnate; rachis ending in a tendril or bristle; stamens diadelphous; pods not jointed, dehiscent.......................................V. **Vicieae**

F. Leaves pinnately 3-foliolate or (1-foliolate as in *Flemingia*) rachis not ending in a tendril or bristle; stamens mon or diadelphous; pods not jointed; dehiscent..VI. **Phaseoleae**

G. Leaves imparipinnate; pods indehiscent, trees ....VII. **Dalbergieae**

Tribe I Genisteae:

Leaves simple, rarely digitately 3-foliolate; Stamens monadelphous; pods turgid..**Crotalaria**

Tribe II Trifolieae:

1. Leaves digitately 3-foliolate ....**Trifolium**

1. Leaves pinnately 3-foliolate:

    2. Pods spirally twisted ....**Medicago**

    2. Pods short, round or oblong ....**Melilotus**

Tribe III Galegeae:

1. Anthers apiculate or gland tipped, hairs centrally attached; flowers pink or purple; pod cylindric fir..**Indigofera**

1. Anthers muticous; hairs basifixed:

    2. Inflorescence axillary raceme, flowers yellow; pod slender, septate..**Sesbania**

    2. Inflorescence terminal or leaf opposed; pod not septate:

        3. Herbs or undershrubs; stamens diadelphous ....**Tephrosia**

        3. Trees, or climbing shrubs; stamens monadelphous ....**Millettia**

Tribe IV Hedysareae:

1. Stamens monadelphous; leaves 2-foliolate ....**Zornia**

1. Stamens diadelphous; leaves not as above:

    2. Stamens in bundles of 9 and 1; plant terrestrials:

        3. Pods one seeded; leaflets exstipellate ....**Lespedeza**

        3. Pods 2 or more seeded, jointed;

            4. Pods twisted inside the calyx..................................**Uraria**

            4. Pods not twisted inside the calyx:

                5. Joints of pods turgid ....**Alysicarpus**

                5. Joints of pods flattened:

6. Trees; racemes fascicled from the old branches ....**Ougeinia**

6. Herbs or shrubs; racemes simple or panicled from the young shoots ....**Desmodium**

2. Stamens in 2 bundles of 5 each; plant hydrophyte ....**Aeschynomene**

Tribe V Vicieae:

1. Climbers; stamen 9, the posterior stamen absent ....**Abrus**

1. Erect or diffuse herbs; stamen 10:

2. Style bearded all round ....**Vicia**

2. Style bearded down one side ....**Lens**

Tribe VI Phaseoleae:

Subtribe 1. **Cajaniae**: Leaves gland dotted beneath; leaflets exstipellate; style beardless.

1. Pods 3 or more seeded:

2. Pods with depressed line between the seeds ....**Atylosia**

2. Pods without depressed lines between the seeds ....**Dunbaria**

1. Pods 1 to 2-seeded:

3. Leaves simple or digitately 3-foliolate; pods turgid ....**Flemingia**

3. Leaves pinnately 3-foliolate; pods compressed ....**Rynchosia**

Subtribe 2. **Phaseolineae**: Leaves not gland-dotted; style bearded: Pods of irritatory bristles; stamens all similar ....**Vigna**

Subtribe 3. **Erythrinae**: Leaves not gland-dotted; leaflets stipellate; flowers conspicuous:

1. Twining shrubs with unequal anthers ....**Mucuna**

1. Dwarf shrubs with uniform anthers ....**Erythrina**

Subtribe 4. **Galactieae**: Leaves not gland-dotted; leaflets stipellate; petals equal:

1. Twining herbs and shrubs:

2. Stamens diadelphous ....**Galactia**

2. Stamens monadelphous ....**Pueraria**

1. Trees or woody climbers ....**Butea**

Tribe VII Dalbergieae:

1. Pods linear-oblong ....Dalbergia
1. Pods suborbicular ....Pterocarpus

Crotalaria Linn.

1. Leaves compound ....C. medicaginea
1. Leaves simple :
   2. Branches winged; stipules decurrent ....C. bialata
   2. Branches not winged; stipules not decurrent:
      3. Trailing herbs ....C. prostrata
      3. Erect herbs or undershrubs:
         4. Pods smaller than calyx:
            5. Flowers yellow ....C. calycina
            5. Flowers blue ....C. sessiliflora
         4. Pods longer than calyx:
            6. Branches slender, densely silky:
               7. Stipules present ....C. mysorensis
               7. Stipules absent ....C. albida
            6. Branches stout; glabrescent ....C. spectabilis

C. medicaginea Lamk. Encycl. 2:201.1790; de Munk in Reinwardtia 6:208.1962; FBI 2:81; FUGP 1:190; P.C. Kanjilal, For. Fl. UP. 2:187.1966; Srivastava, FG. 94.1976.
C. medicaginea var. *luxurians* Baker in FBI 2:81.1876; FUGP 1:191.

Small, perennial herbs with woody base. Leaves 3-foliolate; leaflets variable in shape and size, oblanceolate to ovate. Flowers yellow in long terminal and leaf opposed racemes. Pods obliquely subglobose, adpressed silky outside, 2-seeded.

Local name : Gulabi
Flowers and Fruits : March-September

Commonly found throughout the area in wastelands and in grasslands. Singh 2913, 3073, 7309, 7687, 7741 (CDRI)

C. bialata Schrank, Pl. Raz. Hort. Monac. t. 13.1819; Backer & Bakh.f. Fl. Java 1:579.1963.
C. alata Buch.-Ham. ex D.Don, Prodr. 241.1825; de Munk in Reinwardtia 6:200.1961; P.C. Kanjilal, loc.cit. 186.1966.

Small, hairy, suberect herbs, winged along the internodes by decurrent stipules. Leaves 5-7 cm long, subsessile, obovate to ovate-oblong, mucronate, silky pubescent beneath. Flowers yellow in 2 to 3-flowered racemes on long peduncles. Pods smooth, stalked, beaked.

Local name : Bansani
Flowers and Fruits : March-April

Found frequently in the forest area along forest roads. Singh 2994, 4288, 7863 (CDRI).

C. prostrata Roxb. Hort. Beng. 54.1814, nom. nud. et Fl. Ind. 3:270.1832; FBI 2:67; FUGP 1:186.

Prostrate, perennial herbs, clothed with yellowish-brown silky hairs. Leaves 2-3 cm long, subsessile, obliquely obovate-oblong, obtuse. Flowers yellow in 2 to 4-flowered peduncled raceme. Pods nearly sessile, linear-oblong, glabrous.

Flowers and Fruits : March-May

Found commonly within the forest area and in grasslands. Singh 2967, 4198, 7628, 7841 (CDRI); Inayat 21478 (CAL).

C. calycina Schrank, Pl. Raz. Hort. Monac. t. 12. 1819; FBI 2:72; FUGP 1:188; de Munk in Reinwardtia 6:202.1961; Srivastava, FG. 93.1976.

Erect herbs, branches clothed with silky adpressed hairs. Leaves 3-9 cm long, subsessile, linear, acute, glabrous above, silky beneath. Flowers few in lax racemes. Calyx densely clothed with long silky hairs. Pods sessile, linear-oblong, included, glabrous.

Flowers and Fruits : March-May

Found frequently in the forest area in open situation. Singh 2962, 4201, 7688, 7864 (CDRI).

C. **sessiliflora** Linn. Sp. Pl. ed. 2. 1004.1763; FBI 2:73; FUGP 1:188; de Munk in Reinwardtia 6:214.1961.

Erect, hairy annuals. Leaves 5-13 cm long, linear or lanceolate, narrowed at both ends with setaceous small stipules. Flowers blue, crowded in elongated racemes or occasionally solitary in the axil of upper leaves, deflexed. Pods sessile, oblong, glabrous.

Flowers and Fruits : March-May

Found rarely in the forest area and in grasslands. Singh 7967, 8499 (CDRI); Kanjilal 10660 (LWG).

C. **mysorensis** Roth, Nov. Pl. Sp. 388.1821; FBI 2.70; FUGP 1:188.

Much branched annual herbs, clothed with silky hairs. Leaves sessile, alternate, 3-5 x 0.5-1 cm, linear-oblong, obtuse, hairy. Stipules linear, persistent. Calyx 1.5 cm long, densely hairy. Corolla yellow. Pods sessile, 3 cm long, glabrous, oblong.

Flowers and Fruits : August-October

Found rarely in Dudhwa range of the Park in grassland. Singh 7805 (CDRI).

C. **albida** Heyne ex Roth, Nov. Pl. Sp. 333.1821; FBI 2:71; FUGP 1:189; de Munk in Reinwardtia 6:200.1961; Backer & Bakh.f. Fl. Java 1:580.1963; P.C. Kanjilal, loc.cit. 186.1966.

Erect, perennial, much branched pubescent herbs. Leaves 2-2.5 cm long, linear or oblanceolate, apex mucronate or rounded, gland-dotted, glabrous above and silky pubescent beneath. Flowers yellow in lax usually terminal racemes. Pods oblong, cylindrical, glabrous, 6 to 12-seeded.

Flowers : October-December; Fruits : April-June

Found commonly in the forest area and in grasslands. Singh 2922, 7301, 7675 (CDRI); Inayat 21481 (CAL).

C. spectabilis Roth, Nov. Pl. Sp. 341.1821; Ali in Biol. 226.1966; P.C. Kanjilal, loc.cit. 185.1966; Bennet Nomen. Chang. Fl. Plants 154: 1987.
C. sericea Retz., Obs. Bot. 5.26.1789; FBI 2.75; FUGP 1:189; de Munk in Reinwardtia 6:214.1961; Srivastava, FG. 94.1976.

Erect undershrubs. Leaves 6-12 cm long, oblong-lanceolate, acute, cuneate at the base, with leaf-like persistent stipules. Flowers yellow, tinged with purple, on long terminal racemes with leafy persistent bracts. Pods oblong, glabrous.

Local name : Bansani
Flowers and Fruits : October-December

Occurs rarely in the forest area. Singh 3094 (CDRI).

Trifolium Linn.

T. alexandrinum Linn. Cent. Pl. 1:25.1755; Bailey, Man. Cult. Pl. 581; Hossain in Notes R. Bot. Gdn. Edinb. 23:425.1961; Raizada, Suppl. FUGP 59.1976.

Annuals, 30-50 cm tall. Leaves digitately 3-foliolate; leaflet oblong, obtuse. Flowers greenish-yellow in dense globular or ovate-oblong head provided with involucral bract.

Local name : Bursem
Flowers and Fruits : November-December

Occasionally cultivated in the area for fodder and manure. Singh 7375 (CDRI).

Medicago Linn.

1. Pods reniform, 1-seeded, unarmed .... **M. lupulina**

1. Pods spirally twisted, 2-more seeded, spinous .... **M. polymorpha**

M. lupulina Linn. Sp. Pl. 779.1753; FBI 2.90; FUGP 1:194; Srivastava, FG. 102.1976.

Procumbent, trailing hairy branched herbs. Leaflets obovate, toothed, base deltoid, hairy beneath; stipules lanceolate, acuminate, toothed. Flowers yellow in dense heads, on 2.5 cm long peduncles. Pods unarmed, reniform, faintly veined longitudinally, indehiscent, 1-seeded.

Flowers and Fruits : February-April

Found commonly in moist and muddy places near ponds and in grasslands. Singh 8356 (CDRI).

M. polymorpha Linn. Sp. Pl. 779.1753; van Steenis in Blumea 12:15.1963; Srivastava, FG. 102.1976.
M. denticulata Willd. Sp. Pl. 3:1414. 1803; FBI 2:90; FUGP 1:194.

Glabrous annuals. Leaves 3-foliolate; leaflets 2-2.5 cm long, obovate, cuneate, toothed. Flowers yellow, few umbelled. Pods subglobose of 2-4 spirals, spinous, 6 x 6 mm.

Flowers and Fruits : November-February

Occurs commonly in the winter season as a weed in cultivated fields and in grasslands. Singh 7625, 8355 (CDRI).

Melilotus Juss.

1. Flowers white .... **M. alba**

1. Flowers yellow .... **M. indica**

M. alba Desr. in Lamk. Encycl. 4:63.1797; FBI 2:89; FUGP 1:192; Srivastava, FG. 103. 1976.

Erect or decumbent annuals. Leaves 3-foliolate; leaflet 2-5 cm long, obovate-oblong, distantly serrulate. Flowers white in elongated racemes. Pods 5 mm long, round-oblong.

Flowers and Fruits : November-February

Occurs commonly as a winter weed in the forest plantations. Singh 7608 (CDRI).

M. indica All. Fl. Pedem. 1:308.1785; FUGP 1:191; Santapau in RBSI 16(1):63.1953; Srivastava, FG. 103.1976. *M. parviflora* Desf. Fl. Atlant. 2:192.1798; FBI 2:89.

Erect annuals with slender stems and linear, acuminate stipules. Leaves 3-foliolate; leaflets 2.5 cm long, obovate or oblanceolate, retuse, serrulate. Flowers yellow in dense flowered raceme. Pods glabrous, 1-seeded.

Local name : Mentha
Flowers and Fruits : December-April

Found commonly as a winter weed in the forest plantations and in the fields. Singh 7607 (CDRI).

Indigofera Linn.

1. Leaves simple; pod globose ....**I. linifolia**
1. Leaves compound; pod not globose:
    2. Leaflets 3-5:
        3. Pod angular, winged ....**I. trifoliata**
        3. Pod straight, not winged:
            4. Pod 1 cm long, glabrous, not torulose turgid ....**I. glabra**
            4. Pod 2.5 cm long, compressed pointed at tips..**I. hamiltonii**
    2. Leaflets 5-20:

5. Pods glabrous, 8 to 12-seeded .... **I. cassioides**

5. Pods densely pubescent, 3 to 6-seeded .... **I. hirsuta**

I. linifolia (Linn.f.) Retz. Obs. Bot. 4:29.1786 & 6:33, t.2,1791; FBI 2:92; FUGP 1:229; Gillett in Kew Bull. Add. Ser. 1:35.1958; Ali in Bot. Not.111:546.1958; Srivastava, FG. 101.1976.
Hedysarum linifolium Linn.f. Suppl. Pl. 331.1781.

Perennial trailing or procumbent much branched silvery-pubescent herbs. Leaves simple, 2-3 cm long, linear, silvery mucronate. Flowers bright red, subsessile in dense axillary racemes. Pods silvery white, globose, apiculate, 1-seeded.

Flowers and Fruits : August-January

Found frequently in the grasslands. Singh 4347, 7279, 7789, 7970 (CDRI).

I. trifoliata Linn. Cent. Pl. 2:29.1756; FBI 2:96; FUGP 1:232.

Trailing or suberect branched perennial herbs. Leaves digitately 3-foliolate; leaflets 1.2-2.4 cm long, oblanceolate, cuneate at the base, terminal leaflets sessile; stipules setaceous. Flowers red, 6-12 in short sessile racemes. Pods deflexed, 1.5 cm long, 4-angular and winged.

Flowers and Fruits : August-October

Found rarely in grasslands. Singh 8535 (CDRI).

I. glabra Linn. Sp. Pl. 751.1753; FBI 2:95; FUGP 1:231; Gillet, loc.cit. 48.1958.
I. pentaphylla Murr. in Linn. Syst. Veg. ed. 13, 564.1774.

Slender branched annual herbs. Leaves shortly stalked; leaflets 5, opposite, obovate, slightly hairy. Flowers red in short, axillary, laxly 2 to 4-flowered racemes. Calyx teeth setaceous. Pods turgid, straight, glabrous.

Flowers and Fruits : October-December

Found frequently in waste and open lands; Saran & Party 25778 (LWG).

I. hamiltonii R. Grah. ex Duthie & Prain in Annals R. Bot. Gdn. Calc. 9, pt. 1:22.t.29.1901; FUGP 1:233; P.C. Kanjilal, For. Fl. UP. 2:191.1966.
I. atropurpurea auct. non Buch.-Ham. ex Horn; Baker in FBI 2:101. 1876 pro parte.

Low undershrubs, branches striated with less adpressed hairs. Leaflets usually 3-5, opposite, obovate or oblong, emarginate, sparsely appressed pubescent on both surfaces. Flowers pale-purple, arranged in long racemes. Pods 2-3.5 cm long, linear, ligulate, compressed.

Flowers and Fruits : September-December

Found occasionally in the grasslands. Singh 7183 (CDRI); Inayat 21578 (CAL).

I. cassioides Rottl. ex DC. Prodr. 2.225.1825, Ali in Bot. Notiser 111:569.1958.
I. pulchella sensu Baker in FBI 2:101.1876, pro parte (non Roxb.).

Tall, deciduous shrubs with obscurely hispid branches. Leaves 13 cm long; leaflets 13-21, opposite, elliptic or obovate-oblong, obtuse, thinly adpressed hairy on both surfaces. Flowers pink or violet on axillary raceme. Pods glabrous, turgid, 8 to 12-seeded.

Local name : Girhul
Flowers and Fruits : February-April

Found occasionally in the forest area. The flowers are cooked as vegetable. Singh 4102, 4290, 7635 (CDRI).

I. hirsuta Linn. Sp. Pl. 751.1753; FBI 2:98; FUGP 1:234; Ali in Bot. Notiser 111:559.1958; Srivastava, FG.

100.1976. (Fig. 5).
. astragalina DC. Prodr. 2:228.18225.

Suberect or erect herbs or undershrubs. Stems densely hairy. Leaves imparipinnate; stipules setaceous, plumose; leaflets 7, opposite, 3-4 x 1-1.5 cm, obovate, mucronate. Flowers pink, crowded in erect raceme. Pods 4-gonous, 2 cm long densely hairy, deflexed, 4 to 5-seeded.

Flowers and Fruits : September-October

Found occasionally in grasslands. Singh 7806 (CDRI).

Sesbania Scop. *nom. cons.*

1. Prickly shrub or undershrub .... **S. bispinosa**
1. Unarmed shrub or small tree .... **S. sesban**

S. bispinosa (Jacq.) Fawcett & Rendle, Fl. Jam. 4:24.1923; Santapau in RBSI 16(1):64.1953; P.C. Kanjilal, For. Fl. UP. 2:196.1966; Srivastava, FG. 109.1976.
Aeschynomene bispinosa Jacq. Ic. 3:13,t.564.1793.
Sesbania aculeata Pers. Syn. 2:316.1807, FBI 2:114; FUGP 1:224.

Erect shrubs or undershurbs with prickly branches. Leaves 15-20 cm long; leaflets several on prickly rachis. Flowers yellow on axillary racemes. Pods long, narrowly linear, slightly falcate, beaked.

Local name : Dhaincha
Flowers and Fruits : September-October

Occurs occasionally in the cultivated fields and in swampy places. The leaves and plant is used as green manure. Singh 7454 (CDRI).

S. sesban (Linn.) Merr. in Philip. J. Sci. (Bot.) 7:235.1912; Gillett in Kew Bull. 17:112.1963; P.C. Kanjilal, For. Fl. UP. 2:194.1966; Srivastava, FG. 109.1976.

Aeschynomene sesban Linn. Sp. Pl. 714.1753.
Sesbania aegyptiaca Pers. Syn. 2:316.1807; FBI 2: 114; FUGP 1:223.

Glabrous, soft wooded large shrubs or small trees. Leaves 15-20 cm long; leaflets many 1-2.5 cm long, linear-oblong, mucronate. Flowers pale yellow, often dotted or tinged with purple in lax axillary racemes. Pods 15-25 cm long, twisted, many seeded.

Local name : Jait
Flowers and Fruits : September-October

Commonly cultivated or found as an escape in the waste-lands. The leaves, roots and seeds are used in medicine. Singh 7455 (CDRI).

Tephrosia Pers. *nom. cons.*

1. Calyx lobes deltoid, short .... T. candida
1. Calyx lobes cuspidate, narrow:
    2. Erect glabrescent herbs; pods glabrescent .... T. purpurea
    2. Diffuse downy herbs; pods hairy .... T. pumila

T. candida DC. Prodr. 2.249.1825; FBI 2:111; FUGP 1:225; P.C. Kanjilal, For. Fl. UP. 2:198.1966; Srivastava, FG. 110.1976.
Robinia candida Roxb. Hort. Beng. 56. 1814, nom. nud. et Fl. Ind. 3:327.1832.

Shrubs with grooved branches, clothed with grey persistent velvety pubescence. Leaves about 15 cm long; leaflets 11-12 pairs, elliptic, glabrous above and silky tomentose beneath. Flowers white on long terminal axillary raceme. Pods 7-10 cm long, clothed with brown silky hairs.

Flowers and Fruits : November-December

The plant has been collected from Dudhwa forest nursery. Singh 4235 (CDRI).

T. purpurea (Linn.) Pers. Syn. Pl. 2:329.1807; FBI 2:112; FUGP 1:225; P.C. Kanjilal, For. Fl.UP. 197.1966; Srivastava, FG. 110.1976.
Cracca purpurea Linn. Sp. Pl. 752.1753.

Suberect, much branched, glabrescent or finely downy herbs. Leaflets 7-13, obovate to oblanceolate. Flowers purplish in leaf opposed racemes. Pods slightly recurved, 6 to 8-seeded, glabrescent.

Local name : Sarpokha
Flowers and Fruits : June-October

Commonly found in the wastelands. The root is used against snake bite and in venereal diseases. Singh 7406, 7837 (CDRI).

T. pumila (Lamk.) Pers. Syn. Pl. 2:330.1807; FUGP 1:226; Shah in JBNHS 60(1):296.1963; P.C. Kanjilal, For. Fl. UP. 2:197.1966.
Galega pumila (Lamk. Encycl. 2:599.1788.
T. purpurea (Linn.) Pers. var. *pumila* Baker in FBI 2: 113.1876.

Small, procumbent, perennial, downy herbs. Leaflets 9-11, oblanceolate. Flowers pink, 1-3 on slender peduncles. Pods straight, downy.

Flowers and Fruits : August-October

Occurs occasionally in the grasses near moist places. Singh 7429, 7719 (CDRI).

Millettia W. & A.

1. Woody climbers; pods long, woody, tomentose ....**M. extensa**
1. Trees; pods small, narrow, glabrous ....**M. peguensis**

M. extensa (Benth.) Baker in FBI 2:109.1876; Nasir & Ali in Fl. W. Pak. 413.1972.

Otosema extensa Benth. in Miq. Pl. Jungh. 249.1854.
Millettia auriculata Baker ex Brandis, For. Fl. Ind. 138.1874; FBI 2:108.1876; FUGP 1:227; P.C. Kanjilal, For. Fl. UP. 2:199.1966.

Woody climbers; young parts tomentose. Leaves imparipinnate, crowded at the end of the branches. Leaflets 7-9, opposite, oblong or obovate, cuspidate, entire. Flowers in dense axillary racemes. Pods long, straight, woody, clothed with brown tomentum.

Local name : Gouj
Flowers : April-June; Fruits : Cold season

Common climber in the sal forest. The roots are used for stupefying fish by Tharus. Singh 3099, 4377, 7900 (CDRI).

M. peguensis Ali in Kew Bull. 21(3):489.1969.
M. ovalifolia Kurz in Journ. Asi. Soc. Beng. 42(2):68.1873; FBI 2:107; Benthall, Trees Cal. 145; P.C. Kanjilal, loc. cit. 200.1966; Srivastava, FG. 103.1976.

Small, deciduous trees with greyish bark. Leaves imparipinnate; leaflet usually 7, elliptic-ovate, glabrous, entire. Flowers purple in drooping racemes. Pods linear, oblong, narrowed at both the ends, flattened and slightly curved.

Flowers and Fruits : March-May

Planted in the Forest Rest House campus and on roadside in the district. Singh 7341 (CDRI).

Zornia J.F. Gmel.

Z. gibbosa Span. in Linnaea 15:192.1841; Mohlenbrock in Webbia 16:112, ff.44 & 46.1961; Srivastava, FG. 112.1976.
Z. diphylla auct. plur. (non Pers. 1807).

Slender, diffuse, annual herbs. Leaves stalked; leaflets 2, lanceolate, acute, dotted with black glands beneath. Bracts large, acuminate. Flowers yellow in 2 to 10-flowered raceme. Pods jointed, densely prickly.

Local name : Hirankhuri
Flowers and Fruits : August-October

Occurs commonly in sandy moist places in grasslands. Singh 7469, 7840 (CDRI).

This taxon has been described under the name of *Zornia diphylla* Pers. or *Z. angustifolia* Sm. in Indian Floras. Mohlenbrock loc. cit. points out that the species growing in India belongs to *Z. gibbosa* Span, and not *Z. diphylla* Pers.; the latter being a perennial herb with a limited range of distribution.

Lespedeza Mich.

L. **cuneata** (Du Mont-d. Cours) G.Don, Gen. Syst. 2:307.1832; M.A. Rau in Bull. Bot. Surv. India 10:26.1969; Hara, Fl. E. Himl. 66.1971.

Anthyllis cuneata Du Mont. d. Cours, Bot. Cult. 6:100.1811.

Lespedeza sericea Miq. Ann. Mus. Bot. Lugd. Bot. 3:49.1867; FBI 2:142; FUGP 1:250.

Erect undershrubs. Leaves 3-foliolate, ascending, crowded; leaflets linear-cuneate, glabrous above and adpressed hairy beneath. Flowers 2-4 on short pedicels in the leaf axil. Pods sessile, globose.

Flowers and Fruits : April-May

Occurs occasionally near the edge of the forest in open areas of the Park. Singh 4314 (CDRI).

Uraria Desv.

1. Upper leaves 5 to 9-foliolate .... **U. picta**
1. Upper leaves 1-foliolate, lower 3-foliolate:

2. Racemes dense about 7 cm long, very hairy; lower calyx teeth elongated .... **U. alopecuroides**

2. Racemes more than 7 cm long, elongated; calyx teeth sub-equal:

3. Pods glabrous .... **U. neglecta**

3. Pods hairy .... **U. rufescence**

U. picta (Jacq.) Desv. ex DC. Prodr. 2:324.1825; FBI 2:155; FUGP 1:252; van Meeuwen in Reinwardtia 5:453.1961; P.C. Kanjilal, For. Fl. UP. 202.1966; Srivastava, FG. 110.1976. (Plate XIII).

Hedysarum pictum Jacq. Coll. Bot. 2:262.1788.

Erect, shrubby perennials, with stout pubescent stems. Leaves 20-30 cm long, 5 to 9-foliolate on the upper side; leaflets 16 cm long, linear, lanceolate, glabrous, white above along mid rib. Flowers purple in dense racemes. Pods glabrous, polished, 3 to 6-jointed.

Local name : Ban Makari
Flowers and Fruits : August-October

Found frequently in the sal forest. The plant is used in medicine. Singh 3001, 8485 (CDRI).

U. alopecuroides (Roxb.) Wight Ic. 290.1840; Prain in J. As. Soc. Beng. 66:381.1897; FUGP 1:252; P.C. Kanjilal, loc. cit. 203.1966. *Doodia alopecuroides* Roxb. Fl. Ind. 3:368.1832.

U. repanda Wall. Cat. 5677 nom. nud. 1831-1832; FBI 2.156.

Erect herbs or undershrubs, clothed with small hooked bristles. Leaves 1 to 3-foliolate; leaflets 3, terminal large, lateral small, ovate, elliptic-oblong, mucronate, base cordate. Flowers pink arranged in terminal and axillary very dense racemes. Pods glabrous, 2-jointed.

Flowers and Fruits : September-November

Occurs occasionally in the sal forest as undergrowth. Singh 2987, 7382, 7384, 8461 (CDRI).

U. neglecta Prain in J. Asiat. Soc. Beng. 66:382.1897; FUGP 1.253; P.C. Kanjilal, For. Fl. UP. 204.1966.
U. lagopus Royle Ill. 201.1835 (non DC. 1825).
U. lagopus sensu Baker in FBI 2:156.1876. pro parte (non DC. 1825).

Erect undershrubs with pubescent branches. Leaves 1 to 3-foliolate; stipules long pointed from a broad base; leaflets 3, terminal largest, oblong-elliptic or obovate, apex rounded, mucronate. Flowers purple in dense racemes provided with bristles. Pods 2 to 6-jointed, smooth.

Flowers and Fruits : August-November

Occurs frequently in sal forest in moist shady places. Singh 4320 (CDRI).

U. rufescens (DC.) Schind. in Fedde Rep. 21:14.1925; van Meeuwen in Reinwardtia 5:453.1961.
Desmodium rufescens DC. in Ann. Sc. Nat. 4:101.1825.
Uraria hamosa (Roxb.) Sweet, Hort. Brit. ed. 2:149.1830; FBI 2:156; FUGP 1:253; Prain in J. Asiat. Soc. Beng. 66:382.1897; P.C. Kanjilal, loc.cit. 204.1966.

Straggling, diffuse undershrubs; young parts covered with hooked bristles. Leaves 1 to 3-foliolate; leaflets usually elliptic-ovate or oblong, obtuse or emarginate, glabrous above, downy beneath. Flowers purple on long axillary and terminal lax branched racemes. Pods hairy, 2 to 6-jointed.

Flowers and Fruits : September-November

Found frequently in the sal forest in shady and moist situations. Singh 4154, 4289, 7383, 7871 (CDRI).

Alysicarpus Desv. *nom. cons.*

1. Calyx shorter or equal than the first joint of the pod ....**A. vaginalis**

1. Calyx longer than the first joint of the pod:

2. Flowers in pairs or spike like lax raceme; joints of pods not veined ....**A. bupleurifolius**

2. Flowers in spike like rather dense raceme; joint of pods reticulately veined..**A. longifolius**

A. vaginalis (Linn.) DC. Prodr. 2:353.1825; FBI 2:158; FUGP 1:277; Santapau in RBSI 16(1):69.1953; van Meeuwen in Reinwardtia 6:87.1961; P.C. Kanjilal, For. Fl. UP. 208.1966; Srivastava, FG. 90.1976. *Hedysarum vaginalis* Linn. Sp. Pl. 746.1752.

Erect or prostrate herbs. Leaves 1-foliolate, very variable in size and shape, suborbicular or broad-oblong to oblong-lanceolate, cordate at the base. Flowers yellow, tinged with pink in long raceme. Pods 6 to 8-jointed.

Flowers and Fruits : September-November

Found commonly in the grasslands and in sal forest. The plant is used as fodder. Singh 4126, 4327, 7392, 7424, 7721, 7808 (CDRI).

A. bupleurifolius (Linn.) DC. Prodr. 2:352.1825; FBI 2:158; FUGP 1:255; van Meeuwen in Reinwardtia 6:88.1961; Srivastava, FG. 90.1976.
Hedysarum bupleurifolius Linn. Sp. Pl. 745.1753.

Diffuse or ascending herbs. Leaves linear-lanceolate, acute, glabrous. Flowers pink usually in pairs in spike like loose raceme. Pods stalked, moniliform.

Flowers and Fruits : September-November

Found frequently in moist places amongst grasses. Singh 4348, 7401, 7485 (CDRI).

A. longifolius (Rottl. ex Spreng.) W.&A. Prodr. 233. 1834; FBI 2:159; FUGP 1:256; P.C. Kanjilal, For. Fl. UP. 2:206.1966.
Hedysarum longifolium Rottl. ex Spreng. Syst. 3:319.1826.

Erect glabrous under shrubs. Leaves 1-foliolate, 6-12 cm long, oblong or lanceolate, mucronate, stipules large, lanceolate, acuminate. Flowers reddish in densely elongated racemes. Pods shortly stalked, apiculate, moniliform, glabrous.

Flowers and Fruits : October-November

Occurs occasionally in grasslands. Singh 7444, 7807, 7972 (CDRI).

Ougeinia Benth.

O. oojeinensis (Roxb.) Hochr. in Ann. Cons. Jard. Bot. Geneve 13 & 14: 51.1909; P.C. Kanjilal, For. Fl. UP. 209.1966.

Dalbergia ougeinensis Roxb. Hort. Beng. 53.1814, nom. nud. et Fl. Ind. 3:220.1832 *(D. oojeinensis).*

Ougeinia dalbergioides Benth. in Miq. Pl. Jungh. 216.1852; FBI 2:161; FUGP 1:258.

Desmodium oojeinense (Roxb.) Ohashi in Ginkgoana 1: 117. 1973.

Medium-sized, deciduous trees with grey or light brown bark. Leaves pinnately 3-foliolate; terminal leaflet largest, roundish or obovate, lateral smaller, oblique, ovate or elliptic. Flowers light pink or lilac in short fascicled racemes. Pods long, flat.

Local name : Chajan
Flowers and Fruits : March-May

Occurs occasionally in the sal forest. A good timber tree. Singh 7533, 7657 (CDRI).

Desmodium Desv. *nom. cons.*

1. Pods distinctly jointed and separating into 1-seeded segment:

 2. Leaves 3-foliolate:

  3. Shrubs or undershrubs:

4. Flowers in dense axillary umbellate heads.....**D. triangulare**

4. Flowers in elongated spike or panicled racemes:

5. Bracts foliaceous, concealing the flowers..**D. pulchellum**

5. Bract not concealing the flowers:

6. Joints of pods indehiscent, and more longer than broad ....**D. laxiflorum**

6. Joints of pods dehiscing along the lower uture...........**D. heterocarpon var. heterocarpon**

3. Trailing herbs:

7. Flowers 1-3 in the axil of the leaf ....**D. triflorum**

7. Flowers 6-8 in lax racemes ....**D. microphyllum**

2. Leaves 1-foliolate:

8. Shrubs; pods 4-8-jointed:

9. Petiole winged ....**D. triquetrum subsp. triquetrum**

9. Petiole not winged:

10. Leaflets entire, glabrous above; stem angular...**D. gangeticum**

10. Leaflet repand, scabrous above; stem terete...**D. velutinum**

8. Trailing herbs; pods 2-jointed ....**D. brachystachyum**

1. Pods indistinctly jointed, not separating into segments, but dehiscing in a continuous line along the ventral suture ....**D. motorium**

D. triangulare (Retz.) Merr. in JAA 23:170.1942; Santapau in Kew Bull. 276.1948 (as *D. triangulare* (Retz.) Sant.); van Meeuwen in Reinwardtia 6(3):261.1962; P.C. Kanjilal, For. Fl. UP. 2:211.1966; Srivastava, FG. 98.1976.
Hedysarum triangulare Retz. Observ. Bot. 3:40.1783.
Desmodium cephalotes (Roxb.) W.&A. Prodr. 224.1834; FBI 2:162; FUGP 1:260. (including var. *congestum* (W.&A.) Prain).

Shrubs, about 2 m high, with slender, striated, grey downy branches. Leaves 3-foliolate; leaflets oblong-obovate or elliptic, subacute, glabrous above, grey silky beneath. Flowers pink in dense axillary, erect, many flowered umbels. Pods 2 to 5-jointed, grey silky, slightly curved.

Flowers : July-August; Fruits : November-December

Found commonly in the sal forest as an undergrowth. Singh 3004, 3049, 4161, 7930 (CDRI).

D. pulchellum (Linn.) Benth. Pl. Hongk. 83.1861; FBI 2:162; FUGP 1:260; van Meeuwen in Reinwardtia 6:256.1962; P.C. Kanjilal, loc. cit. 212; Srivastava, FG. 97.1976. (Plate XII).
Hedysarum pulchellum Linn. Sp. Pl. 747.1753.

Shrubs with slender striated, grey pubescent branches. Leaves 3-foliolate; terminal leaflets 12 cm long, oblong-elliptic, undulate, coriaceous, lateral leaflets 6.5 cm long, oblique. Flowers white or cream coloured in clusters on 15-20 cm long erect terminal or axillary racemes, each cluster enclosed between 2-foliolate, orbicular, leafy, persistent bracts. Pods usually 2-jointed covered by bracts.

Local name : Titali
Flowers : July-September; Fruits : November-December

Occurs frequently in sal forest as an undergrowth. The plant is used for making brooms. Singh 3023, 4155, 7391, 8490 (CDRI).

D. laxiflorum DC. in Ann. Sc. Nat. Paris 4:100.1825 (Jan.) et Prodr. 2:325.1825 (Oct.); FBI 2:164; FUGP 1:261; van Meeuwen in Reinwardtia 6:252.1962; P.C. Kanjilal, loc. cit. 213. 1966.

Suberect shrubs with angled pubescent branches. Leaves 3-foliolate; leaflets elliptic-ovate or oblong, obtuse, green, glabrous, rough above, with terminal leaflet largest. Flowers purplish in long clusters of lax axillary and terminal racemes Pods slender, 6 to 8-jointed, slightly downy.

Flowers : July-September; Fruits : November-February

Found frequently in sal forest. Singh 4225, 7512, 8491 (CDRI).

D. heterocarpon (Linn.) DC. Prodr. 2:337.1825.
Hedysarum heterocarpon Linn. Sp. Pl. 747.1753.
Desmodium polycarpum DC. Prodr. 2:334.1825; FBI 2:171 incl. var. *tricaulon* (DC.) Baker; FUGP 1:262.

var. **heterocarpon**

Erect undershrubs with slender, grey-tomentose branches. Leaves 3-foliolate; leaflets elliptic, oblong or obovate, rounded or slightly nothced at the apex. Flowers purple, clustered in dense and axillary racemes. Pods erect, 6 to 8-jointed; ciliate.

Flowers : July-September; Fruits : October-January

Found frequently in sal forest. Singh 2999, 4160, 7915 (CDRI).

D. triflorum (Linn.) DC. Prodr. 2:334.1825; van Meeuwen in Reinwardtia 6:261.1962; FBI 2:173; FUGP 1.263; Srivastava, loc.cit. 98.1976.
Hedysarum triflorum Linn. Sp. Pl. 749.1753.

Much-branched, trailing, perennial herbs, rooting at the nodes. Leaves 3-foliolate; leaflets obovate, obcordate 4-6 x 4-6 mm. Flowers pink or purplish, 3 together in the axil of leaves. Pods 3 to 5-jointed, curved, flat, pubescent.

Flowers : July-September; Fruits : November-January

Found commonly on the sandy soil in grasslands. Singh 7282, 7493 (CDRI).

D. microphyllum (Thunb.) DC. Prodr. 2:337.1825; van Meeuwen in Reinwardtia 6(3):254.1962; Srivastava, FG.

97.1976.
Hedysarum microphyllum Thunb. Fl. Jap. 284.1784.
Desmodium parvifolium DC. in Ann. Sc. Nat. Paris 4:100.1825, et Prodr. 2:334.1825; FBI 2:174; FUGP 1:263.

Procumbent trailing, much branched herbs. Leaves 3-foliolate; leaflets 1.5 cm long, oval or obovate, obtuse, glabrous above, pubsecent beneath; stipules linear, setaceous. Flowers purple in terminal racemes, laxly 6 to 8-flowered. Pods 1.5 cm long, both sutures indented.

Flowers and Fruits : August-November

Found frequently in grasslands. Singh 7932 (CDRI).

D. triquetrum (Linn.) DC. Prodr. 2:326.1825.
Hedysarum triquetrum Linn. Sp. Pl. 746.1753.
D. pseudotriquetrum DC. in Ann. Sci. Nat. Pris 4:100.1825 (Oct.).
Desmodium triquetrum sensu Baker in FBI 2:163.1876 pro parte (quod. syn. *D. pseudotriquetrum* DC.)

Subsp. **pseudotriquetrum**

Erect or decumbent shrubs with grooved triangular branches. Leaves 3-foliolate with winged petiole, oblong-lanceolate, acute, base rounded or cordate. Flowers bluish-purple on long axillary and terminal racemes. Pods compressed, 4 to 6-jointed, densely clothed with silky hairs.

Flowers and Fruits : August-November

Found occasionally in the sal forest along forest roads and in grasslands. Singh 3021, 4156, 7390, 7983, 8498 (CDRI) Sis Ram 772 (DD).

D. gangeticum (Linn.) DC. Prodr. 2:327.1825; FBI 2:168 (including *maculatum* (Linn.) Baker, FUGP 1:264; van Meeuwen in Reinwardtia 6:250.1962.

Hedysarum gangeticum Linn. Sp. Pl. 746.1753.

Erect or decumbent undershrubs upto 1 m high with striate, pubescent branches. Leaves 1-foliolate, 16 x 8 cm, ovate-elliptic to ovate-oblong, glabrous above, grey puberulous beneath. Flowers purple or white in several fascicles, arranged in racemes. Pods falcate, deeply indented on lower suture, indehiscent, covered with minute hooked bristles.

Local name : Sal parni
Flowers and Fruits : August-November

Occurs commonly in sal forests as an undergrowth. Singh 2820, 2839, 3063, 4260, 4301, 7386, 7446, 7676 (CDRI); Sis Ram 2139 (DD).

D. velutinum (Willd.) DC. Prodr. 2:328.1825; Schindl. in Fedde. Repert. 21:6.1925; van Meeuwen in Reinwardtia 6:264.1962; Srivastava, FG. 99. 1976.
Hedysarum velutinum Willd. Sp. Pl. 3:117.1803.
Desmodium latifolium DC. Prodr. 2:328.18225; FBI 2:168; FUGP 1:264.

Erect undershrubs with terete branches, densely clothed with fulvous hairs. Leaves 1-foliolate, densely fulvous-hairy; stipules cordate-ovate at the base, long acuminate, hairy, ciliate leaflets, broadly ovate or suborbicular, obtuse or mucronate, densely clothed with short fulvous hairs beneath. Flowers fascicled, arranged on terminal and axillary long racemes. Pods 4 to 6-jointed, densely clothed with minute hooked bristles.

Flowers and Fruits : August-December

Occurs commonly in the sal forest. Singh 2967, 7385 (CDRI).

D. brachystachyum R. Grah. ex Benth. in Miq. Pl. Jungh. 223.1832; FBI 2:71; FUGP 1:265.

Trailing herbs with slender branches. Leaves 1-foliolate, round or broadly oblong, obtuse, cordate at the base, Flowers violet in short dense axillary and terminal subsessile racemes. Calyx densely clothed with white hairs. Corolla twice the calyx.

Flowers : February-March

Found scarcely in the grasslands. Singh 7494 (CDRI).

D. motorium (Houtt.) Merr. in JAA 19:343.1948; van Meeuwen in Reinwardtia 6:245.1962; P.C. Kanjilal, loc. cit. 216. 1966; Srivastava, FG. 97.1976.
Hedysarum motorium Houtt. Nat. Hist. 2:10,146.1779.
Desmodium gyrans (Linn.f.) DC. Prodr. 2:326.1825; FBI 2:174; FUGP 1:265. var. *roylei* Baker in FBI 2:175. 1876.

Erect, undershrubs, with slender grooved glabrous branches. Leaves 1 to 3-foliolate; leaflets sub-coriaceous, oblong-lanceolate, obtuse, glabrous above, slightly pubescent beneath; side leaflets much smaller or absent, moving by jerks. Flowers pink in lax terminal and axillary racemes, elongating after flowering. Pods 2.5-3 cm long, broadly falcate, 6 to 10-jointed, dehiscing along the ventral sutures.

Flowers and Fruits : August-December

Occurs occasionally in the sal forest along forest roads. Singh 4281, 4283, 7549, 7917, 8500 (CDRI); Inayat 21463 (CAL); Sis Ram 786 (DD).

Aeschynomene Linn.

1. Stems slender; calyx glabrous .... A. indica
1. Stems stout; calyx hispid .... A. aspera

A. indica Linn. Sp. Pl. 713.1753; FBI 2:151; FUGP 1:249; Subramanyam, Aq. Ang. 14,f.9.1962; Srivastava, FG. 89. 1976.

Shruby annuals with slender branches. Leaves pinnate on long rachis; leaflets numerous, close, linear, obtuse, sessile, nerved. Flowers yellowish-white in lax, 1 to 4-flowered, axillary racemes. Pods 2.5-5 cm long, linear, jointed.

Local name : Jhingini
Flowers and Fruits : July-October

Occurs commonly in the wet places, water and rice fields. Singh 2968, 7486, 7711 (CDRI).

A. aspera Linn. Sp. Pl. 713.1753; FBI 2:152; FUGP 1:249; P.C. Kanjilal, For. Fl. UP. 2:254.1966; Srivastava, FG. 89.1976.

I have included this taxon on the authority of P.C. Kanjilal loc. cit. who reports the occurrence of this plant from Kheri district. A brief description is reproduced below.

Tall aquatic perennials with stout, spongy, glabrous stems. Leaves. Leaves imparipinnate compound. Flowers large, 2-4, arranged in axillary corymbose racemes which are much shorter than the leaves.

Abrus Linn.

A. precatorius Linn. Syst. 472.1767; FBI 2:175; FUGP 1:241; Breteler in Blumea 10:613.1960; P.C. Kanjilal, For. Fl. UP. 2:219.1966; Srivastava, FG. 89.1976.

Deciduous climbers with tough slender branches. Leaves paripinnate; leaflets 10-20 pairs, oblong, ligulate, membranous, Flowers pinkish in dense racemes. Pods oblong, turgid. Seeds scarlet with a black hilum.

Local name : Ghumachi, Ratti
Flowers : July-September; Fruits : November-December

Found frequently in the mixed forest climbing on hedges and small trees. The root, leaves and seeds are used in medicine. Singh 3003, 4331, 7423 (CDRI).

Vicia Linn.

1. Flowers white; pod 2-seeded, hairy .... V. hirsuta

1. Flowers blue; pod more than 2-seeded, glabrous .... V. sativa

V. hirsuta (Linn.) S.F. Gray, Nat. Arr. Brit. 21.614. 1821; FBI 2:177; FUGP 1:238; Srivastava, FG. 111.1976.
Ervum hirsutum Linn. Sp. Pl. 738.1753.

Twining, hairy annual herbs. Leaves paripinnate ending in tendrils; leaflets 5-10 pairs. Flowers very small, white, on 3 to 6-flowered long peduncles. Corolla not exserted. Pods oblong, rhomboid, turgid, hairy.

Local name : Masur chana
Flowers and Fruits December-March

Commonly found in the cultivated fields and forest plantations. The leaves are cooked as vegetable Singh 7582 (CDRI).

V. sativa Linn. Sp. Pl. 736.1753; FBI 2:178; FUGP 1:238; Srivastava, FG. 111.1976.

Annuals with slender suberect stems, glabrous or obscurely downy. Stipular nectaries present. leaflets 8 pairs. Flowers reddish, solitary or paired, Corolla twice the calyx. Pods 8 to 10-seeded.

Local name : Akara
Flowers and Fruits : January-March

Found commonly in the cultivated fields and in moist waste places. The leaves are cooked as vegetable and used as fodder. Singh 7136 (CDRI).

Lens Mill. *nom. cons.*

L. culinaris Medik in Vorles, Churpf. Phys. Ges. 1:361. 1787 ("culinare"); Bailey, Man. Cult. Pl. ed. 2:553.1949.

vastava, FG. 102.1976.
Lens esculenta Moench. Meth. 131.1794; FUGP 1:237.

Small erect softly pubescent herbs. Leaflets 4-6 pairs; rachis ending in short bristles or tendril. Flowers pale-purple, arranged in racemes. Pods compressed, smooth. Seeds 2, lenticular.

Local name : Masur
Flowers and Fruits : November-January

The plant is cultivated for its seeds, escape in the area. The grains are edible. The leaves and stem are used as fodder. Saran & Party 26593 (LWG).

Atylosia W.&A.

1. Leaflets ovate, round, cuspidate; flowers in long lax racemes..**A. volubilis**
1. Leaflets elliptic-oblong, obtuse, flowers in short axillary racemes ....**A. scarabaeoides**

A. volubilis (Blanco) Gamble, Fl. Presid. Madras 1:369. 1918; P.C. Kanjilal, For. Fl. UP. 2:232.1966.
Cylista volubilis Blanco, Fl. Filip. 509.1837.
Atylosia crassa Prain in J. As. Soc. Beng. 66(2): 45. 1897; FUGP 1:195.
A. mollis sensu Benth. in FBI 2:213.1876 pro parte (non Benth. 1852).

Twining herbs with firm slender pubescent stems. Leaves 3-foliolate; leaflets 6 cm long, ovate, cuspidate-acuminate, strongly nerved beneath. Flowers yellow on long lax racemes. Pods straight, rounded at both ends.

Flowers and Fruits : March-May

Found rarely in the sal forest. Singh 7656 (CDRI); Inayat 21465 (CAL).

A. scarabaeoides (Linn.) Benth. in Miquel. Pl. Jungh. 245.1852; FBI 2:215; FUGP 1:196; P.C. Kanjilal, For. Fl.

UP. 2:233.1966; Srivastava, FG. 91.1976.
Dolichos scarabaeoides Linn. Sp. Pl. 1020.1753.

Biennial herbs with slender trailing or twining, densely grey downy stems. Leaves subdigitately 3-foliolate; leaflets elliptic-oblong, obtuse, 3-nerved. Flowers yellow on 2 to 6-flowered axillary short raceme. Pods 1-2.5 cm long, 4 to 6-seeded, clothed with fine spreading brown silky hairs.

Flowers and Fruits : Most part of the year. Occurs frequently in the area. Singh 3020, 4232, 7717 (CDRI).

Dunbaria W.&A.

D. glandulosa Prain in J. Asiat. Soc. Bengal 66:433.1897; FUGP 1:197; M.A. Rau in Bull. Bot. Surv. India Suppl. 2.10:25.1969.

Woody climbers, clothed with short hairs. Leaves petiolate, 2-foliolate; leaflets exstipellate, rhomboid, cuspidate, gland dotted beneath. Flowers yellow in pairs arranged in long peduncled lax racemes. Pods 4-6 cm long, linear, clothed with yellow bulbous based bristles.

Flowers and Fruits : September-October

Occurs rarely in open grasslands in Sonaripur range of the Park. Singh 8523 (CDRI).

Flemingia Roxb. ex Rottl.

1. Leaves simple:
    2. Flowers in small cymes, enclosed by a large folded, reniform persistent bract:
        3. Leaves oblong or ovate-lanceolate, acute ....**F. strobilifera**
        3. Leaves broadly ovate-cordate, cuspidate ....**F. chapper**
    2. Flowers in loose panicled racemes, bracts minute, caducous, not enclosing the flowers ....**F. paniculata**
1. Leaves 3-foliolate, compound:

4. Erect shrubs or undershrubs with rather tall stems ....**F. macrophylla**

4. Prostrate shrubs with woody subterranean stems:

 5. Leaflets 5-8 cm long ....**F. nana**

 5. Leaflets 2 cm long ....**F. procumbens**

F. strobilifera (Linn.) R.Br. ex Aiton, Hort. Kew ed. 2,4:350.1812; FBI 2:227.
Hedysarum strobiliferum Linn. Sp. Pl. 746.1753.
Moghania strobilifera (Linn.) St. Hil. ex Jacks. in Index Kew. 2:252.1894; Hui-Lin-Li in Amer. J. Bot. 31:227.1944; Mukerjee in Bull. Bot. Soc. Beng. 6:10.1952; including *M. bracteata* (Roxb.) Li and *M. fruticulosa* (Wall. ex Benth.) Mukerjee, Nooteboom in Reinwardtia 5:433.1961; P.C. Kanjilal, For. Fl. UP. 2:235.1966.
Flemingia bracteata (Roxb.) Wt. Icon. 1:t.68.1840; FUGP 1:199.

Erect shrubs with slender striate, pubescent branches. Leaves subcoriaceous, 7.5-10 cm long, oblong or ovate-lanceolate, acute, rounded at the base. Flowers greenish-white, arranged in short cymes on zig zag terminal and hairy racemes, infolded by a conduplicate, membranous, bract. Pods turgid, 2-seeded.

Local name : Kharpati
Flowers and Fruits : September-December

Found commonly in the sal forest as undergrowth. Singh 3091, 4157, 7643, 7666, 2863, 2923, 7255, 8337 (CDRI).

F. chapper Buch.-Ham. in Wall. Cat. 5757.1832; Benth. in Miq. Fl. Jungh. 244.1853; FBI 2:227; FUGP 1.200.
Moghania chapper (Buch.-Ham. ex Benth.) O. Kuntze, Rev. Gen. Pl. 1:199; 1891; Mukerjee in Bull. Bot. Soc. Beng. 6:13.1952; P.C. Kanjilal, loc. cit. 236.1966; Srivastava, FG. 104. 1976.

Erect shrubs with terete adpressed hairy branches. Leaves broadly ovate-cordate, cuspidate, glabrous or slightly hairy beneath, 3-nerved. Flowers yellowish-white in few flowered cymes which are infolded by conduplicate bracts. Pods hairy.

Local name : Titali
Flowers and Fruits : March-May

Commonly met in sal forest as undergrowth. The plant is used for making basket and brooms. Singh 2833, 4282, 7868 (CDRI).

F. paniculata Wall. ex Benth. in Miq. Pl. Jungh. 245.1852; FBI 2:227; FUGP 1:200.
Moghania paniculata (Wall. ex Benth.) Hui-Lin.-Li in Amer. J. Bot. 31:227.1944; Mukerjee loc.cit. 6:12.1952; P.C. Kanjilal, loc.cit. 237.1966.

Erect shrubs with finely pubescent, slender, terete branches. Leaves simple, subcoriaceous, ovate-lanceolate, acuminate. Flowers pink, fascicled in lax terminal and axillary, densely grey hairy panicles. Pods oblong, thinly hairy, 2-seeded.

Local name : Rahari
Flowers and Fruits : December-April

Found frequently in the sal forest as undergrowth. The leaves are used in diabetes. Singh 2987, 7644 (CDRI).

F. congesta Roxb. Hort. Beng. 98.1814, nom. nud. et Fl. Ind. 3:340.1832; FBI 2:228; including var. *semialata);* FUGP 1:201.
Moghania macrophylla (Willd.) O. Kuntze, Rev. Gen. Pl. 1:199.1891; Hui-Lin.-Li in Amer. J. Bot. 31:226.1944; Mukerjee in Bull. Bot. Soc. Beng. 6:16.1952 including *M. prostrate* (Roxb.) Mukerjee, *M. semialata* (Roxb.) Mukerjee and *M. stricta* (Roxb.) O. Ktze., Nooteboom in Reinwardtia 5.434.1961.

Erect shrubs. Young branches silky pubescent. Leaves 3-foliolate; leaflets very variable in size and shape, subcoriaceous, puberulous, rough above, thinly grey-silky beneath. Flowers bracteate, pink, arranged in dense spike like fascicled racemes. Pods hairy, gland dotted.

Flowers and Fruits : September-December

Found commonly in the sal forest areas as undergrowth. Singh 2844, 2859, 3015, 4153, 4101, 4185, 7374, 7935, 8336 (CDRI).

Note: This species is very variable as regards the habit, shape and size of leaflets, inflorescence, the indument of calyx and the size and shape of calyx lobes etc.

F. congesta var. **nana** Baker in FBI 2:229.1876.
Moghania nana (Roxb.) Mukerjee in Bull. Bot. Soc. Beng. 6:20.1952; P.C. Kanjilal, loc.cit. 241.1966.
Flemingia nana Roxb. Hort. Beng. 56.1814, nom.nud. et Fl. Ind. 3:339.1832; Prain in J. Asiat. Soc. Beng. 444; FUGP 1:202.

Dwarf shrubs. Leaves 3-foliolate on long winged petiole; leaflets large, 7-15 cm long, terminal obovate, cuneate, mucronate, the laterals obliquely-oblong. Flowers small, pale-green tinged with pink, arranged on elongated, shortly peduncled racemes. Pods 7-8 mm long, downy covered with glands.

Flowers and Fruits : March-April

Occurs occasionally in the area. Inayat 21507 (DD).

F. procumbens Roxb. Fl. Ind. 3:338.1832; FUGP 1:202.
Moghania procumbens (Roxb.) Mukerjee in Bull. Bot. Soc. Beng. 6:20.1952; P.C. Kanjilal, loc.cit. 242.1966.

Dwarf undershrubs with trailing angled branches, softly grey-hairy when young. Leaves 3-foliolate; leaflets 1.5 cm to 2.5 cm long, terminal obovate, laterals oblique-ovate, oblong, villous, prominently veined. Flowers purple, small, in short dense long peduncled, axillary racemes. Pods sessile, oval, downy densely provided with orange coloured glands.

Flowers and Fruits : April-May

Occurs occasionally in grasslands. Singh 7180 (CDRI); Inayat 21515 (CAL).

Rhynchosia Lour. *nom. cons.*

1. Leaflets small, glabrous .... **R. minima**
1. Leaflets large, softly pubescent .... **R. rothii**

R. minima (Linn.) DC. Prodr. 2:385.1825; FBI 2:223; FUGP 1:204; Nooteboom in Reinwardtia 5:1961; Srivastava, FG. 108.1976.
Dolichos minimus Linn. Sp. Pl. 726.1753.

Twining annuals with slender, much branched stems. Leaves 3-foliolate; leaflets 1-2.5 cm long, obtuse, rhomboid-ovate or obovate, glabrous above, gland dotted beneath. Flowers yellow, arranged in axillary, 6 to 12-flowered, short peduncled racemes. Pods 1 cm long, glabrescent, turgid, slightly recurved.

Flowers and Fruits : September-January

Occurs commonly in the area in open places on small bushes and in grasslands. Singh 2934, 7785 (CDRI).

R. rothii Benth. ex Ait. Cat. Pl. Pb. & Sindh 50.1869; Nooteboom in Reinwardtia 5:441.1961.
R. sericea Span. in Linnaea 15:195.1841; non Gill. ex W.&A. 1833; FBI 2:225; FUGP 1:205; P.C. Kanjilal, For. Fl. UP. 2:243.1966.

Slender woody climbers clothed with short spreading glandular hairs. Leaves 3-foliolate; leaflet 5-10 cm long, softly hairy on both surfaces, gland dotted beneath, terminal rhomboid, laterals oblique, broad ovate, acuminate. Flowers purple, arranged in many flowered racemes. Pods 2.5-4 cm long, pubescent, recurved, beaked.

Flowers and Fruits : August-February

Found frequently in sal forest in moist shady places. Singh 4345, 7441 (CDRI).

Vigna Savi

1. Flowers yellow .... **V. aconitifolia**

1. Flowers reddish-purple .... **V. vexillata**

V. aconitifolia (Jacq.) Marechal in Bull. Jard. Bot. Nat. Belg. 39: 160. 1969. Verdc. in Kew Bull. 23. 464. 1969.
Phasseolus aconitifolius Jacq. Obs. 3:2, t. 52. 1768; FBI 2:202; FUGP 1:207; Srivastava, FG. 106.1976.

Diffuse, slender trailing, hairy herbs. Leaflets 2 to 7-lobed in upper part or 3-lobed; lobes linear-lanceolate; stipules lanceolate. Flowers yellow arranged in heads on axillary peduncles. Pods 2.5 cm long, cylindrical, 7 to 8-seeded.

Flowers and Fruits : August-October

Found frequently in grassland. Singh 7922 (CDRI).

V. vexillata (Linn.) A. Rich. in Sagra Ess. Fl. Cub. 440. 1845.
Phaseolus vexillatus Linn. Sp. Pl. 1017.1753.
Vigna capensis Walp. in Linnaea 13:533.1839; Santapau in RBSI 16(1):80.1953.

Perennial twining herbs. Leaves 3-foliolate; 5-15 cm long, broadly ovate or lanceolate, acute or short acuminate, clothed with appressed pubescent. Flowers reddish-purple, crowded at the ends of axillary peduncles which are longer than leaves. Pods 9-12 cm long, slender.

Flowers and Fruits : September-November

Found occasionally in the forest areas in open places. Singh 7478, 8495 (CDRI).

Mucuna Adans. *nom. cons.*

M. prurita Hook. Bot. Misc. 2:348.1830-1831; Santapau in RBSI 16(1):74.1953; P.C. Kanjilal, For. Fl. UP. 2:219.1966; Srivastava, FG. 105.1976.
M. pruriens Baker in FBI 2:187.1876; FUGP 1:218 (non DC.).

This plant has been included here on the basis of working plan of the South KheriForest division. Hence, a brief description is reproduced below.

Slender climbing annuals. Leaves 3-foliolate; leaflets ovate-rhomboid, glabrous above, grey silky beneath. Flowers dark purple in large, drooping racemes. Pod 5 to 6-seeded, turgid, recurved, abruptly hooked at the tips, clothed with dense pale-brown, irritating bristles.

Local name : Kewanch
Flowers and Fruits : Sept.-Dec.

Erythrina Linn.

E. resupinata Roxb. Pl. Cor. t. 220.1795 et Fl. Ind. 3:257.1832; FBI 2:189; FUGP 2:219; P.C. Kanjilal, For. Fl. UP. 2:221.1966.

Dwarf beautiful undershrubs producing herbaceous stems from a perennial root-stock. Flowers bright scarlet, arranged in racemes which is produced from rootstock directly. Calyx campanulate, 2-lipped. Pods 6.5 x 1.2 cm, flat, curved, 3-seeded.

Flowers and Fruits : March-May

Found frequently in the grassland in Sonaripur range of the Park. Singh 7685, 8350 (CDRI).

Galactia R.Br.

G. tenuiflora (Klein ex Willd.) W.&A. Prodr. 206.1834; FBI 2:192; Noteboom in Reinwardtia 5:431.1961.
Glycine tenuiflora Klein ex Willd. Sp. Pl. 3:1959.1800.
Galactia villosa W.&A. Prodr. 207.1834; FUGP 1:220; Benth. in Miq. Pl. Jungh. 332.1852.

Herbaceous twiners with slender, slightly pubescent stems. Leaves 3-foliolate; leaflets 2.5-3 cm long, oval or elliptic-oblong, obtuse slightly silky pubescent beneath. Flowers purple in elongated racemes. Pods slightly recurved, pubescent afterwards glabrescent.

Flowers and Fruits : October-November

Found occasionally in the open area. Singh 2996 (CDRI).

Pueraria DC.

1. Woody climbers; flowers purplish-blue ....**P. tuberosa**
1. Slender climbers; flowers white or pale-blue ....**P. phaseoloides**

P. tuberosa (Roxb. ex Willd.) DC. Prodr. 2:240.1825; FBI 2:197; FUGP 1.215; P.C. Kanjilal, For. Fl. UP. 2:228.1966.
Hedysarum tuberosum Roxb. Hort. Beng. 57.1814, nom. nud. et Fl. Ind. 3:363.1832; ex Willd. Sp. Pl. 3:1197.1803.

Climbing shrubs with large tuberous roots and tubercled woody stems. Leaves 3-foliolate; leaflets subcoriaceous, ovate-oblong, subacute, base rounded, laterals slightly smaller than the terminal, oblique. Flowers purplish-blue in dense often panicled leafless racemes. Pods 5-8 cm long, flat, clothed with soft brown bristles.

Local name : Patal Kohadr, Vidari Kand.
Flowers : April-May (leafless condition)

Found rarely in Sonaripur range near Chandan Chowki area climbing on trees. The tubers are edible. Singh 8359 (CDRI).

P. phaseoloides Benth. in JLS 9:125.1867; FBI 2:199; FUGP 1:216; P.C. Kanjilal, loc.cit. 229.1966.

Slender climbers with depressed pubescent branches. Leaves 3-foliolate; leaflets membranous, sparsely hairy above, velvety-tomentose beneath, oblique-ovate, cuneate. Flowers pale-blue fading white, fascicled, arranged on long peduncled, axillary puberulous racemes. Pods 5-7 cm long, slender, thinly clothed with stiff adpressed hairs.

Flowers and Fruits : September-November

Found frequently in the sal forest in shady places and in grasslands. Singh 7378, 7727, 8504 (CDRI); Inayat 21464 (a) (DD).

Butea Roxb. ex Willd. *nom. cons.*

1. Flowers large showy, orange red, a tree ....**B. monosperma**

1. Flowers rather small, cream coloured, a climber ....**B. parviflora**

B. monosperma (Lamk.) Taub. in Engl. & Prantl, Nat. Pfam. 3(3):366.1894; P.C. Kanjilal, For. Fl. UP. 2:224.1966; Srivastava, FG. 91.1976.
Erythrina monosperma Lamk. Encycl. 2:391.1786.
Butea frondosa Roxb. Pl. Cor. 1:21.1795; FBI 2:194; FUGP 1:221.

Small deciduous trees with crooked trunk; young parts silky pubescent. Leaves 3-foliolate; leaflets coriaceous, glabrous above, silky tomentose beneath, terminal rhomboid, lateral obliquely ovate. Flowers orange-red on rigid axillary and terminal racems. Pods pendulous, silky tomentose, 1-seeded.

Local name : Paras, Dhak
Flowers and Fruits : March-June

Found frequently in grasslands. The root bark used for making rope Yellow dye is prepared from flowers. The seed is anthelmintic. Singh 7671 (CDRI).

B. parviflora Roxb. Hort. Beng. 54. 1814, nom. nud. et Fl. Ind. 3:248.1832; Santapau in RBSI 16(1):76.1953; P.C. Kanjilal, For. Fl. UP. 2:225.1966.

Spatholobus roxburghii Benth. in Miq. Pl. Jungh. 283.1852; FBI 2:193; FUGP 1:222.

S. parviflorus O. Ktze. Rev. Gen. Pl. 1:205.1891; Hara in Fl. East. Him. 1:164.1966.

Large woody climbers. Young parts, inflorescence silky pubescent. Leaves 3-foliolate; leaflets obovate or elliptic, acute or shortly acuminate, subcoriaceous, glabrous above, finely hairy beneath. Flowers white or cream coloured, small, fascicled on large terminal, pubescent panicles. Pods 7-9 x 2.5-3 cm, clothed with densely brown velvety tomentum.

Local name : Godari
Flowers : September-December; Fruits : March-April

Found commonly climbing on sal and other trees. Singh 2366, 3090, 8360 (CDRI).

Dalbergia Linn.f. *nom. cons.*

1. Stamens 9, in one bundle; leaflets acuminate .... **D. sissoo**

1́. Stamens 10, in two bundles of 5 each; leaflets obtuse or emarginate:

2. Trees .... **D. lanceolaria**

2. Climbing shrubs .... **D. volubilis**

D. sissoo Roxb. Hort. Beng. 53. 1814, nom. nud. et Fl. Ind. 3.223.1832; FBI 2:231; FUGP 1:243; P.C. Kanjilal, For. Fl. UP. 2:247.1966; Srivastava, FG. 95.1976.

Large deciduous trees. Leaves alternate, imparipinnate, 3-5-foliolate; leaflets rhomboid, obliquely ovate, acuminate, glabrous. Flowers yellowish-white in panicled raceme. Pods 3-6 cm long, strap-shaped, glabrous.

Local name : Shisham
Flowers : March-April; Fruits : September-November

Found commonly in the area constituting Shisham-Khair community. A good timber tree. Singh 2850, 7209 (CDRI).

D. lanceolaria Linn.f. Suppl. 316.1781; FBI 2:235; FUGP 1:243 P.C. Kanjilal, For. Fl. UP. 2:248.1966; Srivastava, FG. 95.1976.

I have included this plant on the authority of P.C. Kanjilal loc. cit. who reports the occurrence of this species from Kheri forest area. Hence, a brief descriptions is reproduced below.

Large deciduous trees with glabrous branches. Leaflets 7-17, elliptic or oblong, often emarginate, glabrous above, puberulous beneath. Flowers pinkish-white, unilateral on the spreading branches of large lax axillary and terminal panicles. Pods narrowed at both ends.

Flowers and Fruits : April-June

D. volubilis Roxb. Pl. Cor. 2:48, t. 191. 1798 et Fl. Ind. 3:231.1832; FBI 2:235; FUGP 1:244; P.C. Kanjilal, For. Fl. UP. 2:250.1966.

Large scandent shrubs with green glabrous branches. Leaves 10-20 cm long; leaflets 11-13, thick coriaceous, oblong or ovate-oblong, mucronate. Flowers pale-blue in axillary and terminal leafless panicles bearing dense corymbose cymes. Pods linear, oblong, thin.

Local name : Hariyara
Flowers and Fruits : March-May

Found frequently in sal forest and in open situation. Singh 4175, 8395 (CDRI).

Pterocarpus Linn.

P. marsupium Roxb. Pl. Cor. 2:9 t.116.1798; et Fl. Ind. 3:234.1832; FBI 2:239; FUGP 1:245; P.C. Kanjilal, loc. cit. 251.1966.

I have included this plant on the basis of the working plan of the North Kheri Forest division. Hence, a brief description is reproduced below.

Large deciduous trees. Leaves imparipinnate; leaflets 5-7, coriaceous, elliptic or ovate-lanceolate, acute or shortly acuminate. Flowers yellow in large, much branched, brown pubescent terminal panicles. Pods orbicular, compressed.

Local name : Bijasal

Occurs rarely in Fort range of the Park. The heartwood is used in the treatment of diabetes.

39. CAESALPINIACEAE

1. Leaves unipinnate:
   2. Petals 5; stamens free:
      3. Leaflets many; anthers basifixed ....**Cassia**
      3. Leaflets combined to form one leaflet; anthers versatile. **Bauhinia**
   2. Petals 3; stamens monadelphous ....**Tamarindus**
1. Leaves bipinnate:
   4. Pinnae distant; pods slightly winged ....**Caesalpinia**
   4. Pinnae crowded; pods moniliform ....**Parkinsonia**

Cassia Linn.

1. Fertile stamens 10; staminodes absent:
   2. Pods cylindrical, indehiscent ....**C. fistula**
   2. Pods flat, dehiscent:
      3. Leaflets 4-6 pairs, elliptic-oblong ....**C. surattensis**

3. Leaflets 9-12 pairs, elliptic, obtuse....**C. surattensis** var. **suffruticosa**

1. Fertile stamens 5-7, staminodes present:

4. Trees ....**C. siamea**

4. Herbs or shrubs:

5. Leaves rachii with single gland at the base:

6. Fertile stamens 5; glands stalked ....**C. pumila**

6. Fertile stamens 7; glands not stalked:

7. Leaflets 3-5 pairs ....**C. occidentalis**

7. Leaflets 5-12 pairs ....**C. sophera**

5. Leaves rachii with glands between one or more pairs of leaflets:

8. Leaflets 3 pairs, deltoid; flowers yellow ....**C. tora**

8. Leaflets 2 pairs, oblong, obtuse; flowers pink ......**C. absus**

C. fistula Linn. Sp. Pl. 377. 1753; FBI 2:161; FUGP 1:268; Santapau in RBSI 16(1):90.1953; de Wit in Webbia 11:207.1956; P.C. Kanjilal, For. Fl. UP. 2:1966; Srivastava, FG. 118.1976.

Small trees with greenish-grey or yellowish bark. Leaflets 4-5 pairs, glabrous, 8-15 x 5-6 cm, ovate to oblong-lanceolate. Flowers bright yellow in drooping racemes. Pods 25-60 cm long, cylindrical, dark brown. Seeds flat, embedded in pulp.

Local name : Amaltas
Flowers : April-July; Fruits : October-November

Found occasionally in the mixed forests. The root, leaf and pods are used in medicine. Singh 4380 (CDRI).

C. surattensis Burm. f. Fl. Ind. 97.1768; de Wit in Webbia 11:269.1956; P.C. Kanjilal, loc. cit. 259.1966; Srivastava, FG. 120.1976.
C. glauca Lamk. Encycl. 1:647.1785; FBI 2:265; (exl.

syn. *C. fastigiata* (Vahl); FUGP 1:268. Senna surattensis (Burm. f.) Irvin.

Large shrubs or small trees. Leaflets 4-6 pairs, 6 x 2 cm, elliptic-oblong, pale green, glaucous, glabrous. Flowers yellow in axillary, 7-9 cm long corymbose racemes. Pods straight, flat, thin.

Flowers and Fruits : September-November

Commonly planted in the district along roadside; Singh 4210 (CDRI).

C. surattensis Burm.f. var. **suffruticosa** (Baker) Chatterjee in JBNHS 57:695.1960; Raiz. Indian For. 92.304.1966. *C. glauca* Lamk. var. *suffruticosa* Baker in Hook. f. FBI 2: 265.1878; FUGP 1:269.

Large shrubs. Leaflets 9-12 pairs, 4 x 1 cm, elliptic, obtuse. Flowers yellow. Pods 9 cm long, flat, thin, hirsute.

Flowers and Fruits : March-May

Occurs rarely in Dudhwa range of the Park along railway lines. Singh 2926 (CDRI).

C. siamea Lamk. Encycl. 1:648.1785; FBI 1:2:264; Baily, Man. Cult. Pl. 586; P.C. Kanjilal, For. Fl. UP. 2:263.1966; Srivastava, FG. 119.1976; Raizada, Suppl. FUGP 63.1976.

Middle-sized or large trees with grey bark. Leaflets 6-14 pairs, elliptic-oblong. Flowers yellow in corymbose-raceme, clustered at the end of the branches. Pods flat, slightly curved at the posterior end.

Flowers : April-July; Fruits : October-December

Planted on roadside and in the Forest Rest House campus. Singh 4371 (CDRI).

C. pumila Lamk. Encycl. 1:651.1784; FBI 2:266; FUGP 1: Santapau in RBSI 16(1):91.1953; P.C. Kanjilal, For. Fl. UP. 2:263.1966; Srivastava, FG. 119.1976.

Prostrate ascending or suberect pubescent much branched herbs. Leaflets small, 10-30 pairs, linear, acute. A stipitate gland is present at the base of pinna. Flowers yellow, usually solitary. Pods 2.5-3 cm long, straight, flat, 6 to 12-seeded.

Flowers and Fruits : September- November

Found frequently in sal forest in moist places. Singh 7377, 7843 (CDRI).

C. occidentalis Linn. Sp. Pl. 377.1753; FBI 2:262; FUGP 1:269; de Wit in Webbia 11:256.1956; P.C. Kanjilal, loc. cit. 260.1966; Srivastava, FG. 119.1976.

Undershrubs upto 1 m tall. Leaves provided with an ovoid gland at the base of the petiole. Leaflets 3-5 pairs, ovate-oblong or elliptic-oblong, acuminate. Flowers orange-yellow in axillary corymbose racemes. Pods 9-13 cm long, compressed, slightly curved and transversely septate.

Local name : Kasundha
Flowers and Fruits : July-November

Found commonly in wastelands along roadsides. Singh 4236 (CDRI).

C. sophera Linn. Sp. Pl. 379.1753; FBI 2:262; FUGP 1:269; de Wit in Webbia 11:256.1956; P.C. Kanjilal, loc. cit. 261.1966; Srivastava, FG. 120.1976.

Shrubs or undershrubs. Leaves provided with a gland at the base of the petiole. Leaflets 8-12 pairs, opposite, oblong-lanceolate, acute. Flowers pale-yellow in corymbose racemes. Pods 7-10 cm long, slightly curved, transversely septate, 30 to 40-seeded.

Local name : Kasondhi
Flowers and Fruits : August-December

Found occasionally in wastelands near villages. The seeds and leaves are used in dysentery, piles and stomach disorder. Singh 4168, 4381 (CDRI).

C. tora Linn. Sp. Pl. 376.1753; FBI 2:263, in part; Santapau in RBSI 16(1):90.1953; P.C. Kanjilal, loc.cit. 262. 1966; Srivastava, FG. 120.1976.

Annual herbs upto 90 cm tall. Leaflets 3 pairs, opposite, 3-5 x 1.5-2 cm, obovate-oblong with a gland between the lowest pair. Flowers bright yellow in subsessile pairs, in the leaf axils. Pods stout, obliquely septate.

Local name : Chakwad
Flowers and Fruits : July-November

Found commonly in wastelands along roadsides and other places. The young leaves are cooked as vegetable. Singh 7410 (CDRI).

C. absus Linn. Sp. Pl. 376.1753; FBI 2:265; FUGP 1:271; Santapau in RBSI 16(1):91.1953; de Wit in Webbia 11:279.1956; P.C. Kanjilal, loc.cit. 263.1966; Srivastava, FG. 117.1976 (Plate XIV).

Erect, viscous hairy annual herbs, 30-40 cm high. Leaflets in 2 pairs, 2.5-3 cm long, elliptic-obovate or rhombic-ovate. Flowers reddish, in long racemes. Pods 5 cm long, oblique with grey bristly hairs.

Flowers and Fruits : August-November

Found occasionally in the wastelands. Singh 7530 (CDRI).

Bauhinia Linn.

de Wit in Reinwardtia 3(4):381-539.1956 has reduced the Linnaean genera and the various section recognised under

Bauhinia proper have been restored to their generic status by him. Thothathri Bull. Bot. Soc. Bengal 10(2):130-134.1965. has proposed a number of new combinations affecting the Indian and Burmese species. However, I have followed de Candolle (1825), Bentham & Hooker (1865), Santapau (1967) and Ali (1967) in treating *Bauhinia* in a much wider sense.

1. Fertile stamens 10:

   2. Trees with white flowers; pod dehiscent .... **B. linnaei**

   2. Trees with cream or pale-yellow flowers; pods indehiscent:

      3. Leaves acidic; pods straight, marked with diagonal veins .... **B. malabarica**

      3. Leaves not acidic, pods falcate, reticulately veined.. **B. racemosa**

1. Fertile stamens 3-5:

   4. Woody climbers; stamens 3 .... **B. vahlii**

   4. Trees; stamens 5 .... **B. purpurea**

B. linnaei Ali in Portug. Acta Biol. (B)8:243.1967.
B. acuminata auct. non. Linn. Sp. Pl. 376.1753; Baker in FBI 2:276.1878; FUGP 1:277; P.C. Kanjilal, For. Fl. UP. 2:273.1966.

Small deciduous trees. Leaves 13 x 11 cm, with 2-subacute lobes, coriaceous. Flowers white in short terminal and lateral racemes.

Local name : Safed Kachnar
Flowers : March-April

Planted in Forest Rest House campus. Singh 7690, 8379 (CDRI).

B. malabarica Roxb. Hort. Beng. 31.1814, nom. nud. et Fl. Ind. 2:321.1824; FBI 2:277; FUGP 1:275; Santapau in RBSI 16(1):92.1953; P.C. Kanjilal, loc. cit. 268.1966.
Piliiostigma malabaricum (Roxb.) Benth. Pl. Jungh. 262.1852; Thothathri, Bull. Bot. Soc. Beng. 19:130-134.1965; Srivastava, FG. 122.1976.

Medium-sized trees with dark brown bark. Leaves 5-7 x 6-9 cm, divided into two rounded lobes, coriaceous, glabrous, glaucous beneath, acidic. Flowers cream coloured in dense terminal and axillary corymbose pubescent racemes. Pods straight, 15-22 cm long, glabrous, tapering into a beak.

Local name : Amlosa
Flowers : August-October; Fruits : January-March

Occurs commonly in the mixed forest of the Park. Singh 2867, 4333, 7495, 7909 (CDRI).

B. racemosa Lamk. Encycl. 1:390.1704; FBI 2:276; FUGP 1:274; Santapau in RBSI 16(1):92.1953; P.C. Kanjilal, loc.cit. 267.1966.

Small trees with dark scabrous bark and drooping branches. Leaves 4 x 5 cm, deeply cleft, lobes cordate at the base, obtuse. Flowers pale-yellow in terminal, leaf opposed, 5-10 cm long racemes. Pods slightly falcate, 12-20 cm long, indehiscent.

Local name : Mahuli
Flowers : March-June; Fruits : October-December

Found frequently in mixed forests and in grasslands. The stem bark is used for making fibre. Singh 2861, 7213, 7946 (CDRI).

B. vahlii W.&A. Prodr. 297.1834; FBI 2:279; FUGP 1.276; Santapau in RBSI 16(1):93.1953; P.C. Kanjilal, loc. cit. 269.1966; Srivastava, FG. 114.1976.
Phanera vahlii (W.&A.) Benth. Pl. Jungh. 263.1851-55.

Large woody climbers with opposite, circinate tendrils; young parts clothed with rusty-tomentum. Leaves 10-18 cm long, suborbicular, divided into 2-obtuse lobes, glabrous above, densely tomentose beneath, base cordate. Flowers cream coloured in long peduncled terminal corymbose racemes.

Local name : Mahulain bel
Flowers : March-April

Found frequently in the forest areas climbing on sal trees. The seeds are edible. A strong fibre is obtained from the stem bark. Singh 7335 (CDRI).

B. variegata Linn. Sp. Pl. 375.1753; W.&A. Prodr. 296.1834; FBI 2:284; FUGP 1:277; P.C. Kanjilal, loc.cit. 270.1966.

Middle-sized trees. Leaves divided into 2 obtuse lobes, subcoriaceous, 10 to 15-nerved. Flowers purple veined, variegated, white, fragrant, arranged in lateral corymbs. Pods hard, flat, slightly curved, 30 cm. long 10-20 seeded.

Local name : Kachnar
Flowers and Fruits : November-January

Planted in Rest House campus. The young leaves and flowers are cooked as vegetable. Singh 7987 (CDRI).

Tamarindus Linn.

T. indica Linn. Sp. Pl. 34. 1753; FBI 2:273; FUGP 1:277; P.C. Kanjilal, For. Fl. UP. 2: 274. 1966; Srivastava, FG. 123. 1976.

Large trees with spreading branches. Leaflets 10-12 pairs, subsessile, oblong, obtuse, glabrescent. Flowers yellow in lax few flowered racemes. Pods 7-18 cm long, falcate, indehiscent. Seeds many, dark brown, compressed.

Local name : Imli
Flowers and Fruits : May-June

Commonly planted on roadsides. The fruits are edible. Singh 7997 (CDRI).

Caesalpinia Linn.

1. Pods densely covered with prickles ....**C. bonduc**

1. Pods unarmed, glabrous ....**C. decapetala**

C. bonduc (Linn.) Roxb. Fl. Ind. 2:362.1824; Brenan in Milne-Redhea & Polhill. J. Trop. Afr. 37:1967.
Guilandiorua bonduc Linn. Sp. Pl. 381.1753.
Caeselpinia bonducella (Linn.) Flem. in As. Res. 11: 159.1810; FBI 2:254; FUGP 1:278.

Scandent, prickly shrubs. Leaflets on prickly, 30 cm long rachis, opposite, 6-18 pairs, elliptic-oblong, mucronate, membranous, glabrous. Flowers yellow in dense axillary long peduncled racemes. Pods oblong, 5-7 x 4.5 cm, densely armed with spines.

Local name : Kanja, Karang.
Flowers and Fruits : September-December

Found frequently on the edge of sal forest and open areas. The seeds are used in fever. Singh 3053 (CDRI).

C. decapetala (Roth) Alston in Trimen Handb. Fl. Ceylon 6 (suppl.)89.1931; Santapau & Wagh in Bull. Bot. Surv. India 5(2):105.1963; Srivastava, FG. 116.1976.
Reichardia decapetala Roth, Nov. Pl. Sp. 212.1821.
Caesalpinia sepiaria Roxb. Hort. Beng. 32.1814, nom. nud. et Fl. Ind. 2:360.1832; FBI 2:256; FUGP 1:279; P.C. Kanjilal, For. Fl. UP. 2:277.1966.

I have included this species on the basis of the Working Plan of the North Kheri Forest division. Hence, a brief description is reproduced below.

Climbing shrubs with prickly branches. Leaflets on long prickly rachis, 8-12 pairs, subsessile, oblong, rounded at the apex. Flowers bright yellow in simple terminal and axillary, racemes. Pods oblong, smooth, beaked.

Parkinsonia Linn.

P. aculeata Linn. Sp. Pl. 375.1753; FBI 2:260; FUGP 1:280; P.C. Kanjilal, loc.cit. 279.1966; Srivastava, FG. 121.1976.

Shrubs or small trees with smooth green bark. Branches green and armed with sharp woody spines. Leaves bipinnate; leaflets numerous on flattened rachis, oblanceolate, obtuse. Flowers yellow in lax axillary racemes. Pods 6-9 cm long, turgid, moniliform.

Local name : Bilayati Babool.

Flowers : November-February; Fruits : March-May

Occasionally cultivated near village side and along roads. Singh 7907 (CDRI).

40. MIMOSACEAE

1. Stamens definite (4-8) .... **Mimosa**

1. Stamens indefinite:

   2. Stamens free .... **Acacia**

   2. Stamens united:

      3. Pods straight .... **Albizzia**

      3. Pods twisted or curved .... **Pithecellobium**

Mimosa Linn.

1. Prickly spreading undershrubs; stamens 4 .... **M. pudica**

1. Prickly erect shrubs; stamens 8 ....**M. rubicaulis subsp. himalayana**

M. pudica Linn. Sp. Pl. 518.1853; FBI 2:191; FUGP 1:286; Santapau in RBSI 16(1):95.1953; Brenan in Kew Bull.

184.1955; P.C. Kanjilal, For. Fl. UP. 2:291.11966; Srivastava, FG. 127.1976.

Spreading, undershrubs, provided with prickles and bristles. Pinnae 2 pairs, digitately arranged; pinnules 12-20 pairs, sensitive. Flowers purplish in oblong-globular heads. Pods flat, 3 to 5-jointed, spinous, bristly.

Local name : Chuee-muee
Flowers and Fruits : August-December

Found frequently in open places. Singh 7407 (CDRI); Sis Ram 893 (DD).

M. rubicaulis Lamk. subsp. **himalayana** (Gamble) Ohasi in EFPN 2: 216. 1979; Bennet Nomen. Chang. Fl. Plants 372. 1987. P.C. Kanjilal, For. Fl. UP. 2:291.1966; Raizada, Suppl. FUGP 66.1976 (Plate XV).
M. rubicaulis Baker in FBI 2:291.1878. pro parte, not of (Lamk. 1783); FUGP 1:286.

Erect or straggling shrubs about 2 m high. Branchlets reddish, ribbed, covered with prickles. Pinnae pinnately arranged, 8-12 pairs; pinnules 10-20 pairs on each pinnae. Flowers pink in fascicled heads, forming terminal panicles. Pods slightly curved, smooth, 6-jointed, sutures persisting.

Local name : Aila
Flowers : July-September; Fruits : October-December

Occurs commonly in grasslands and in open places. Singh 3031, 4264, 7947 (CDRI).

Acacia Willd.

1. Climbing shrubs; flowers pink in panicled globose heads ....**A. sinuta**
1. Shrubs or trees; flowers yellow or white:
    2. Flowers in globose heads; spines straight:
        3. Pods cylindric, thick, turgid ....**A. farnesiana**

3. Pods moniliform ....**A. nilotica subsp. indica**

2. Flowers in cylindric spikes; spines hooked:

4. Leaves bipinnate, pinnae 20 pairs ....**A. catechu**

4. Leaves reduced to phyllods ....**A. auriculiformis**

Acacia sinuata (Lour.) Merr. in Trans. Am. Phil. Soc. n.s. 24(2):186.1935; Bennet Nomen. Chang. Fl. Plants 4. 1987.

A. rugata (Lamk.) Merr. in Philipp. J. Sc. Bot. 5:28.1910; Bahadur & Gaur, Acta Bot. Indica 4:66.1976.

Mimosa rugata Lamk. Encycl. 1:20.1783.

A. concinna (Willd.) DC. Prodr. 2:464.1825; FBI 2:296; FUGP 1:292, Santapau in RBSI 16(1):96.1953; P.C. Kanjilal, loc. cit. 303.1966.

Large prickly, climbing shrubs; branches brown dotted with white. Pinnae 4-8 pairs, 3-6 cm long; pinnules 13-20 pairs, oblong, subsessile. Flowers pink in globose, panicled heads. Pods straight, thick and succulent.

Local name : Aila, Shikakaee
Flowers : March-May; Fruits : November-December

Occurs commonly near the edge of the forest. The pods are used medicinally and detergent. Singh 2918, 7189, 8352 (CDRI).

A. farnesiana (Linn.) Willd. Sp. Pl. 4:1083.1806; FBI 2:292; FUGP 1:288; P.C. Kanjilal, For. Fl. UP. 2:294.1966; Srivastava, FG. 125.1976.

Mimosa farnesiana Linn. Sp.. Pl. 521. 1753.

Thorny shrubs or small trees; branches marked with grey dots. Pinnae 2-7 pairs; pinnules 10-20 pairs, minute. Flowers yellow, fragrant, in axillary fascicled globose heads. Pods 4-6 x 1.2 cm, thick, cylindric, hooked at the tips.

Flowers : March; Fruits : April-September

Occurs rarely in the area. Singh 7769 (CDRI).

A. nilotica (Linn.) Willd. ex Del. subsp. **indica** (Benth.) Brenan in Kew Bull. 12:84.1957 et Fl. E. Trop. Afr. 109. 1959; Srivastava, FG. 125.1976.
Mimosa nilotica Linn. Sp. Pl. 521.1753.
Acacia arabica auct. non Willd. Sp. Pl. 4:1085.1806; FBI 2:293; FUGP 1:288; P.C. Kanjilal, loc.cit. 295.1966.

Moderate-sized trees with black bark and straight white spines. Pinnae 2-6 pairs, pinnules 10-20 pairs, 5 x 1 mm, membranous. Flowers yellow in axillary fascicled heads. Pods 7-16 x 2 cm, flat, moniliform, fleshy, grey downy.

Local name : Babul
Flowers : August-October; Fruits : January-April

Found frequently near village side and in open areas. The gums and tannin materials is obtained. The fruits are used as fodder. Singh 3062, 8458 (CDRI).

A. catechu (Linn.f.) Willd. Sp. Pl. 4:1079.1806; FBI 2:295; FUGP 1:290; P.C. Kanjilal, For. Fl. UP. 2:299. 1966; Srivastava, FG. 125.1976.
Mimosa catechu Linn.f. Suppl. Pl. 439.1781.

Moderate-sized trees with rough dark coloured bark and hooked, paired spines. Pinnae 15-20 pairs or more; pinnules numerous, 5 x 1 mm, linear-oblong. Flowers white, fragrant, in cylindrical axillary pubescent spikes. Pods 5-5.6 cm long, flat, acute, dark brown.

Local name : Khair
Flowers and Fruits : August-November

Found commonly in the area alongwith Shisham. The heart-wood is used for making catechu. Singh 3032, 7981 (CDRI).

A. auriculiformis A. Cunn. in Hook. Lond. J. Bot. 1:377.1842; P.C. Kanjilal, loc cit. 305.1966; Srivastava, FG. 124.1976; Raizada, Suppl. FUGP 64.1976.

Medium-sized evergreen trees. Phyllodia curved, coriaceous, parallel nerved, glabrous, narrowed at both ends. Flowers yellow in slender axillary, lax spikes. Pods moniliform, dehiscent, brown.

Local name : Bilayati Babul.
Flowers : September-November; Fruits : February-March

Grown as ornamental trees in Forest Rest House campus and along roads. Singh 7497, 8564 (CDRI).

Albizia Durazz.

1. Stipules foliaceous; leaflets 20-40 pairs ....**A. chinensis**

1. Stipules not foliaceous; leaflets 6-20 pairs:
   2. Flowers stalked; pods straw-coloured ....**A. lebbeck**
   2. Flowers sessile; pods brown or reddish-brown ....**A. procera**

A. chinensis (Osb.) Merr. in Amer. J. Bot. 3:575.1916; P.C. Kanjilal, For. Fl. UP. 2:309.1966.
Mimosa chinensis Osbeck, Dagbok, Ostind. Resa, 233.1757.
Albizia stipulata Boiv. Encycl. 19(2):33.1838; FBI 2: 300; FUGP 1:294.

Large trees with dark-grey rugose bark. Young parts and inflorescence pubescent. Stipules foliaceous, obliquely-cordate, acuminate, velvety beneath. Pinnae 6-20 pairs; pinnules 20-45 pairs, sessile, narrowly falcate, acute, membranous, glabrous. Flowers cream coloured, arranged in globose axillary and terminal panicled heads. Pods thin, glabrous, light brown.

Local name : Kala Siris
Flowers : March-June; Fruits : October-December

Found occasionally in the forest areas and in open places. A timber tree. Singh 2896 (CDRI).

A. lebbeck (Linn.) Benth. in Hook. Lond. J. Bot. 3:87.1844; FBI 2:298; FUGP 1:295; Santapau in RBSI 16(1):97.1953; P.C. Kanjilal, loc.cit. 306.1966; Srivastava, FG. 126.1976.
Mimosa lebbeck Linn. Sp. Pl. 516.1753.

Large or medium-sized deciduous trees with grey-bark. Leaves paripinnate with a large gland at the base. Pinnae 2-3 pairs; pinnules 5-10 pairs, 3.5 x 1.5 x 2 cm. Flowers pale-yellow, fragrant, arranged in globose, umbellate heads. Pods 18-20 x 3.5 cm, flat, dry, straw-coloured with brown spots.

Local name : Siris
Flowers : March-June; Fruits : September-January

Found frequently in the forest areas. A timber tree. Singh 3092 (CDRI).

A. procera (Roxb.) Benth. in Hook. Lond. J. Bot. 89.1844; FBI 2:299; FUGP 1:296; Santapau in RBSI 16(1):97.1953; P.C. Kanjilal, loc.cit. 308.1966.
Mimosa procera Roxb. Pl. Cor. 2:12, t.121.1798.

Tall trees with yellowish or greenish-white bark. Leaves bipinnate with a large gland at the base. Pinnae 2-6 pairs; pinnules 6-16 pairs, oblong-ovate, obtuse, oblique. Flowers yellowish-white in small heads. Pods reddish-brown, shining.

Local name : Safed Siris
Flowers : August-October; Fruits : December-February

Found occasionally in open areas and near the edge of the mixed forests. Singh 7525, 7873 (CDRI).

Pithecellobium Mart. *nom. cons.*

P. dulce (Roxb.) Benth. in Hook. Lond. J. Bot. 3:199.1844; FBI 2:302; FUGP 1:297; Santapau in RBSI 16(1):98.1953; P.C. Kanjilal, For. Fl. UP. 2:311.1966; Srivastava, FG. 127.1976.
Mimosa dulcis Roxb. Pl. Cor. 1:67,t.99.1795.

Medium-sized thorny trees with grey-white or slate coloured bark. Leaves bipinnate. Pinnae 2; leaflets 2, oblique-rhomboid or elliptic-oblong, coriaceous, glabrescent. Flowers white in globose, pubescent, panicled heads. Pods spirally twisted, reddish brown.

Local name : Jangal Jalebi
Flowers and Fruits : March-June

Occurs commonly in the area in villages. The fruits are edible. Singh 2847 (CDRI).

41. ROSACEAE

1. Ripe carpels adnate tc the calyx tube .... **Pyrus**
1. Ripe carpels not enclosed within calyx tube:
    2. Leaves simple; carpel one .... **Prunus**
    2. Leaves compound; carpel many:
        3. Climbing shrubs; calyx ebracteolate .... **Rubus**
        3. Herbs; calyx bracteolate .... **Potentilla**

Pyrus Linn.

P. pashia Buch.-Ham. ex D.Don, Prodr. Fl. Nep. 236.1825; FBI 2:374; FUGP 1:302; P.C. Kanjilal, For. Fl. UP. 2:315.1966.

I have included this species on the authority of P.C. Kanjilal (loc.cit.) who reports its occurrence from Kheri forest area. Hence, a brief description is reproduced below.

Medium-sized, deciduous trees; young parts woolly. Branches dark brown, smooth and dotted with lenticels. Leaves ovate or ovate-lanceolate, crenulate, acuminate. Flowers white in corymbose or subumbellate cymes at the end of short branches. Fruits pome, globose, rough with whitish dots.

Prunus Linn.

P. persica (Linn.) Stokes, Bot. Mat. Med. 3:100.1812; FBI 2:313; FUGP 1:298; P.C. Kanjilal, For. Fl. UP. 2:316.1966; Srivastava, FG. 129.1976.
Amygdalus persica Linn. Sp. Pl. 472.1753.

Small deciduous trees. Leaves 5-7 x 2 cm, oblong-lanceolate, serrate, acute. Flowers pink, sessile, appearing before the leaves. Drupes 4 x 3 cm, pubescent.

Local name : Aru
Flowers and Fruits : March-June

Cultivated in forest nursery for its edible fruits. Singh 2875 (CDRI).

Rubus Linn.

R. niveus Thunb. Diss. Rubo 9, f.3.1813; Focke in Bibl. Bot. 17(72): 182.1911; P.C. Kanjilal, For. Fl. UP. 2:313.1966.
R. lasiocarpus Smith in Rees Cycl. 30, 6.1819; FBI 2:339; FUGP 1:299.

I have included this species on the authority of Kanjilal loc. cit. who reports its occurrence from Kheri forest areas. hence, a brief description is reproduced below.

Straggling prickly shrubs with reddish-brown and whitish tomentose branches. Leaves imparipinnate; leaflets ovate-elliptic or ovate-lanceolate, serrate, glabrous above, densely white tomentose beneath. Flowers pink in terminal and axillary panicled lax corymbs. Fruits globose.

Potentilla Linn.

1. Carpels on a fleshy receptacle ....P. indica
1. Carpels on a dry receptacle ....P. supina

P. indica (Andr.) Wolf. in Asch. & Graebn. Syn. Mitt. Eur. Fl. 6(1):661.1904; et Bibl. Bot. Heff. 71:664.1908.
Fragaria indica Andr. Bot. Rep. 7 t.479.1807; FBI 2:343; FUGP 1:300.
Duchesnea indica (Andr.) Focke in Engl., Pfl.-fam.III 3:33.1888; Hara and Kurosaw in J. Jap. Bot. 34:165.1959.

Perennial hairy herbs with many prostrate stems and root stocks. Leaves long petioled; leaflets 3, stalked, about 3 cm long, obovate, double crenate, membranous. Flowers yellow, solitary on long axillary peduncles. Fruits globose, oblong, bright red.

Flowers and Fruits : September-October

Occurs occasionally on moist shady places in sal forests. The fruits are edible. Singh 7342, 8365 (CDRI).

P. supina Linn. Sp. Pl. 497.1753; FBI 2:359; FUGP 1:301; Srivastava, FG. 128.1976.

Suberect much branched leafy, hairy herbs. Leaves pinnately 3 to 9-foliolate; leaflets lobulate, obovate-oblong, cuneate. Flowers yellow, solitary on slender pedicels. Achenes minute, numerous in globose, villous receptacle.

Flowers and Fruits : January-April; September-October

Occurs commonly in the muddy and moist sandy places. Singh 2889, 7103, 8402 (CDRI).

42. COMBRETACEAE

1. Trees; petiolar glands present ....**Terminalia**
1. Shrubs; petiolar glands absent ....**Combretum**

Terminalia Linn. *nom. cons.*

1. Fruits coriaceous, 5-7 winged:
    2. Leaves glabrous; bark pale, smooth ....**T. arjuna**

2. Leaves tomentose; bark rough, dark-grey ....**T. alata**

1. Fruits fleshy, not winged:

3. Fruits glabrous, ribbed ....**T. chebula**

3. Fruits tomentose, not ribbed ....**T. bellirica**

T. arjuna (Roxb. ex DC.) W.&A. Prodr. 314.1834; FBI 2:447; P.C. Kanjilal, For. Fl. UP. 2:321.1966; Srivastava, FG. 131.1976.
Pentaptera arjuna Roxb. Hort. Beng. 34.1814; nom. nud; DC. Prodr. 3:114.1823.
Terminalia glabra W.&A. Prodr. 314.1834; FUGP 1:307.

Medium-sized, glabrous trees with pale or white smooth bark. Leaves opposite, 10-17 cm long, oblong or elliptic-oblong, coriaceous with 2 glands at the base. Flowers yellowish-white in pendulous, axillary and terminal spikes. Fruits ovoid or obovoid-oblong, 4-5 winged.

Local name : Arjun, Kahua.
Flowers and Fruits : March-June

Found rarely in the forest area. Planted along roadside. The bark is valued for the treatment of heart diseases and diabetes. Sis Ram 1000 (DD).

T. alata Heyne ex Roth, Nov. Pl. Sp. 379.1821; P.C. Kanjilal, loc. cit. 322.1966; Srivastava, FG. 131.1976.
T. tomentosa W.&A. Prodr. 314.1834; FBI 2:447; FUGP 1:310.

Large deciduous trees with dark, deeply furrowed bark; young parts rusty tomentose. Leaves sub-opposite, 25 x 9 cm, oblong-obovate, coriaceous, entire, emarginate. Flowers dull yellow in panicled erect spikes. Fruits 5 winged, crenulate.

Local name : Asana
Flowers : March-May; Fruits : October-November

Occurs commonly in the area associated with sal trees. The ash of the bark is used as detergent. Timber tree. Singh 2943, 4335 (CDRI).

T. chebula Retz. Obs. 5:31.1789; FBI 2:446; FUGP 1:309; Santapau in RBSI 16(1):102.1953; P.C. Kanjilal, loc. cit. 320.1966; Srivastava, FG. 132.1976.

Medium-sized, deciduous trees with dark brown bark. Leaves 10-18 cm long, subopposite, elliptic-oblong, acute, coriaceous, short petiolate. Flowers cream-coloured in terminal panicled cymes. Drupes 3.5 cm long, ovoid or obovoid, ribbed.

Local name : Har
Flowers : April-May; Fruits : August-September

Found occasionally in the forest area. The fruits are used medicinally. Singh 3026, 7908 (CDRI) Sis Ram 993 (DD).

T. bellirica (Gaertn.) Roxb. Pl. Cor. 2.54, t.198.1798; FBI 2:445; FUGP 1:308; Santapau in RBSI 18(1):102.1953; Exell in Fl. Males. 4:569.1954; P.C. Kanjilal, loc.cit. 319.1966; Srivastava, FG. 132.1976.
Myrobalanus bellirica Gaertn. Fruct. 2:9 t,97.1790.

Large deciduous trees with dark grey bark. Leaves alternate, 9-18 x 6-10 cm, elliptic, coriaceous, long petiolate, crowded at the end of the branches. Flowers greenish-yellow, foetid, in slender interrupted spikes. Drupes ovoid or ellipsoid, grey velvety.

Local name : Bahera
Flowers : March-May; Fruits : August-September

Found frequently in the forest areas. The fruits are used medicinally and kernels are eaten. Singh 7910 (CDRI) Sis Ram 985 (DD).

Combretum Linn. *nom. cons.*

C. nanum Buch.-Ham. ex D.Don, Prodr. Fl. Nep. 219.1825; FBI 2:457; FUGP 1:311; P.C. Kanjilal, For. Fl. UP. 2:325.1966.

Woody undershrubs from which several slender stems are given out annually. Leaves 8-9 x 5.4 cm, obovate or elliptic-ovate, coriaceous, glabrous, apex pointed. Flowers cream-white arranged in racemes. Fruits 3.5 cm long with papery wings.

Flowers and Fruits : March-June

Found commonly in grasslands. Singh 7300, 7629 (CDRI); Sis Ram 1020 (DD).

43. MYRTACEAE

1. Fruits berry .... **Syzygium**
1. Fruits capsule .... **Eucalyptus**

Syzygium Gaertn. *nom. cons.*

1. Lateral nerves of leaves distant .... **S. cerasoides**
1. Lateral nerves very fine and close:
   2. Leaves elliptic, ovate-oblong .... **S. cumini**
   2. Leaves narrowly lanceolate-oblong .... **S. heyneanum**

S. cerasoides (Roxb.) Chatt. & Kanjilal. f. JBNHS 55 n.3, 1958 (Dec.); P.C. Kanjilal, For. Fl. UP. 2:331.1966.
Eugenia cerasoides Roxb. Fl. Ind. 2:482.1824.
E. operculata Roxb. Hort. Beng. 37.1814, nom. nud. et Fl. Ind. 2:486.1824; FBI 2:498; FUGP 1:314.

Medium-sized trees with pale-brown or grey bark. Leaves 12-14 x 8-8.5 cm, ovate-oblong or suborbicular, acuminate or rounded, subcoriaceous, nerves prominent beneath. Flowers greenish-white in 3-flowered cymes, arranged in lateral panicles, tetramerous. Berry globose, ovoid.

Local name : Piyaman
Flowers : April-May; Fruits : July-August

Found frequently in the mixed forest. Singh 4136, 7234, 7310 (CDRI).

S. cumini (Linn.) Skeels in U.S. Dept. Agric. Bur. Pl. Ind. Bull. No. 248:2.1912; Santapau in RBSI 16(1):105.1953; P.C. Kanjilal, loc. cit. 333.1966; Srivastava, FG. 134.1976.
Myrtus cumini Linn. Sp. Pl. 471.1753.
Eugenia jambolana Lamk. Encycl. 3:198.1789: FBI 2:499; FUGP 1:314.

Large evergreen trees with smooth, grey bark. Leaves about 15 x 6 cm, elliptic-oblong, bluntly acuminate or acute, coriaceous, glabrous. Flowers white-green, arranged in threes in lateral or terminal trichotomous panicles. Berry oblong-ovoid or subglobose, black.

Local name : Kath Jamun
Flowers and Fruits : March-July

Found commonly in the forest areas. The fruits are edible. The bark and seeds are used in diabetes. Singh 2879, 2949, 7203, 7233 (CDRI).

S. heyneanum Wall. ex Duthie in FBI 2:500.1879; P.C. Kanjilal, loc.cit. 335.1966; Srivastava, FG. 135.1976.
Eugenia heyneana Duthie in FBI 2:500.1879.

Small evergreen trees. Leaves 10-13 x 3 cm, elliptic-lanceolate, slightly acuminate, finely and closely nerved. Flowers white in lax panicled cymes. Berries oblong-obovoid, crowded by the calyx lobes.

Local name : Kathjamuni
Flowers and Fruits : April-September

Found occasionally along river banks and in swamp. The fruits are edible. Singh 7352 (CDRI).

Eucalyptus L'Herit

E. citriodora Hook. in Mitch. J. Trop. Austr. 235.1848; Benthall, Trees Cal. 246; P.C. Kanjilal, For. Fl. UP. 2:338.1966; Raizada, Suppl. FUGP 73.1976.

Tall trees with shining whitish or reddish-grey bark. Leaves 10-15 cm long, narrow, lanceolate, lemon-scented. Flowers white in short panicles. Calyx tube short, operculum double, inner transparent; outer hemispheric, pointed. Capsules ovoid, urceolate.

Local name : Safeda, liptus.
Flowers and Fruits : January-March

Commonly planted in the Forest Rest House campus and along forest roads. A timber wood. Singh 4224, 7571 (CDRI).

44. LECYTHIDACEAE

1. Flowers in slender drooping racemes; fruits angled, 1-seeded..**Barringtonia**
1. Flowers in erect racemes; fruits globose, many seeded ....**Careya**

Barringtonia Forst. *nom. cons.*

B. acutangula (Linn.) Gaertn. Fruct. 2:97,t.1011.1791; FBI 2:508; FUGP 1:316; P.C. Kanjilal, For. Fl. UP. 2:341.1966; Srivastava, FG. 135.1976.
Eugenia acutangula Linn. Sp. Pl. 673.1753.

Small glabrous trees. Leaves 12-16 x 4-5 cm, short-petiolate; obovate-oblong or elliptic, cuneate, crenate-denticulate, apex broadly subacute, glabrous. Flowers dark pink in 15-20 cm long, slender racemes. Fruits 2.5-3 cm long, quadrangular, crowned by the persistent calyx.

Local name : Panyala
Flowers : May-September; Fruits : October-November

Found frequently in swampy areas and near river banks. Singh 4342 (CDRI).

Careya Roxb.

1. Trees; flowers sessile .... **C. arborea**
1. Undershrubs; flowers stalked .... **C. herbacea**

C. arborea Roxb. Hort. Beng. 54.1814; nom. nud. et Pl. Cor. 3:14.t.218.1819; FBI 2:511; FUGP 1:317; Santapau in RBSI 16(1):107.1953; P.C. Kanjilal, For. Fl. UP. 2:343.1966.

Large deciduous trees with rough dark grey bark. Leaves 13-30 x 7-14 cm, obovate-oblong, rounded at apex, shortly acuminate, crenate-dentate glabrous. Flowers yellowish or pinkish-white, sessile, 2-4 together in short terminal spikes. Fruits globose with persistent calyx and styles.

Local name : Kumbhi
Flowers and Fruits : March-July

Found frequently in the forest area. Singh 4374, 7355 (CDRI).

C. herbacea Roxb. Pl. Cor. 3:13 t.217.1819. et Fl. Ind. 2:638.1824; FBI 2:510; FUGP 1:316; P.C. Kanjilal, loc. cit. 343.1966.

Dwarf beautiful undershrubs with a woody root stock and annual shoots. Leaves 17-19 x 8 cm, cuneate, oblong, obovate, obtuse, serrulate, glabrous. Flowers stalked, purple, 1-2 on each shoot, terminal, jointed within the bracts. Fruits globose with persistent calyx and styles.

Local name : Bhumi Kumbhi
Flowers and Fruits : March-May

Occurs commonly on the burnt grasslands. Singh 7302 (CDRI) Sis Ram 1074 (DD).

45. LYTHRACEAE

1. Herbs:
    2. Erect aquatics with woody stem; flowers in whorled cymes............ .... **Ammannia**

2. Prostrate aquatics with herbaceous stem; flowers axillary, sessile, solitary or in spike ....**Rotala**

1. Shrubs or trees:

3. Stamens 8-12:

4. Calyx tube curved; stamens 12 ....**Woodfordia**

4. Calyx tube straight; stamens 8 ....**Lawsonia**

3. Stamens numerous ....**Lagerstroemia**

Ammannia Linn.

1. Leaves cuneate at the base; flowers sessile, clustered in the leaf axils ....**A. baccifera**

1. Leaves subauriculate at the base; flowers stalked in compound cymes..... ....**A. multiflora**

A. baccifera Linn. Sp. Pl. 120.1753; FBI 2:569; FUGP 1:321; Santapau in RBSI 16(1):111.1953; Srivastava, FG. 136.1976.

Glabrous, erect annual herbs. Leaves opposite, 2-4 cm long, sessile, linear-oblong or lanceolate, obtuse. Flowers in dense axillary racemose clusters. Capsules red, globose, depressed. Seeds black, subhemispheric.

Flowers and Fruits : Greater part of the year

Occurs commonly in moist and marshy places throughout the area. Singh 4149, 7219 (CDRI).

A. multiflora Roxb. Hort. Beng. 11.1814, nom. et Fl. Ind. 1:447.1820; FBI 2:570; FUGP 1:322; Santapau in RBSI 16(1):111.1953; Srivastava, FG. 137.1976.

Erect glabrous herbs with 4-gonous woody stems. Leaves opposite, 2-3.5 cm long, sessile, linear-oblong, subauriculate at the base. Flowers stalked, axillary solitary and in 3-flowered simple or compound cymes. Capsules globose, red.

Flowers and Fruits : October-January

Occurs commonly in the moist and marshy places throughout the area. Singh 4148 (CDRI).

Rotala Linn.

1. Leaves obovate or spathulate; calyx campanulate ....**R. indica**

1. Leaves linear; calyx hemispheric ....**R. mexicana**

R. indica (Willd.) Koehne in Engl. Bot. Jahrb. 1:172.1880; Santapau in RBSI 16(1):112.1953; Srivastava, FG. 139.1976.
Peplis indica Willd. Sp. Pl. 2:244.1799.
Ammannia peploides Spreng. Syst. 1:444.1825; FBI 2:566; FUGP 1:320.

Prostrate creeping herbs, rooting at the nodes and then ascending. Leaves opposite, about 2 cm long, obovate or spathulate, rounded at the apex. Flowers sessile in short axillary, branches, solitary in the axil of reduced leaves. Capsules 2-valved, ellipsoid.

Flowers and Fruits : October-January

Found commonly in moist and muddy places along the forest roads. Singh 4180, 7859 (CDRI).

R. mexicana Cham. & Schlect in Linnaea 5:567.1830.
Ammannia pygmaea Kurz in Seem. J. Bot. 5:376.1867; FBI 2:568; FUGP 1:321.

Prostrate annual herbs with tufted stems. Leaves opposite, 1-1.5 cm long, linear or narrow oblong. Flowers minute, sessile and axillary. Capsules 2-valved, globose, acute.

Flowers and Fruits : October-December

Found commonly on moist shady places along forest roads. Singh 4203, 7857 (CDRI).

Woodfordia Salisb.

W. fruticosa (Linn.) Kurz in J. Asiat. Soc. Beng. 4(2): 56.1871; P.C. Kanjilal, For. Fl. UP. 2:345.1966; Srivas-

tava, FG. 139.1976.
Lythrum fruticosum Linn. Sp. Pl. ed. 2, 1:641.1762.
Woodfordia floribunda Salisb. Parad. Lond. t. 42.1806; FBI 2:572; FUGP 1:323.

Large spreading shrubs with reddish-brown bark. Leaves opposite, 10-12 x 2.5 cm, ovate-lanceolate, acuminate, entire. Flowers scarlet in panicled dense axillary clusters. Capsules ellipsoid, enclosed within the persistent calyx tube.

Local name : Dhawaru
Flowers and Fruits : March-May

Found occasionally in the open areas. The flowers are collected and sold in the market. Singh 2932, 7129, 7648, 8347 (CDRI).

Lawsonia Linn.

L. inermis Linn. Sp. Pl. 349.1753; Santapau in RBSI 16(1):114.1953.
L. alba Lamk. Encycl. 3:106.1789; FBI 2:573; FUGP 1.323.

Shrubs with grey bark and spinous branches. Leaves opposite, 1.5-2 cm long, elliptic, subsessile, coriaceous, mucronate. Flowers greenish-white in large terminal panicles. Capsules globose, depressed.

Local name : Mehandi
Flowers and Fruits : May-August

Commonly planted near village side and in Forest Rest House campus. The leaves are used medicinally and for dye. Singh 7283 (CDRI).

Lagerstroemia Linn.

1. Flowers purple; calyx ribbed .... **L. speciosa**

1. Flowers white; calyx not ribbed .... **L. parviflora**

L. speciosa (Linn.) Pers. Syn. 2:72.1806; Benthall, Trees Calc. 260.

Munchausia speciosa Linn. in Muench. Hausv. 1:357, t. 2. 1770.

Lagerstroemia flos-regine Retz. Obs. 5:25.1788; FBI 2:57 pro parte; FUGP 1:325.

Large trees with smooth pale-brown bark. Leaves 14-20 x 6-9 cm; elliptic or oblong-lanceolate, acuminate, coriaceous, lateral nerves prominent. Flowers purple in terminal panicles. Capsules 1.5-2 cm, oblong-ovoid or subglobose.

Flowers and Fruits : July-September

Planted in forest nursery in Dudhwa range of the Park. Singh 7566 (CDRI).

L. parviflora Roxb. Pl. Cor. 1:47, t.66.1795; FBI 2:575; FUGP 1:324; P.C. Kanjilal, For. Fl. UP. 2:346.1966; Srivastava, FG. 138.1976.

Medium-sized deciduous trees with grey or light brown smooth bark. Leaves 5-9 x 3-3.4 cm; opposite, shortly petiolate; elliptic-oblong or ovate, acute, coriaceous, glaucous beneath. Flowers white in axillary and terminal panicles. Capsules 2-3 cm long, ovoid or ellipsoid.

Local name : Asidh
Flowers : April-May; Fruits : October-December

Found frequently in sal forests. Used as fuelwood. Singh 7901 (CDRI).

46. ONAGRACEAE

Ludwigia Linn.

1. Plants prostrate creeping or floating:
2. Flowers white with cream centre ....**L. adscendens**
2. Flowers yellow ....**L. prostrata**

1. Plants erect, not floating:
   3. Capsules linear-cylindric ....**L. octovalvis** subsp. **sessiliflora**
   3. Capsules oblong, inflated ....**L. perennis**

L. adscendens (Linn.) Hara in J. Jap. Bot. 28:291.1953; Raven in Reinwardtia 6(4): 387. 1963; Shah in JBNHS 66(1): 232. 1969; Srivastava, FG. 140.1976.
Jussiaea adscendens Linn. Mant. Pl. 1:69.1767.
J. repens Linn. Sp. Pl. 388.1753; FBI 2:587; FUGP 1:326. (non *Ludwigia repens* Forst. 1771).

Herbs with creeping and floating stems, rooting at the nodes and possessing a number of spongy vescicles. Leaves 2-6 x 1-1.5 cm, obovate to oblanceolate, obtuse. Flowers cream or white, solitary. Capsules cylindric.

Flowers and Fruits : September-November

Commonly found in ponds during rainy season. Singh 3083, 4368, 7409 (CDRI).

L. prostrata Roxb. Hort. Beng. 11.1814, nom. nud. et Fl. Ind. 1:441.1820; FBI 2:588; FUGP 1:327; Santapau in RBSI 16(1):116.1953; Raven loc.cit. 374, f.7,1963; Srivastava, FG. 141.1976.

Prostrate or decumbent herbs. Leaves 4-6 cm long, linear or lanceolate, tapering to the base. Flowers yellow, solitary axillary, tetramerous. Capsules 2-3 cm long, 4-angled.

Flowers and Fruits : August-November

Found frequently on moist and shady places along forest roads. Singh 7538, 7413, 4142, 8482 (CDRI).

L. octovalvis (Jacq.) Raven subsp. **sessiliflora** (Mich.) Raven in Reinwardtia 6:362.1963; Hara, Fl. E. Himal. 225.1966; Srivastava, FG. 141.1976.
Oenothera octovalvis Jacq. Enum. Syst. Pl. 19.1760.
Jussiaea suffruticosa Linn. Sp. Pl. 388.1753; FBI 2;587; FUGP 1:327; P.C. Kanjilal, For. Fl. UP. 2:350.1966.

Erect, perennial herbs. Leaves 8-9 cm long, sessile, linear-lanceolate, slightly hairy, acute. Flowers yellow, solitary in leaf axils, tetramerous. Capsules 2-3 cm long, linear, cylindric, 8-ribbed. Seeds minute, reddish-brown.

Flowers and Fruits : October-December

Found frequently near muddy and water-logged areas. Singh 4261, 7487 (CDRI).

L. perennis Linn. Sp. Pl. 119. 1753; Raven in Reinwardtia 6:367.1963; Srivastava, FG.141.1976.
L. parviflora Roxb. Hort. Beng. 11.1814, nom. nud. et Fl. Ind. 1:440.1820, FUGP 1:327.

Erect herbs. Leaves 10-13 x 8 cm, oblong-lanceolate or linear-lanceolate. Flowers yellow, solitary axillary, tetramerous. Capsules 10 x 2 mm, oblong, inflated, many seeded.

Flowers and Fruits : September-November

Occurs commonly on moist shady places along forest roads. Singh 8532 (CDRI).

47. TRAPACEAE

Trapa Linn.

T. natans Linn. var. **bispinosa** (Roxb.) Makino in Linuma Somuko-Dzeusetz (ed.3) 1:137.1907; Santapau & Wagh in Bull. Bot. Surv. India 5(2):109.1965; Srivastava, FG. 142.1976.
T. bispinosa Roxb. Pl. Cor. 234.1798; et Fl. Ind. 3:428.1832; FBI 2:590; FUGP 1:3299; P.C. Kanjilal, For. Fl. UP. 2:351.1966.

Aquatic herbs, ascending in water. Floating leaves 4 x 6 cm, arranged in rosettes, rhomboidal; submerged leaves dissected; petioles spongy near the top. Flowers white, solitary. Fruits angled nuts, turbinate, provided with a sharp spine on either side.

Local name : Singhara
Flowers and Fruits : August-November

Commonly cultivated in ponds for its edible fruits. Singh 7694 (CDRI).

48. CUCURBITACEAE

1. Female flowers in racemes; stamens 5, free with straight anthers .... **Actinostemma**

1. Female flowers usually solitary; stamens usually 3, rarely 2:

 2. Anther-cells conduplicate or flexuose:

  3. Corolla divided to the base:

   4. Petals fimbriate .... **Trichosanthes**

   4. Petals entire:

    5. Calyx tube of male flowers elongate ....**Gymnopetalum**

    5. Calyx tube of male flowers short:

     6. Stamens inserted near the mouth of calyx tube .... **Luffa**

     6. Stamens inserted below the mouth of calyx tube ...**Momordica**

  3. Corolla divided only to the middle .... **Coccinia**

 2. Anther cells straight or recurved, not flexuose:

 7. Style with an annular disk at the base:

  8. Male flowers in clusters; fruits globose .... **Mukia**

  8. Male flowers in corymbs; fruits oblong .... **Solena**

 7. Style without a disk at the base . Seeds betted in the middle........ .... **Bryonopsis**

Actinostemma Griff.

A. tenerum Griff. Pl. Cantor. 24, 5.3.1837; FBI 2:633 *in part;* Cogn. DC. Mon. Phan. 3:919.1881; FUGP 1:351.

Weak climbing herbs with 2-fid tendrils. Leaves 10 x 6 cm long, deeply cordate or hastate, 3-lobed, acute, glabrous, denticulate. Flowers unisexual in lax axillary panicles. Fruits ovate-elliptic, conical.

Flowers and Fruits : September-November

Found occasionally in the open areas. Singh 7833 (CDRI); Inayat 21429, 23657 (DD).

Trichosanthes Linn.

1. Male flowers solitary .... **T. dioca**
1. Male flowers in racemes:
   2. Bracts of male flowers large:
      3. Bracts sheathing at the base; fruits globose:
         4. Bracts laciniate .... **T. bracteata**
         4. Bracts entire .... **T. cordata**
      3. Bracts not sheathing at the base; fruits cylindric ...**T. himalensis**
   2. Bracts of male flowers minute or absent; fruits ovoid, conical........ .... **T. cucumerina**

T. dioica Roxb. Hort. Beng. 70.1814, nom. nud. et Fl. Ind. 3:701.1832; FBI 2:609, FUGP 1:334; P.C. Kanjilal, For. Fl. UP. 2:355.1966; Srivastava, FG. 151.1976.

Slender twining herbs. Leaves *ca* 8 x 3 cm, ovate-oblong, cordate, acute, sinuate-dentate, scabrous. Male flowers solitary, white. Fruits ellipsoid, 3-5 cm long.

Local name : Parwal
Flowers and Fruits : August-November

Found scarcely in the areas, probably an escape. The fruits are used as vegetable. Singh 7344 (CDRI).

T. bracteata (Lamk.) Voigt, Hort. Sub. Calc. 58.1845; Cogn. in DC. Mon. Phaner. 3:375.1881; Santapau in RBSI 16(1):118.1953; P.C. Kanjilal, For. Fl. UP.2:355.1966. *Modeca bracteata* Lamk. Encycl. 4:210.1797; DC. Prodr.

3:337.1828.
Trichosanthes palmata Roxb. Fl. Ind. 3:704.1832; FBI 2:606; FUGP 1:332.

Scandent or climbing herbs with 3-fid tendrils. Leaves palmately 3 to 5-lobed, 13 x 10 cm, ovate-cordate, serrate, scabrous. Male racemes few flowered provided with fringed bracts. Female flowers axillary, solitary. Fruits globose, bright-red.

Flowers and Fruits : September-November

Found occasionally in open areas. The fruits are used in medicine. Singh 7465 (CDRI).

T. cordata Roxb. Fl. Ind. 3:703.1832; FBI 2:608; FUGP 1:333.

Climbers with 3-fid tendrils. Leaves *ca* 13 x 9 cm, ovate-cordate, obscurely angular, shortly acuminate, scabrous. Male racemes few flowered provided with large, obovate entire bracts. Fruits globose, orange streaked.

Flowers and Fruits : September-November

Found occasionally in open areas. Singh 3055, 7834, 8542, 8547 (CDRI).

T. himalensis Clark in FBI 2:603.1879; Cogn. in DC. Mon. Phan. 3:379.1881; FUGP 1:333.

Climbers with 3-fid tendrils. Leaves 10-13 cm long, palmately 3-lobed, deeply cordate, irregularly serrate, softly pubescent. Fruits long, cylindric.

Flowers and Fruits : September-November

Found rarely in the area. Inayat 21443 (DD).

T. cucumerina Linn. Sp. Pl. 1008.1753; FBI 2:609; FUGP 1:333; Chakravarty in RBSI 17:31.1959; Srivastava, FG. 151.1976.

Twining or climbing annuals. Leaves 11 x 7 cm, cordate, 3-5 lobed, obtuse, denticulate, glaucous. Male flowers in axillary racemes, white, solitary. Fruits 3-5 cm long, ovoid, conical.

Flowers and Fruits : August-November

Occurs frequently on the open areas climbing on trees and hedges. Singh 7363, 8454 (CDRI).

Gymnopetalum Arn.

G. cochinchinense Kurz in Flora 295.1871; FBI 2:611; Raizada, Suppl. FUGP 77.1976.

Twining herbs with simple or rarely 2-fid tendrils. Leaves 5-6 cm in diameter, scabrous on both surfaces, 5 angular, deeply lobed. Male flowers white, solitary, bracteate. Fruits ovate-oblong, acute, orange red with ten longitudinal ribs.

Flowers and Fruits : September-October

Found occasionally in open areas twining on grasses. Singh 7759, 7921, 8521 (CDRI).

Luffa Cav.

1. Flowers yellow .... **L. graveolens**
1. Flowers white .... **L. echinata**

L. graveolens Roxb. Fl. Ind. 3:716.1832; FBI 2:614; Chakravarty in RBSI 17:74.1959; Raizada, Suppl. 78.1976.

Extensive climbers with 2 to 5-fid tendrils. Leaves 12 x 13 cm, reniform, orbicular angled, denticulate, scabrous above. Flowers yellow, unisexual. Fruits 4 x 2 cm, covered with spinous papillae.

Local name : Ghosiari, Ghosalan
Flowers and Fruits : August-November

Found commonly on the waste ground near village side. The plant is medicinally used in rheumatism and fever by local people. Singh 4351, 7362, 7693 (CDRI).

L. echinata Roxb. Hort. Beng. 104.1814, nom. nud. et Fl. Ind. 3:716.1832; FBI 2:615; FUGP 1:337; Srivastava, FG. 149.1976.

Procumbent trailing herbs with angled stems. Tendrils 2-fid. Leaves 3-5-lobed, ovate, deeply cordate, denticulate, scabrous. Flowers white, dioecious without bracts. Capsules ovoid, densely covered with stiff ciliate bristles, except at the operculate apex. Seeds many, scabrous.

Local name : Bindal
Flowers and Fruits : August-October

Found occasionally in open and wastelands. The fruits are used in sinusites. Singh 7875 (CDRI).

Momordica Linn.

1. Plants monoecious; fruits tubercled ....**M. charantia**
1. Plants dioecious; fruits spinous ....**M. dioica**

M. charantia Linn. Sp. Pl. 1009. 1753; FBI 2:616; FUGP 1;338; Chakravarty in RBSI 17:88.19589; Srivastava, FG. 149.1976.

Climbing annuals with simple slender tendrils. Leaves 6 x 3 cm, cordate, orbicular, deeply 5 to 7-lobed. Flowers unisexual. Male flowers on 6 cm long peduncle, solitary, yellow. Fruits 2.5-4 cm long, fusiform, tubercled.

Local name : Jangali Kareli
Flowers and Fruits : May-September

Found occasionally on waste places near village side: The fruits are edible and juice given in diabetes. Singh 7404, 7603, 8305 (CDRI).

M. dioica Roxb. ex Willd. Sp. Pl. 4:605.1805; FBI 2:617; FUGP 1:339; Chakravarty in RBSI 91.1959; Srivastava, FG. 149.1976.

Dioecious perennials with simple tendrils. Leaves 5-13 x 5-6 cm, cordate, ovate, entire or denticulate, membranous, acute. Male flowers solitary on 3-5 cm long, peduncles, embraced by large bracts. Fruits 2.5-3 cm long, ellipsoid, spinous.

Local name : Khekasa
Flowers and Fruits : August-October

Occurs occasionally on the edge of the forest near village side. The fruits are edible. The tubers are used medicinally and edible. Singh 2970, 8453 (CDRI).

Coccinia W.&A.

C. grandis (Linn.) Voigt, Hort. Sub. Calc. 59.18445; Bhandari, Fl. Indian Desert 163.1977.
Bryonia grandis Linn. Mant. 1:126.1767.
Coccinia cordifolia Cogn. in DC. Mon. Phan. 3:529.1881; Chakravarty in RBSI 17:117.1959; Srivastava, FG. 145.1976. non *Bryonia cordifolia* Linn. 1753.
Coccinia indica W.&A. Prodr. 347.1834; FUGP 1:345; P.C. Kanjilal, For. Fl. UP. 2:356.1966.

Twining or spreading annual herbs. Leaves 3-5-lobed or angled, 5-7 cm in diameter, cordate, shining, margin gland tipped. Flowers white. Fruits ovoid-oblong, 5 cm long, bright red when ripe.

Local name : Kundaru
Flowers and Fruits : August-November

Found frequently in the open areas and on bushes. The fruits and tubers are edible useful in diabetes. Singh 7998 (CDRI).

Mukia Arn.

M. maderaspatana (Linn.) M. Roem. syn. Pep. 47.1846.
Cucumis maderaspatanus Linn. Sp. Pl. 1912.1753.
Mukia scabrella (Linn.f.) Arn. in Hook. J. Bot. 3:276.1841; FBI 2:623.
Melothria maderaspatana (Linn.) Cogn. in DC. Mon. Phan. 3:623.1881; Chakravarty loc. cit. 141.1959.

Scabrous climbing herbs with tendrils. Leaves 3-7 cm long, ovate or deltoid, 3 to 4-lobed, acute, dentate. Flowers unisexual. Male flowers yellow in clusters. Female flowers solitary axillary. Fruits globose, 1 cm in diameter, bright red.

Flowers and Fruits : September-November

Found frequently in scrub forest in open areas. Singh 3034, 4265, 4373, 8515, 8517 (CDRI).

Solena Lour.

Solena heterophylla Lour. Fl. Cochinch. 1:514.1790; Raizada, Suppl. FUGP 80.1976.
Melothria heterophylla Cogn. in DC. Mon. Phan. 3:618.1881; FUGP 1:347; Santapau in RBSI 16(1):120.1953.
Zehneria umbellata Thwait; Enum. 125.1858; FBI 2:625 pro parte.

Perennial climbing herbs with tendrils. Leaves cordate, hastate, 9-12 cm long, acute or acuminate, dentate. Flowers unisexual. Male flowers cream coloured, in corymbs. Female flowers solitary. Fruits oblong, cylindric, red.

Flowers and Fruits : September-November

Found frequently in the open areas on hedges. Singh 7372, 7476, 8308, 8480 (CDRI).

Bryonopsis Arnott.

B. laciniosa (L.) Naud. Ann. Sci. Nat. 4.12:141,1859 FBI 2:262; FUGP 1:349.

Bryonia laciniosa Linn. Sp. Pl. 1013.1753; p.p. FBI 2:622.

Diplocyclos palmatus (Linn.) Jeffrey in Kew Bull. 15(3):352.1962; Maheshwari in Bull. Bot. Surv. India 5(2):128.1963; Srivastava, FG. 147.1976.

Bryonopsis laciniosa Sensu Naud; FBI 2:622; FUGP 1:349; Chakravarty loc.cit. 17:135.1959. pp.

Glabrous climbing annual herbs with bifid tendrils. Leaves 11-13 cm long, petiolate, palmately 5-lobed, glabrous, dentate, acuminate. Flowers yellowish. Fruits subsessile, 2-3 together, globose, smooth, white-green streaked with vertical lines.

Local name : Shivlingi

Flowers and Fruits : September-November

Found commonly in the open areas. The seeds are given for regulating fertilily in women. Singh 3972, 4273 (CDRI).

49. PASSIFLORACEAE

Passiflora Linn.

P. foetida Linn. Sp. Pl. 959.1753; Bor & Raiz. Beaut. Ind. Climb. & Shrubs 258, f. 151.1954; Srivastava, FG. 143.1976.

Twining viscid annuals herbs, with tendrils. Leaves 10 cm long, 3-lobed, hispid, membranous, margins provided with glandular hairs. Flowers white, solitary axillary, 3 cm long, clothed with glands. Fruits ovoid, globose, orange-yellow, enfolded by large bracts.

Flowers and Fruits : September-December

Found only in Dudhwa Forest nursery. Singh 4239, 8555 (CDRI).

50. CACTACEAE

Opuntia Mill.

O. stricta var. **dilleni** (Ker-Gawl.) Benson. Cact. & Succ. U. America 41: 26. 1969.
O. elatior Mill. Gard. Dict. ed. 8. n.4.1768; Suppl. FUGP. 83.1976.
O. dillenii (Ker-Gawl). Haw. Pl. Succ. 79.1812; FBI 2:657; P.C. Kanjilal, For. Fl. UP. 2:359.1966; Srivastava, FG.152.1976; FUGP. 1:384.

Thorny shrubs up to 2 m tall. Phylloclades, jointed, obovate, provided with long yellow spines. Leaves represented by tufts of spines. Flowers yellow, solitary, sessile. Berries purple, pear shaped.

Local name : Nagphani
Flowers : March-April

Occasionally met near village side in an open areas. Singh 7284 (CDRI).

51. MOLLUGINACEAE

1. Flowers in axillary fascicles; seeds appendaged ....**Glinus**

1. Flowers in terminal cymes; seeds slightly appendaged or not ......**Mollugo**

Glinus Linn.

1. Plants hairy; leaves spathulate ....**G. lotoides**

1. Plants glabrous; leaves elliptic ....**G. oppositifolius**

G. lotoides Linn. Sp. Pl. 463.1753; Srivastava, FG. 152.1976.
Mollugo hirta Thunb. Prodr. Pl. Cap. 24.1794; FBI 2:662; FUGP 1:354.

Prostrate or ascending, stellately hairy annual herbs. Leaves about 2 cm long, opposite or in whorls of 3-5, suborbicular or spathulate, obtuse. Flowers green, 2-6 in sublateral fascicles. Petals absent. Capsules 4 x 3 mm, with enlarge calyx. Seeds dark brown, appendaged.

Flowers and Fruits : April-August

Found commonly in dry ponds, puddles, and moist sandy places. Singh 2927 (CDRI).

G. oppositifolius (Linn.) DC. in Bull. Herb. Boiss. II, 1:559.1901; Backer in Fl. Males. 4(1):270.1951; Srivastava, FG. 153.1976.

Mollugo oppositifolia Linn. Sp. Pl. 89. 1753; FUGP 1:355.

M. spergula Linn. Syst. 881.1759; FBI 2:662.

Trailing much branched glabrous herbs with long internodes. Leaves 10-25 x 5-9 mm, usually in whorls, obovate or elliptic, apiculate. Flowers white, clustered at the nodes. Capsules trilocular. Seeds brown, tuberculate, appendaged.

Flowers and Fruits : July-September

Found frequently on moist, sandy soils near ponds and puddles. Singh 4107 (CDRI).

Mollugo Linn.

M. pentaphylla Linn. Sp. Pl. 89.1753; FUGP 1:355; Santapau in RBSI 16(1):124.1953; Backer in Fl. Males. Vol. 4.(1):268.1951; Srivastava, FG. 153.1976.

M. stricta Linn. Sp. Pl. ed. 2.131.1762; FBI 2:663.

Erect, glabrous annual herbs. Leaves whorled, 3 cm long, subsessile, lanceolate, acute. Flowers white in terminal cymes. Capsules globose. Seeds dark brown, not appendaged.

Flowers and Fruits : August-October

Found occasionally near cultivated fields and forest nursery. Singh 7467, 8494 (CDRI).

52. AIZOACEAE

Trianthema Linn.

T. portulacastrum Linn. Sp. Pl. 223.1753; Backer in Fl. Males. 4(1):274.1951; Srivastava, FG. 153.1976.
T. monogyna Linn. Mant. 69.1767; FBI 2:660; FUGP 1:353.

Prostrate glabrous or puberulous succulent herbs. Leaves 2-3 cm long, obovate, unequally paired; petiole with a sheath at its base. Flowers pink, solitary in pouch like sheath. Stamens many, with pink anthers. Capsules 5 x 3 mm, dehiscent.

Local name : Pathari
Flowers and Fruits : June-November

Found frequently in the cultivated fields as well as in waste moist places. Leaf juice given in Jaundice. Singh 7999 (CDRI).

53. APIACEAE nom. alt. UMBELLIFERAE

1. Umbels simple:
   2. Leaves orbicular, lobed; petals acute ....**Hydrocotyle**
   2. Leaves reniform; petals obtuse ....**Centella**
1. Umbels compound:
   3. Leaves pinnately compound:
      4. Fruits subglobose, distinctly ridged ....**Seseli**
      4. Fruits globose, ovoid, thinly ridged ....**Cnidium**
   3. Leaves pinnately or ternately decompound ....**Peucedanum**

Hydrocotyle Linn.

H. sibthorpoides Lamk. Encycl. 3:153.1789; BUW. in Fl. Males. 4(1):115.1949; Hara, Fl. E. Himal. 230.1966.

H. rotundifolia Roxb. ex DC. Prodr. 4:64.180; FBI 2:668; FUGP 1:358.

Prostrate herbs with filiform stems, rooting at the nodes. Leaves long petiolate, orbicular 5-1 cm in diameter, lobed, crenate, hirsute. Flowers 8-15, minute, sessile, pink, in simple umbels. Fruits orbicular, smooth.

Flowers and Fruits : September-November

Found rarely in moist shady places in South Kheri Forest division. Singh 7651 (CDRI).

Centella Linn.

C. asiatica (Linn.) Urban in Mart. Fl. Bras. 11:287.1879; Santapau in RBSI 16(1): 124. 1953; Buw. in Fl. Males. 4(1): 117. 1949; Srivastava, FG. 155. 1976. (Fig.6). *Hydrocotyle asiatica* Linn. Sp. Pl. 234.1753; FBI 2:669; FUGP 1:358.

Prostrate creeping herbs, rooting at nodes. Leaves many at each node, 2-2.5 cm in diameter, long petiolate, reniform, glabrous, crenate. Flowers small, pink in clusters of umbels. Fruits 2-seeded, indehiscent, laterally compressed.

Local name : Brahmi
Flowers and Fruits : April-August

Occurs commonly in moist shady places and along river banks. The plant is used in medicine and for improving memory. Singh 2931, 4223, 7137, 7252, 7961 (CDRI).

Seseli Linn.

S. diffusum (Roxb. ex Sm.) Santapau & Wagh in Bull. Bot. Surv. India 5(2):108.1963; Srivastava, FG. 157.1976. *Ligusticum diffisum* Roxb. ex Sm. in Rees Cyclop. 21: No. 11.1812. *Seseli indicum* W.&A. Prodr. 371.1834; FBI 2:693; FUGP 1:362.

Erect much branched annual herbs. Leaves 2-pinnate; pinnae ovate, lobed, hispid. Flowers white in compound umbels, with several bracts and bracteoles. Fruits subglobose, ridges thick.

Flowers and Fruits : October-December

Found frequently in open moist places. Singh 7133 (CDRI) Inayat 21659 (DD).

Cnidium Cusson.

C. **monnieri** (Linn.) Cusson. Mem. Soc. Med. Par. 280.1782; Mathials & Constance in North Amer. Flora 286(2): 190. 1945.
Selinum monnieri Linn. Cent. Pl. 1:9.1755.
Seseli daucifolium Clarke in FBI 2:693.1979.
Ligusticum monnieri Calest. Webbia 1:211.1905.

Erect, annual herbs. Leaves 2-3-pinnate; pinnae linear-lanceolate, crowded and more leafy upwards. Flowers whitish-pink in leaf opposed peduncled compound umbels, with numerous bracts and bracteoles. Fruits globose, ridged, glabrous.

Flowers and Fruits : April-May

Found frequently in open areas on moist muddy places. Singh 2938, 7162 (CDRI).

Peucedanum Linn.

P. **dhana** Buch.-Ham. ex Clarke in FBI 2:709.1879; FUGP 1:363.

Glabrous perennial herbs. Leaves twice 3-partite; leaflets 3-5 cm long, linear-oblong, sessile, entire or slightly toothed. Flowers yellow in compound umbels. Fruits ellipsoid or oblong, dorsally much compressed, lateral ridges winged.

Flowers and Fruits : March-May

Found commonly in grasslands in open areas. The roots are used in medicine. Singh 7297 (CDRI) Inayat 21658 (CAL).

54. ALANGIACEAE

Alangium Lamk. *nom. cons.*

A. salvifolium (Linn.f.) Wang. subsp. **salvifolium** Mukerjee in Bull. Bot. Surv. India 10 (3&4):330.1968; P.C. Kanjilal, For. Fl. UP. 2:363.1966; Srivastava, FG. 157.1976.
Grewia salvifolia Linn.f. Suppl. Sp. Pl. 409. 1781.
Alangium lamarckii Thw. Enum. 133.1859; Clarke in FBI 2:741.1897; FUGP 1:368.
A. salvifolium (Linn.f.) Wang. subsp. *decapetalum* Wang. in Eng. Pflanz. 4(220 b):11.1910.

Deciduous small trees with pale-brown bark. Leaves alternate, 15 x 5 cm, elliptic-oblong, acute, glabrous above pubescent beneath. Flowers white, solitary or fascicled from the axils of fallen leaves. Fruits 2 cm long, ellipsoid, black.

Local name : Akohar, Dhera
Flowers and Fruits : March-August

Found occasionally in the open areas and near village side. Ripe fruits are edible. Singh 7393, 7595, 8319 (CDRI); Sis Ram 1119 (DD).

55. RUBIACEAE

1. Flowers in dense globose heads:
   - 2. Shrubs; ovaries confluent .... **Cephalanthus**
   - 2. Trees; ovaries free:
     - 3. Calyx five lobed .... **Adina**
     - 3. Calyx entire, truncate .... **Mitragyna**

1. Flowers not in globose heads:

4. Seeds winged .... **Hymenodictyon**

4. Seeds not winged; angled:

5. Erect or prostrate herbs:

6. Flowers in axillary, terminal panicled cymes .... **Oldenlandia**

6. Flowers in whorls or in axillary heads .... **Borreria**

5. Trees and shrubs:

7. Corolla lobes twisted in buds:

8. Fruits a 2-celled capsule .... **Wendlandia**

8. Fruit fleshy, indehiscent:

9. Flowers white, usually solitary or in pairs:

10. Ovaries 2-celled; placentas on septum .... **Xeromphis**

10. Ovaries 1-celled; placentas parietal............ .... **Gardenia**

9. Flowers greenish, in sessile axillary clusters....... .... **Hyptianthera**

7. Corolla lobes valvate in buds:

11. Herbs; fruits of 2-indehiscent cocci .... **Knoxia**

11. Trees; fruits indehiscent, globose, fleshy pyrenes .... **Morinda**

Cephalanthus Linn.

C. occidentalis Linn. Sp. Pl. 95.1753; FBI 3:24; FUGP 1:374; P.C. Kanjilal, For. Fl. UP. 3:366.1966.

Large evergreen about 2 m tall shrubs. Leaves opposite or whorled, stipulate, 8-14 x 3-6 cm, ovate-elliptic or lanceolate, acuminate, entire. Flowers white, sessile in globose heads. Fruits forming a globose fleshy mass which consists of numerous indehiscent, 1-seeded cocci.

Local name : Jal Kaim
Flowers and Fruits : March-July

Occurs occasionally in marshy area along river banks. Inayat 22381, 23668 (DD).

Adina Salisb.

A. cordifolia (Roxb.) Hook.f. ex Brandis, Gen. Pl. 2:30.1873; FBI 3:24; FUGP 375; Santapau in RBSI 16(1):128.1953; P.C. Kanjilal, For. Fl. UP. 3:367.1966; Srivastava, FG. 159.1976.

Nauclea cordifolia Roxb. Pl. Cor. 1:40, t.53.1795.

Haldina cordifolia (Roxb.) Ridsdale in Blumea 24: 361. 1978.

Large trees with grey, rough bark. Leaves opposite, petiolate, 15-25 x 7-13 cm, ovate-cordate, acuminate, pubescent. Flowers yellow in globose peduncled heads. Ovaries free; stigma clavate. Fruits of numerous tiny free capsules, which split into 2 dehiscent cocci. Seeds winged.

Local name : Haldu, Karam
Flowers : May-July; Fruits : October-December

Found commonly in the areas associated with sal. The stem bark is used in fever. Singh 3008, 7982 (CDRI).

Mitragyna Korth. *nom. cons.*

M. parvifolia (Roxb.) Korth. Obs. Naucl. Ind. 19.1839; FUGP 1:376; Santapau in RBSI 16(1):129.1953; P.C. Kanjilal, For. Fl. UP. 3:368.1966; Srivastava, FG. 164.1976.

Nauclea parvifolia Roxb. Pl. Cor. 1:40, t. 52. 1795.

Stephegyne parvifolia Korth. in Verh. Nat. Ges. 161.1840; FBI 3:25.

Medium-sized, deciduous trees with smooth, brown or grey bark. Leaves opposite, stipulate, 7-12 x 5-6 cm, ovate, elliptic or suborbicular, obtuse or acute, cuneate to subcordate at the base, entire. Flowers greenish-yellow in globose heads; stigma mitriform. Fruits of numerous ovoid, ribbed capsules, each splitting into 2-cocci. Seeds winged.

Local name : Phaldu, Karam
Flowers : July-August; Fruits: October-December

Occurs frequently in the sal forests and in grasslands. The plant is worshipped and used in religious ceremonies. Singh 4361, 7225, 7904, 7991, 7992, 8463 (CDRI); Saran & Party 26548 (LWG).

Hymenodictyon Wall.

H. excelsum (Roxb.) Wall. in Roxb. Fl. Ind. 2:149.1824; FBI 3:35; FUGP 1.377; Santapau in RBSI 16(1):129.1953; P.C. Kanjilal, For. Fl. UP. 3:370.1966.
Cinchona excelsa Roxb. Pl. Cor. 2:t.106.1798.

Large deciduous trees with brown or grey deeply furrowed, corky bark. Leaves opposite, 12-18 x 6-9 cm, ovate, elliptic, acuminate, pubescent, membranous. Flowers white in terminal, drooping panicles with 5-8 cm long floral leaves. Fruits ellipsoid, loculicidal capsules.

Local name : Bhurkur
Flowers : May-August; Fruits : September-December

Found occasionally in the Park. The bark is used in fever. Singh 7527, 7948 (CDRI).

Oldenlandia Linn. Sensu amplo.

1. Seeds angled, smooth:
    2. Flowers white .... **O. corymbosa**
    2. Flowers pale-brown .... **O. gracilis**
1. Seeds subglobose, coarsely pitted:
    3. Leaves opposite, elliptic-oblong .... **O. paniculata**
    3. Leaves whorled, elliptic-ovate .... **O. nudicaulis**

O. corymbosa Linn. Sp. Pl. 119.1753; FBI 3:64; FUGP 1:380; Santapau in RBSI 16(1):131.1953; Srivastava, FG. 165.1976.
Hedyotis corymbosa (L.) Lamk. Tab. Ency. 1: 272. 1791.

Prostrate, slender, ascending herbs. Leaves 3-5.4 cm long, bristly stipulate, sessile, linear-lanceolate, acute, with recurved margins. Flowers white on filiform pedicels. Capsules 2 x 2 mm. Seeds angular.

Flowers and Fruits : September-October

Occurs frequently in moist places in fields and grasslands. Singh 4186, 4209, 4322, 8538 (CDRI).

O. gracilis Hook.f. in FBI 3:68.1880; FUGP 1:382; Srivastava, FG. 165.1976.
Hedyotis gracilis Wall. in Roxb. Fl. Ind. 1: 371. 1820.

Slender erect herbs. Leaves 6-7 cm long, with truncate stipules, linear-lanceolate, 3-nerved, acuminate. Flowers pale-brown on cymes. Capsules pedicelled, obovoid with teeth.

Flowers and Fruits : March-May

Occurs frequently in grasslands in open areas. Singh 2911, 7298, 8374 (CDRI).

O. paniculata Linn. Sp. Pl. ed. 2:1667.1763; FBI 3:69; FUGP 1:328; Srivastava, FG. 166.1976.

Glabrous, annual herbs. Leaves 3-5 x 1 cm, elliptic-oblong with truncate and bristly stipules, entire, acute. Flowers white in panicled cymes. Capsules didymous or hemispheric, ribbed with teeth.

Flowers and Fruits : August-October

Found occasionally in the forest area in moist places. Singh 3039, 7955 (CDRI).

O. nudicaulis Roth, Nov. Sp. 95.1821; FBI 3:70; FUGP 1.382; Srivastava, FG. 165.1976.

Small, erect, hairy, annual herbs. Leaves opposite below, whorled above, 2-3.5 x 2 cm, elliptic-ovate, obtuse, membranous with distinct nerves. Flowers white, on long peduncled, terminal cymes. Capsules hemispheric with teeth.

Flowers and Fruits : August-October

Found occasionally on moist sandy soil. Singh 7402 (CDRI).

Borreria G.F.W. Mey. *nom. cons.*

1. Flowers white .... **B. stricta**

1. Flowers pale-blue .... **B. articularis**

B. stricta (Linn.f.) K. Schum. in Engl. & Prantl, Nat. Pfam. 4(4):143.1891; Blatt. in JBNHS 36:794.1933; Srivastava, FG. 161.1976. (non *B. stricta* G.F.W. Mey 1818).
Spermacoce stricta Linn. f. Suppl. 120.1781; FBI 3.200; FUGP 1:394. (non *B. stricta* G.F.W. Mey 1818).

Erect, annual, herbs. Leaves 3-4 cm long, subsessile, whorled, ovate-oblong or linear, acute, stipules united into a tube, filiform bristly. Flowers white, numerous in dense, axillary and terminal clusters. Capsules elliptical, provided with teeth.

Flowers and Fruits : August-November

Occurs occasionally on the edge of the forest in moist places. Singh 7427, 7842, 7855 (CDRI).

B. articularis (Linn.f.) F.N. Wils. in Bull. Herb. Boiss. 11, 5:956.1905; Santapau & Merchant in Bull. Bot. Surv. India 3:107.1961; Srivastava, FG. 160.1976.
Spermacoce articularis Linn.f. Suppl. 119.1781.
Spermacoce hispida Linn. Sp. Pl. 102.1753; FBI 3:200; FUGP 1:394. (non *B. hispida* Spruce ex K. Schum. 1888).
Borreria hispida (Linn.) K. Schum Pflan. 4.4:144.1891. (non Spruce ex Shum 1888).

Procumbent hispid herbs. Leaves 2-3 x -1.5 cm, decussate, ovate-oblong to elliptic-oblong, coriaceous, acute; stipules hispid bristly. Flowers pale-blue in nodal whorls, tetramerous. Capsules 4 x 2 mm, oblong, hairy, dehiscing ventrally.

Flowers and Fruits : August-October

Occurs commonly on sandy moist soil in forest and in wastelands. Singh 3012, 7434 (CDRI).

Wendlandia Bartl.

W. heynei (R.&S.) Santapau & Merchant in Bull. Bot. Surv. India 3(2):110.1961; Srivastava, FG. 167.1976.
Rondeletia heynei R.&S. Systs. 234.1819.
Wendlandia exserta DC. Prodr. 4:411.1830; FBI 3:37; FUGP 1:384; P.C. Kanjilal, For. Fl. UP. 3:372.1966.

Small deciduous trees with rough, brown bark and grey-tomentose young branches. Leaves 10-20 x 3-6 cm, opposite, ovate-lanceolate, coriaceous, acute, recurved with persistent stipules. Flowers white, fragrant, sessile, arranged in panicled cymes. Capsules globose, grey tomentose.

Flowers and Fruits : April-July

Occurs occasionally in the Park in mixed forests. Singh 4202, 7232, 8358 (CDRI).

Xeromphis Rafin.

1. Thorn short; corolla glabrous outside ....**X. uliginosa**
1. Thorn long; corolla hairy outside ....**X. spinosa**

X. uliginosa (Retz.) Maheshwari in Bull. Bot. Surv. India 3:1:91.1961.
Gardenia uliginosa Retz. Obs. 2:14.1781.
Randia uliginosa DC. Prodr. 4:386.1830; FBI 3:110.1880; FUGP 1:385; P.C. Kanjilal, For. Fl. UP. 3:373.1966.
Catunaregum uliginosa (Retz.) Sivrajan, Fl. Calicut 132. 1982.

Small deciduous trees with reddish-brown bark and quadrangular branches. Leaves crowded at the end of the branches, 6-15 x 3-7 cm, obovate-oblong, cuneate with triangular stipules. Flowers white, solitary, sessile or shortly peduncled. Fruits ovoid, 3 cm long, with calyx teeth.

Local name : Pindara, Pedar
Flowers and Fruits : April-June

Found commonly in both the Forest divisions, in open grasslands. The fruits are eaten as vegetable by local people. Singh 7215, 7445, 7949, 7980, 8376 (CDRI).

X. spinosa (Thunb.) Keay, Bull. Jard. Bot. Etat. Brux. 28:37.1958; Maheshwari in Bull. Bot. Surv. India 3:91.1961.
Gardenia spinosa Thunb. Diss. Gard. No. 7.1780.
Randia dumetorum. Lamk. Tab. Encycl. 2:227.1812; FBI 3:110; FUGP 1:386; P.C. Kanjilal, For. Fl. UP. 3:374.1966.
Catunaregum spinosa (Thunb.) Trivengadum Taxon 27:515.1978.

Small trees or shrubs with long, straight, opposite, axillary spines. Leaves opposite and fascicled, 6-9 x 3-3.5 cm, obovate, subacute, hispid or glabrous with ovate, acuminate stipules. Flowers cream coloured, sub-sessile, solitary or in pairs, fragrant. Fruits berry, 2.5 cm long yellow, globose.

Local name : Mainphal
Flowers : March-June; Fruits : October-December

Found frequently in scrubs in open areas. The fruits are used for stupefying fish. Singh 3024, 4376, (CDRI).

Gardenia Ellis *nom. cons.*

G. turgida Roxb. Hort. Beng. 15.1814, nom. nud. et Fl. Ind. 1:711.1820; FBI 3:118; FUGP 1:388; P.C. Kanjilal, For. Fl. UP. 3:378.1966.

Small deciduous trees or shrubs with smooth, yellowish bark. Leaves 3-5 x 1.5-2 cm, elliptic-obovate, coriaceous, obtuse, crowded at the end of branches. Flowers dimorphic, polygamous, fragrant, the male fascicled, the female solitary. Fruits globose, 2-3 cm in diameter, pale-brown.

Local name : Thanera, Kharhar
Flowers : March-April; Fruits : November-December

Occurs occasionally in the scrubs of the Park. The dwarf plants with vegetative growth is also seen in the field. The plant root and fruits used in medicine. Singh 7343 (CDRI); Sis Ram 1186, 1188 (DD).

Hyptianthera W.&A.

H. stricta W.&A. Prodr. 399.1834; FBI 3:121; FUGP 1:389; P.C. Kanjilal, For. Fl. UP. 3:379.1966; Srivastava, FG. 163.1976.

Small, glabrous, evergreen shrubs. Leaves opposite, 15 x 3 cm, linear or ovate-lanceolate, acuminate, coriaceous with triangular persistent stipules. Flowers white, sessile in axillary fascicles. Fruit a berry, ovoid to globose.

Flowers and Fruits : March-May

Occurs rarely in the swampy and shady area of the Park in Bankati range. Singh 7140, 7195 (CDRI).

Knoxia Linn.

K. sumatrensis (Retz.) DC. Prodr. 4:569.1830; Backer & Bakh.f. Fl. Java 2:318.1965; Srivastava, FG. 163.1976.
Spermacoce sumatrensis Retz. Obs. 4:23.1786.
Knoxi corymbosa Willd. Sp. 1:582.1797; FBI 3:128; FUGP 1:393.
K. mollis R.Br. in Wall. Cat. 820. 1828.

Erect, softly villous, slender annual herbs. Leaves opposite 8-9.5 x 2.5 cm, ovate-lanceolate, entire, acute with bristly stipules. Flowers white-purplish in dichotomous cymes. Fruits shortly stalked, perforate at the base with closely united mericarps.

Flowers and Fruits : September-November

Occurs commonly in both the Forest divisions, in moist places. Singh 2998, 4194, 7396, 7782, 7966 (CDRI).

Morinda Linn.

M. pubescens Sm. in Rees, Cyclop. 24n. 3. 1813; Verdcourt in Kew Bull. 37: 543; 1983.
M. tomentosa Heyne ex Roth, Nov. Pl. Sp. 147.1821. Backer & Bakh.f. Fl. Java 2:351.1965; Saldanha & Nicolson in Fl. Hass. Distt. 581.1976.
M. tinctoria Roxb. Hort. Beng. 15.1814, nom. nud. et Fl. Ind. 2:197.1824; FBI 3:156; FUGP 1:392; P.C. Kanjilal. For. Fl. UP. 3:369.1966.

Small trees with 4-angled tomentose branches. Leaves opposite with bifid stipules, 10-13 x 5-7 cm, elliptic, tomentose on both surfaces, acuminate. Flowers white in globose heads, fragrant in night. Fruits globose, fleshy, 2.5 cm in diameter, composed of achenes.

Local name : Achchi
Flowers and Fruits : April-June

Planted near villages and in Forest Rest House campus. Saran & Party 26692 (LWG).

56. ASTERACEAE (nom. alt. COMPOSITAE)

Group I Tubuliflorae: Disc florets tubular; sap watery:

Tribe 1. Vernonieae: Heads homogamous, florets blue, bisexual and tubular; anthers cleft at the base; leaves usually alternate:

1. Heads many flowered; pappus long copious ....Vernonia

1. Heads single or few flowered, surrounded by leafy bracts; pappus chaffy .... **Elephantopus**

Tribe 2. Eupatorieae: Heads homogamous, bisexual; anthers subentire at the base; leaves opposite or upper alternate:

1. Florets white with truncate anther tips .... **Adenostemma**

1. Florets blue with appendiculate, anther tips .... **Ageratum**

Tribe 3. Astereae: Heads with all similar flowers or the outer ligulate; anthers subentire at the base:

1. Heads with distinct ray florets; pappus present:

2. Leaves linear-lanceolate, entire; ray florets tubular .... **Conyza**

2. Leaves obovate-oblong, auriculate at the base; ray florets ligulate .... **Erigeron**

1. Heads without ray florets; pappus absent:

3. Receptacle usually elongate; achenes with thickend border .... **Dichrocephala**

3. Receptacle conic or convex; achenes without thickend border:

4. Flowers rose-purple .... **Cyathocline**

4. Flowers yellow .... **Grangea**

Tribe 4. Inuleae: Heads with the flowers all similar or the outer ligulate; anther cells tailed (except Laggera); florets yellow:

1. Heads homogamous, sessile with two large bracts; pappus absent .... **Caesulia**

1. Heads heterogamous; pappus present:

2. Heads radiate; flowers yellow:

3. Pappus in a single row .... **Pentanema**

3. Pappus in double row, outer scaly .... **Pulicaria**

2. Heads not radiate:

4. Invol.-bracts dry or herbaceous; style arms of bisexual flowers filiform:

5. Heads corymbose or panicled; pappus present:

6. Invol.-bracts narrow; herbs:

7. Anther-cells tailed ....Blumea

7. Anther-cells not tailed ....Laggera

6. Invol.-bracts broad; shrubs ....Pluchea

5. Heads in dense globose or ovoid masses; pappus absent ....Sphaeranthus

4. Invol.-bracts scarious or hyaline; style arms of bisexual flowers truncate...Gnaphalium

Tribe 5. Heliantheae: Heads heterogamous, usually radiate; receptacle with chaffy bracts; anther bases not tailed; style arm truncate or appendiculate:

Sub tribe 1. Melampodieae: Heads heterogamous, many flowered; female florets jug-shaped; white ....**Parthenium**

Sub tribe 2. Ambrasieae: Heads monoecious; leaves alternate; capsule armed with spines ....**Xanthium**

Sub tribe 3. Verbesineae: Heads hetero or homogamous; leaves usually opposite:

1. Inner Invol.-bracts embracing the achenes; outer invol.-bracts 5. angular. ....**Sigesbeckia**

1. Inner Invol.-bracts not embracing the achenes; outer invol.-bracts many ....**Eclipta**

Sub tribe 4. Coreopsideae: Heads hetero or homogamous; flowers all fertile or neuter; achenes dorsally compressed; 2 to 4-awned or naked:

1. Leaves opposite, pinnatifid; achenes 2 to 4-awned ....**Bidens**

1. Leaves alternate, cauline, pinnatifid; achenes not awned...**Chrysanthellum**

Sub tribe 5. Galinsogae: Heads as above; flowers all fertile; receptacle pale-aceous; pappus scale awned; plumose ....**Tridax**

Tribe 6. Anthemideae: As in Helianthae; but the receptacle is bractless; the invol.-bracts with membranous tips; pappus absent:

1. Heads solitary:

 2. Leaves pinnatifid; heads disciform on long peduncles ....Cotula

 2. Leaves simple; heads disciform, subsessile ....Centipeda

1. Heads in panicles ....Artemisia

Tribe 7. Senecionideae: Heads heterogamous; rayed or disciform or homogamous and rayless; receptacle usually naked; Invol.-bracts usually 1-seriate, often forming a tube; pappus of fine hairs; heads homogamous; flowers pink ....**Emilia**

Tribe 8. Cynareae: Leaves often spinous; flowers all tubular; Invol.-bracts often with spinous tips; receptacle mostly bristly:

1. Erect robust herbs; heads pink-purple:

 2. Leaves and invol.-bracts spinous; pappus plumose ....Cirsium

 2. Leaves and invol.-bracts unarmed, cottony; pappus; pappus 1-2 seriate, inner plumose ....Saussurea

1. Straggling herbs; heads pale-purple ....Amberboa

Group II. Liguliflorae: Sap milky; flowers all ligulate:

Tribe 1. Cichorieae: Leaves radical and alternate; heads homogamous; corolla all ligulate; anthers base sagittate, rarely tailed; pappus setose, paleaceous or absent:

1. Florets blue; pappus minute, scaly ....Cichorium

1. Florets yellow; pappus of simple hairs:

 2. Achenes usually beaked, narrowed at both ends:

  3. Achenes fusiform, beak short ....Youngia

  3. Achenes ovoid, compressed or flattened, beaked ....Lactuca

 2. Achenes not beaked:

  4. Achenes columnar, truncate at both ends; heads cylindric....... ....Launaea

  4. Achenes compressed, ovoid or obovoid; heads ovoid, campanuate or cylindric ....Sonchus

Vernonia Schreb. *nom. cons.*

1. Heads 0.5 cm in diameter; achenes not ribbed ....**V. cinerea**

1. Heads 1 cm in diameter:

   2. Achenes ribbed:

      3. Invol.-bracts subulate-lanceolate subsquarrose; perennial herbs ....**V. squarrosa**

      3. Invol.-bracts all linear with leafy tips; annual herbs...**V. anthelmintica**

   2. Achenes terete not ribbed ....**V. revoluta**

V. cinerea (Linn.) Less. in Linnaea 4:291.1829 ; FBI 3:233; FUGP 1: 404; Santapau in RBSI 16(1):139.1953; Srivastava, FG.181.1976.
Conyza cinerea Linn. Sp. Pl. 862.1753.

Erect, pubescent, annuals or perennials, 15-120 cm high herbs. Leaves 4-5 x 1 cm, ovate to lanceolate, crenate-serrate, acute, hairy. Heads pink, rosy-pink or lilac, in corymbs, 1 x 0.5 cm. Invol.-bracts linear-lanceolate, mucronate. Achenes terete, hairy. Pappus white.

Local name : Sahadevi
Flowers and Fruits : Most part of the year

Occurs commonly in open places and in moist places, in sal forests. Plant is used in medicine. Singh 2912, 3081, 74802, 7630 (CDRI) Saran 26482 (LWG).

V. squarrosa (D.Don) Less. in Linnaea 6:627.1831. (Fig.7)
Acilepis squarrosa D.Don, Prodr. 169.1825.
Vernonia teres Wall. Cat. 2926.1830; nom. nud; FBI 3:229.1881.

Erect, rigid perennial herbs. Leaves 6-8 x 1-2 cm, subsessile, obovate-lanceolate, acute, serrate, rough on both surfaces, nerves prominent beneath. Heads purple, 1 cm in diameter, solitary axillary at the end of the branches. Invol.-bracts in several series, subulate-lanceolate, subsquarrose, sparsely hairy. Achenes ribbed, appressedly hairy. Pappus hairs yellowish-brown.

Flowers and Fruits : October-December

Occurs occasionally as undergrowth in sal forests. Singh 4286, 7918 (CDRI). Inayat 22222 (DD).

V. anthelmintica (Linn.) Willd. Sp. Pl. 3:1634.1800; FBI 3:236; FUGP 1:402.
Conyza anthelmintica Linn. Sp. Pl. 1207. 1753.
Centratherum anthelminticum (L.) Kuntze, Rev. Gen. Pl. 1: 320. 1891; Babu Herb. Fl. Dehra Dun 25.1.1977.

Pubescent annual herbs, 1.5 m high. Leaves shortly petiolate, 8-10 x 2.5-3 cm, elliptic-lanceolate, membranous, acute, coarsely serrate. Heads purple, 1 cm in diameter, subcorymbose on long peduncles. Invol.-bracts linear-oblong, apiculate. Achenes truncate, ribbed, hairy. Pappus hairs reddish or dirty-white.

Local name : Kalajiri
Flowers and Fruits : September-October

Found rarely near village side in wastelands. The leaves juice is given in fever. Singh 7570 (CDRI).

V. revoluta Buch.-Ham. in TLS 15.218.1827; FBI 3:233; FUGP 1:403.

Prostrate, dwarf perennial herbs. Leaves 4-5 cm long, linear, sessile, apiculate, margins revolute with prominent midrib. Heads 1-2 cm in diameter, pink-purple on long striate peduncles. Invol.-bracts subulate, recurved. Achenes terete. Pappus hairs white.

Local name : Lampati
Flowers and Fruits : October-March

Commonly found in the grasslands and in waste places. The plant paste applied on cuts and boils. Singh 7291, 7673, 8363 (CDRI) Inayat 22248 (DD).

Elephantopus Linn.

E. scaber Linn. Sp. Pl. 814.1753; FBI 3:242; FUGP 1:404; Santapau in RBSI 16(1):140.1953; Srivastava, FG. 176.1976.

Erect, strigose, dichotomously branched, biennial or perennial herbs. Leaves radical, 10-12 x 2-3 cm, obovate-oblong, crenate, scabrous. Heads violet, sessile, 1.5 cm in diameter, surrounded by three cordate leafy bracts. Invol.-bracts linear, acuminate. Achenes truncate and ribbed. Pappus white.

Local name : Bishari
Flowers and Fruits : September-December

Commonly found in shady places as undergrowth in sal forests. The roots are used in filaria. Singh 4256, 7425, 7817, 8462 (CDRI).

Adenostemma Forst.

A. lavenia (Linn.) O. Kuntze, Rev. Gen. Pl. 1:304.1891; Santapau in RBSI 16(1):140.1953; Panigrahi in Kew Bull. 30(4):649.1975; Srivastava, FG. 170.1976. *Verbesina lavenia* Linn. Sp. Pl. 902.1753.
Adenostemma viscosum Forst. Char. Gen. Pl. 90.1776; FBI 3:242; FUGP 1:405.

var. **lavenia**

Erect, glabrous, annual herbs. Leaves opposite, 9-11 x 4-5 cm, ovate-cordate, serrate, acute. Heads white, discoid, arranged in lax, panicles or corymbs. Invol.-bracts oblong, obtuse or acute, scabrid. Achenes angled with glandular tubercles, crowned with 3-5 short pappus hairs.

Flowers and Fruits : September-February

Found commonly in moist shady places in sal forests. Singh 4171, 7938 (CDRI).

Ageratum Linn.

A. conyzoides Linn. Sp. Pl. 839.1753; FBI 3:242; FUGP 1:405; Santapau in RBSI 16(1): 141.1953; P.C. Kanjilal, For. Fl. UP. 3:392.1966; Srivastava, FG. 171.1976.

Erect hispidly hairy annual herbs with purplish stem. Leaves 4-8 x 1.5-2.5 cm, ovate, cuneate or subcordate at the base, crenate, hairy with ciliate margins. Heads pale-blue or white, discoid, arranged in terminal corymbs. Achenes ribbed, black. Pappus of 5-awned scales.

Local name : Kukrona
Flowers and Fruits : Most part of the year.

A common weed found in waste places, in fruit orchard and other moist shady places. Singh 2805, 4141 (CDRI); Saran 26153 (LWG).

Conyza Less. *nom. cons.*

1. Heads with 6 mm across, arranged in panicles; achenes glabrous .... **C. canadensis**
1. Heads about 12 mm across, arranged in corymbs; achenes sparsely hairy .... **C. bonariensis**

C. canadensis (Linn.) Cronquist in Bull. Torr. Bot. Club 70(6): 632. 1943.
Erigeron canadensis Linn. Sp. Pl. 863.1753; FBI 3:254; FUGP 1:408.

Erect much branched hairy annual herbs. Leaves 3-9 x 0.3 x 0.4 cm, narrowly linear or linear or linear-lanceolate, hairy, acuminate. Heads yellowish, heterogamous, peduncled, arranged in branched panicles. Invol.-bracts narrowly-lanceolate. Achenes flat, glabrous. Pappus white becoming red.

Flowers and Fruits : May-August

Occurs frequently in wastelands and in cultivated fields. Singh 3030, 4220, 7606 (CDRI).

C. bonariensis (Linn.) Cronquist in Bull. Torr. Club 70(6): 632. 1943.
Erigeron bonariensis Linn. Sp. Pl. 863.1753.
E. linifolius Willd. Sp. Pl. 3:1955.1803; FBI 3:254.

Erect much branched hirsute, annual herbs with crowded ascending leaves. Leaves 4-10 x 0.4 cm, linear-lanceolate, entire, acute. Heads many arranged in corymbose panicles, heterogamous. Ray florets white. Disc florets yellow. Achenes oblong, brown, slightly curved, loosely hairy.

Flowers and Fruits : May-August

Found frequently in wastelands, cultivated fields and lawns etc. Singh 2990 (CDRI).

Erigeron Linn.

E. asteroides Roxb. Fl. Ind. 3:432.1832; FBI 3:254; FUGP 1:408.

Erect, villous, much branched annual herbs. Leaves sessile or shortly petiolate, 3-4 x 1 cm, obovate-oblong, lobulate and toothed, hairy on both sides, auriculate at the base. Heads 1 cm in diameter, peduncled solitary or sub-corymbose. Ray florets ligulate, violet, spreading. Disc florets yellow, hermaphrodite. Achenes ellipsoid, flattened, glabrous. Pappus grey or reddish.

Flowers and Fruits : September-March

Found frequently in moist shady places and on the edge of the forests. Singh 2899, 7113, 8408 (CDRI).

Dichrocephala DC.

D. integrifolia (Linn.f.) O. Ktze. Rev. Gen. Pl. 33.1891; Hara, Fl. E. Himal. 338.1966.
Hippia integrifolia Linn.f. Suppl. Pl. 389.1781.
Dichrocephala latifolia DC. in Wight, Contrib. 11.1834; FBI 3.245; FUGP 1:406.

Pubescent, annual herbs. Leaves 2-8 cm long, obovate, alternate, serrate. Heads 0.3 cm in diameter, globose, arranged in panicles. Invol.-bracts linear, minute with scarious margins. Achenes compressed, smooth. Pappus of 2 minute bristles.

Flowers and Fruits : August-October

Occurs occasionally in moist shady places in sal forests. Singh 4133, 7327 (CDRI).

Cyathocline Cass.

C. purpurea (Buch.-Ham. ex D.Don). O. Kuntze in Rev. Gen. Pl. 333.1891; Hara, Fl. E. Himal. 337.1966.
Tanacetum purpureum Buch.-Ham. ex D.Don, Prodr. Fl. Nep. 181.1825.
Cyathocline lyrata Cass. in Ann. Sc. Nat. Ser. 1, 17:420.1829; FBI 3:246; FUGP 1:407.

Erect, hairy, branched annual herbs. Leaves 2-6 cm long, sessile, alternate, pinnatifid, segments crenate. Heads rose-purple, 0.5 cm in diameter, arranged in corymbose panicles. Invol.-bracts linear with ciliate margin. Achenes oblong, smooth.

Flowers and Fruits : November-March

Occurs commonly in moist places and along river banks. Singh 7106, 7227 (CDRI); Inayat 22283 (DD).

Grangea Forst.

G. maderaspatana (Linn.) Poir. Encycl. Suppl. 2:825.1812; FBI 3:247; FUGP 1:407; Srivastava, FG. 178.1976.
Artemisia maderaspatana Linn. Sp. Pl. 849.1753.

Prostrate, tufted, pubescent, annual herbs. Leaves 2-4 cm long, sessile, sinuately pinnatifid, pubescent, lobes crenate. Heads yellow, disciform, 0.5 cm in diameter, heterogamous in terminal leaf opposed peduncles. Invol.-bracts elliptic, obtuse, hirsute. Achenes glandular. Pappus cupular.

Flowers and Fruits : October-March

Found commonly in marshy places and other moist places like river banks, puddles etc. Singh 2884, 7182, 7589, 8328 (CDRI); Inayat 22272 (CAL); Saran 26625 (LWG).

Caesulia Roxb.

C. axillaris Roxb. Pl. Cor. 1:64. t.93.1798; FBI 3:291; FUGP 1:422; Srivastava, FG. 173.1976.

Succulent, glabrous, prostrate or suberect, marshy herbs. Leaves sessile, 5-11 x 1-1.5 cm, narrowly lanceolate, acute, serrulate, base auricled. Heads purplish-white, homogamous, axillary sessile; bracts 2. Achenes flat, one ribbed on each side.

Flowers and Fruits : September-April

Found frequently in rice fields and marshy places. Singh 4255, 4324, 7803 (CDRI); Inayat 22253 (DD); Saran 26358 (LWG).

Pentanema Cass.

1. Invol.-bracts villous with recurved tips; plants woolly ....**P. vestitum**
1. Invol.-bracts not villous, erect; plants glabrous ....**P. indicum**

P. vestitum (Wall. ex DC.) Ling in Actr. Phyty. Sin. 10:180.1965. *Inula vestita* Wall. Cat. 2962.1831, nom. nud. ex DC. Prodr. 5:470.1824; FUGP 1:424. *Vicoa vestita* (Wall. ex DC) Benth ex Hook f. in FBI. 3: 297. 1881. Srivastava, FG 182. 1976.

Erect woolly annual herbs. Leaves 4-7 x 1-1.5 cm, oblong or linear oblong, acute, serrate, membranous, base auriculate, villous. Heads bright yellow, heterogamous, axillary and terminal, arranged in corymbs. Invol.-bracts filiform. Achenes oblong-cylindric.

Flowers and Fruits : February-April

Found commonly in wastelands on sandy soil. Singh 2852, 7131, 7150, 7261, 8326 (CDRI) Inaya 22219 (a) (DD).

P. indicum (L.) Ling. in Acta Phyt. Sin 10: 179. 1965.
Inula indica Linn. Sp. Pl. ed. 2.1763; FUGP 1:425.
Vicoa indica (Linn.) DC. in Wt. Contrib. 10. 1834; Santapau in RBSI 16(1):149.1953; Srivastava, FG. 182.1976.
Vicoa auriculata Cass. in Ann. Sc. Nat. ser. 1.17:418.1829; FBI 3:297.

Erect, branched, rigid annual herbs. Leaves 3-6 cm long, sessile, lanceolate or oblong-lanceolate, base auriculate, acuminate, scabrous, entire or serrulate. Heads bright yellow, heterogamous, solitary in large corymbose, loose, terminal panicles. Invol.-bracts erect. Achenes terete. Pappus of ray florets absent, pappus of disc florets few, of slender hairs.

Flowers and Fruits : October-April

Occurs frequently in wet and dry waste places. Singh 7167, 7295, 7677, 7831 (CDRI).

Pulicaria Gaertn.

P. crispa Sch.-Bip. in Webb. & Berth. Phyt. Canar. 2:223.1836-1847; FBI 3:299; FUGP 1:425.

Stout, diffuse, white-woolly herbs. Leaves sessile, 2-3 cm long, linear-oblong or spathulate, half amplexicaul, undulate, crisped. Heads yellow, heterogamous, rayed. Invol.-bracts slender, setaceous. Achenes glabrous, oblong. Pappus longer than achenes.

Flowers and Fruits : February-March

Occurs occasionally in waste grounds and cultivated fields. Saran & Party 26710 (LWG).

Blumea DC. nom. cons.

1. Heads solitary peduncled on the end of branches:

 2. Prostrate herbs with radiating branches from the root..**B. oxyodonta**

 2. Erect herbs, dichotomously branched from the base ....**B. eriantha**

1. Heads numerous:

 3. Heads clustered to form spiciform, panicled cymes:

  4. Corolla purple ....**B. mollis**

  4. Corolla yellow :

   5. Plants woolly; achenes ribbed, hairy ....**B. hieracifolia** var. **hamiltonii**

   5. Plants hairy; achenes not ribbed, glabrous ....**B. lacera**

 3. Heads arranged in corymbs or panicles:

  6. Receptacle pubescent ....**B. laciniata**

  6. Receptacle glabrous ....**B. membranacea**

B. oxyodonta DC. in Wight, Contr. Bot. Ind. 15.1834; FBI 3:266; FUGP 1:413; Randeria in Blumea 10:280.1960; Kitamura in Act. Phytotax. Geobot. 23:1968.

Prostrate or decumbent, pubescent, annual herbs with dichotomously forked branches. Leaves 1.5-2 x 0.5-1 cm, alternate, elliptic-oblong or obovate-oblong, spinulose-crenate. Heads yellow, solitary or corymbose on slender peduncles. Invol.-bracts linear, acute, pubescent. Achenes minute. Pappus white.

Local name : Kukrondha
Flowers and Fruits : March-May

Occurs frequently on sandy soil along the edge of the sal forests. Singh 7324 (CDRI); Inayat 22240 (CAL & DD); Saran 26522 (LWG).

B. eriantha DC. in Wight, Contr. Bot. Ind. 15.1834; FBI 3:266; FUGP 1:413.

Erect, dichotomously branched, pubescent or glabrous annual herbs. Leaves 2.5-3 x 1-1.5 cm, obovate or oblanceolate, alternate, crenate, villous. Heads 0.5 cm in diameter, arranged in slender peduncled dichotomous cymes. Invol.-bracts linear, clothed with long silky hairs. Achenes minute, angled.

Flowers and Fruits : November-March

Occurs frequently in the wastelands and in the sal forests. Singh 2807 (CDRI); Saran 26321,26697 (LWG).

B. mollis (D.Don) Merr. in Philip. J. Sci. Bot. 5:395.1910; Randeria in Blumea 10:261.1960; Srivastava, FG. 172.1976.

Erigeron molle D.Don, Prodr. Fl. Nepal. 172.1825.

Blumea wightiana DC. in Wight, Contr. 14.1838; FBI 3:261; FUGP 1:413.

Erect, villous leafy herbs. Leaves 4-6 x 1.5-3 cm, ovate, obovate or elliptic-lanceolate, serrate. Heads purplish, many in terminal or axillary spike like cymes or panicles. Invol.-bracts linear, acuminate. Achenes terete or angular, sparsely hairy.

Flowers and Fruits : January-May

Found commonly in waste grounds. Singh 7176, 7177, 7198 (CDRI).

B. hieracifolia DC. var. **hamiltonii** Hook.f. in FBI 3:283.1882; FUGP 1:415; Randeria loc. cit. 249.1960.

Erect, woolly, annual herbs. leaves radical, 4-6 cm long, petiolate; upper sessile, elliptic or obovate, serrate. villous. Heads yellow in spiked fascicles. Invol.-bracts lanceolate, acuminate. Achenes strongly ribbed, hairy.

Flowers and Fruits : February-May

Occurs occasionally in open areas near grasslands or on roadsides. Singh 7289, 7292 (CDRI); Inayat 22244, 22228, 23635 (DD).

B. lacera (Burm.f.) DC. in Wight, Contr. 14.1834; FBI 3:263; FUGP 1:414; Randeria in Blumea 10(1):264.1960.
Conyza lacera Burm.f. Fl. Ind. 180, t. 59. Fig. 1.1768.

Erect, glandular, pubescent, turpentine smelling annual herbs. Leaves 5-8 x 2-3 cm, sessile or the lower petiolate, oval or obscurely dentate or serrate, silky pubescent. Heads yellow, numerous, arranged in short axillary cymes or terminal spiciform panicles. Achenes not ribbed, glabrate.

Local name : Kukrondha
Flowers and Fruits : February-June

Occurs occasionally in moist shady places and near river banks. Singh 7596, 8327 (CDRI).

B. laciniata (Roxb.) DC. Prodr. 5:436.1836; FBI 3:264; FUGP 1:415; Kitamura in Act. Phytotax. Geobot. 22:68.1968; Srivastava, FG. 172.1976.
Conyza laciniata Roxb. Hort. Beng. 61. 1814, nom. nud. et Fl. Ind. 3:427.1832.

Erect tall, hairy, corymbosely branched annual herbs. Leaves 7-11 cm long, runcinate-lyrate or pinnatifid, toothed, hairy on both surfaces. Heads many 0.3 cm in diameter, yellow, arranged in close corymbose panicles. Invol.-bracts lanceolate, acuminate, tinged with purple. Achenes ribbed, silky hairy.

Local name : Kukrondha
Flowers and Fruits : February-May

Found frequently in wastelands and unused fields etc. The leaves juice is applied on cuts to stop bleeding. Singh

2821, 7299, 7597, 7595, 7639 (CDRI); Inayat 22220 (CAL & DD); Saran 26232 (LWG).

B. membranacea DC. Prodr. 5:440.1836; FBI 3.265; FUGP 1:415; Randeria in Blumea 10:269.1960.

Slender, much branched, glandular pubescent, annual herbs. Leaves 4-6 cm long, ovate or often lyrate or runcinate, thinly membranous, acute or obtuse, serrulate. Heads numerous, yellow, 0.5 cm in diameter, arranged in much branched open panicles. Invol.-bracts slender, granular pubescent. Achenes 8-10 ribbed, hairy; pappus white.

Flowers and Fruits : January-May

Found frequently as a weed in waste places and near the edge of the forest. Singh 7123, 7154 (CDRI) Inayat 22241 (CAL); Saran 26177, 26421 (LWG).

Laggera Sch.-Bip.

1. Heads pink; leaves shortly decurrent ....**L. aurita**

1. Heads golden yellow; leaves amplexicaul at the base ....**L. falcata**

L. aurita (Willd.) Sch.-Bip. in Schweinf. Beitr. Fl. Aeth. 151.1867; FBI 3:271; FUGP 1:417; Santapau in RBSI 16(1):147.1953.
Conyza aurita Willd. Sp. Pl. 3:1929.1804.

Perennial, villous, branched herbs. Leaves 2-2.5 cm long, sessile, oblanceolate, decurrent, dentate or subpinnatifid. heads pink, 0.4 cm in diameter, arranged in terminal panicled corymbs. Invol.-bracts linear. Achenes reddish, hairy.

Flowers and Fruits : May-July

Found occasionally in waste places and along roadsides. Singh 7615 (CDRI).

L. falcata (D.Don) O. Kuntze, Rev. Gen. Pl. 334.1891.
Erigeron falcatum D.Don, Prodr. Fl. Nepal, 172.1825.

Laggera flava Benth. ex Clarke, Comp. Ind. 90.1876; FBI 3:270; FUGP 1:417.

Blumeopsis falcata Merr. in JAA 19:70.1938.

Erect, annual herbs. Leaves 5-8 x 1-2 cm, lower amplexicaul, ovate-lanceolate or obovate, incised, crenate-serrate, membranous, glabrous. Heads golden yellow, 0.4 cm in diameter, arranged in panicled cymes. Achenes minute, glabrous. Pappus white.

Flowers and Fruits : September-January

Occurs frequently in moist shady places in sal forests. Singh 7634, 7861 (CDRI); Inayat 22282 (CAL).

Pluchea Cass.

P. tomentosa DC. in Wight, Contr. 16.1834; FBI 3:272; FUGP 1:418.

Pubescent, much branched shrubs up to 1 m high. Leaves 7-11 x 1-2 cm, ovate-lanceolate, acute, coarsely toothed, pubescent. Heads purple, 0.5 cm in diameter, arranged in compound, terminal corymbs. Invol.-bracts ovate, acute. Achenes brownish, angled. Pappus white.

Flowers and Fruits : January-March

Found occasionally in sal forest of the Park. Singh 7653 (CDRI).

Sphaeranthus Linn.

S. indicus Linn. Sp. Pl. 927.1753; FBI 3:275; FUGP 1:419; Srivastava, FG. 179.1976.

Much branched, glandular hairy herbs. Branches ascending, winged with the bases of decurrent leaves. Leaves sessile, 2-3.5 x 0.5 cm, ovate-oblong, subacute, serrate, glandular hairy, narrowed at the base. Heads pink-purple, globose, ovoid, ebracteate on solitary glandular peduncles. Achenes glabrous.

Local name : Lal múndi
Flowers and Fruits : November-April

Occurs commonly on moist ground in wastelands and in fields. The plant is used in medicine, and ophthalmic diseases. Singh 2872, 7591, 8339 (CDRI).

Gnaphalium Linn.

1. Heads in leafless corymbose clusters ....**G. luteo-album subsp. affine.**
1. Heads leafy:
    2. Heads arranged in spikes:
        3. Leaves linear-obovate; pappus hairs not coherent at the base ....**G. indicum**
        3. Leaves spathulate; pappus hairs coherent at the base.. ....**G. peregrinum**
    2. Heads in rounded axillary or terminal clusters ....**G. pulvinatum**

G. luteo-album Linn. subsp. **affine** (D.Don) Koster in Blumea 4:484.1941.
G. affine D.Don, Prodr. Fl. Nep. 173.1825; Srivastava, FG. 177.1976.
G. luteo-album Linn. var. *multiceps* Hook.f. in FBI 3:288.1881; FUGP 1:421.
Pseudognaphalium luteo-album subsp. *affine* (D. Don) Hillard & Burtt in Bot. J. Linn. Soc. 82: 206. 1981.

Softly white woolly annual herbs. Leaves sessile, 5 x 0.4 cm, linear-oblong or lanceolate, woolly on both surfaces, obtuse, entire. Heads golden yellow, 0.6 cm in diam; arranged in dense corymbose clusters. Achenes oblong, pappilose.

Flowers and Fruits : February-April

Occurs commonly in moist places in sal forests and in wastelands. Singh 2838, 2936, 7111, 7236, 7659 (CDRI); Inayat 22257 (a) (DD).

G. indicum Linn. Sp. Pl. 852.1753; FBI 3:289; FUGP 1:421; Srivastava, FG. 177.1976.

Slender much branched, white woolly herbs. Leaves 1.5-2.5 cm long, linear-obovate, apiculate, white woolly on both surfaces. Heads pale-brown, 0.3 cm in diam. sessile, arranged in leafy spikes or ovoid spiciform clusters. Invol.-bracts linear-oblong, brown. Achenes ovoid, papillose. Pappus hairs not cohering at the base.

Flowers and Fruits : February-April

Found frequently on the edge of the forests and in waste moist shady places. Singh 2829, 7605, 7115, 7187 (CDRI); Inayat 22259, 22260 (DD).

G. peregrinum Fernald in Rhodora 45.1943; Backer & Bakh.f. Fl. Java 2:394.1965; N.C. Nair in JBNHS 63(3):777-78. 1966; Raizada, Suppl. FUGP 119.1976.
G. purpureum auct. plur. non Linn. 1753. *G. pensylvanicum* Willd. Enum. Hort. Berol. 867. 1809.

Erect, branched, sparsely cottony, annual herbs. Leaves 6-8 cm long, spathulate, tips broadly rounded, hairy beneath. Heads pale-brown, 0.5 cm in diam; arranged in spicate, globose, leafy clusters. Invol.-bracts linear, subacute. Achenes minutely hairy. Pappus hairs coherent at the base.

Flowers and Fruits : December-April

Found frequently on moist shady waste places. Singh 2808, 7115, 7584 (CDRI); Inayat 22254 b.c., 22259 a.b. (DD).

G. pulvinatum Del. Fl. Egypte 266, t.44.f.1.1813; FBI 3:289; FUGP 1:422; Srivastava, FG. 177.1976.

Small prostrate, woolly, annual herbs. Leaves 1-1.5 cm long, obovate, or spathulate, mucronate, woolly. Heads minute, arranged in sub-globose, terminal clusters, covered in white wool. Invol.-bracts linear-lanceolate, acute. Achenes oblong, minutely papillose.

Flowers and Fruits : January-March

Found frequently on moist shady waste places. Singh 7682 (CDRI); Inayat 22258 (CAL).

Parthenium Linn.

P. hysterophorus Linn. Sp. Pl. 988.1753; Maheshwari in Curr. Sci. 35:181-183. 1966; Raizada, Suppl. FUGP 127.1976. (Plate XVI).

Erect, tufted, branched leafy herbs up to 1 m tall, with hairy scabrid angular stems. Leaves 2.5-7 x 2-5 cm, pinnately dissected; segments linear. Inflorescence 4-5 mm in diam. heterogamous, numerous, arranged in corymbose cymes. Involucral bracts in 2 series of 5. Inner involucral bracts subtending a white female floret with 2 male florets on either side. Achenes obovate, black, crowned by corolla.

Local name : Gajar ghas.
Flowers and Fruits : September-December

Occurs frequently near Food Corporation building in Lakhimpur-Kheri. Singh 8413 (CDRI).

Xanthium Linn.

X. strumarium Linn. Sp. Pl. 987.1753; FBI 3:303; FUGP 1:426; Srivastava, FG. 183.1976.
X. indicum Koenig in Roxb. Fl. Ind. ed. 2, 3:601.1832.

Scabrous, erect herbs or undershrubs. Leaves 17-22 cm long, poetiolate, triangular-cordate, irregularly crenate, cuneate at the base. Heads bisexual and female. Bisexual heads in the upper axil, globose, many flowered. Female heads in axillary racemes, 2-flowered, fertile. Involucre of fertile heads ovoid in fruit with 2 erect, mucronate beaks, clothed with hooked prickles.

Local name : Gokharu
Flowers and Fruits : September-June

Found commonly in wastelands and along road sides. The seeds are used for oil. Singh 4144, 7562 (CDRI).

Sigesbeckia Linn.

S. orientalis Linn. Sp. Pl. 900.1753; FBI 3:304; FUGP 1:426.

Erect, pubescent, annual herbs with purple tinged, dichotomous branches. Leaves opposite, 7-9 x 2.3 cm, triangular-ovate, pubescent, acute or acuminate, irregularly crenate, cuneate at the base. Heads yellow in leafy panicles, viscid. Invol.-bracts in 2 series; outer with viscous hairs, inner with glandular hairs.

Flowers and Fruits : October-December

Occurs commonly, in moist waste places and near the edge of the forests in the Park. Singh 4147, 7777, 7943 (CDRI).

Eclipta Linn. *nom. cons.*

E. alba Hassk. Pl. Jav. Rar. 528.1848; FBI 3:304.
E. prostrata (Linn.) Linn. Mant. 286.1771; Maheshwari, Fl. Del. 197.1963; Srivastava, FG. 175.1976.
Eclipta erecta Linn. Mant. 286.1771; FUGP 1:427.

Erect or prostrate branches hirsute annual herbs. Leaves sessile, 2.5-3 x 0.4-0.8 cm, linear or oblong-lanceolate, acute, hirsute on both surfaces. Heads white, solitary or 2-3 together. Invol.-bracts ovate, appressed with white hairs. Achenes winged on the margins. Pappus absent.

Local name : Kala bhangara
Flowers and Fruits : Most part of the year

Occurs commonly in moist shady places in the forests and in wastelands. The plant is used in medicine and liver disorder. Singh 7168 (CDRI).

Bidens Linn.

B. biternata (Lour.) Merr. & Scherff in Bot. Gaz. 88:293. 1929; Santapau in RBSI 16(1):151.1953; Srivastava, FG. 171.1976.
Coreopsis biternata Lour. Fl. Cochinch. 508.1790.
Bidens pilosa auct. pl. non Linn.

Erect, annual herbs with 4-angular stems. Leaves variable, 3-foliolate or 1-2 pinnatifid; the terminal segments, 5-6 x 2 cm, ovate, serrate, acute, glabrous. Heads yellow, heterogamous, arranged in corymbose panicles. Achenes black, 7-11 x 1 mm, quadrangular, ribbed, glabrous. Pappus 2-awned.

Flowers and Fruits : March-December

Found commonly in moist shady waste places. Singh 2917, 4247 (CDRI).

Chrysanthellum Rich.

C. indicum DC. Prodr. 5:631.1836; FBI 3:310; FUGP 1:432; Srivastava, FG. 174.1976.

Procumbent, diffuse, glabrous annual herbs. Leaves alternate, pinnatifid, 2-6 cm long with oblong segments. Heads yellow, axillary and terminal, peduncled. Invol.-bracts oblong, obtuse. Achenes linear-oblong, dorsally compressed. Pappus a minute corona.

Flowers and Fruits : September-November

Found occasionally in grasslands in moist places. Singh 4340, 7888 (CDRI).

Tridax Linn.

T. procumbens Linn. Sp. Pl. 900.1753; FBI 3:311; FUGP 1:433; Srivastava, FG. 181.1976.

Erect or straggling hispid herbs. Leaves opposite, 3-4 x 1 cm, ovate or lanceolate, dentate or pinnatisect. Heads cream coloured, long peduncled. Invol.-bracts few, broadly ovate. Achenes brown. Pappus of feathery bristles.

Local name : Phulani
Flowers and Fruits : Most part of the year

Found commonly in wastelands and along railway lines. The leaf Juice is used for healing cuts and wounds. Singh 7366, 7632 (CDRI).

Cotula Linn.

1. Leaf segments decurrent lobed; achenes ovate, thick winged .... C. anthemoides
1. Leaf segments linear-mucronate; achenes angled, not winged .... C. hemisphaerica

C. anthemoides Linn. Sp. Pl. 891.1753; FBI 3:316; FUGP 1:434; Srivastava, FG. 175.1976.

Weak, diffuse much branched, annual herbs. Leaves alternate, 1-5 cm long, pinnatisect; segments obtuse. Heads yellow, solitary on filiform peduncles. Achenes of disc florets with thick narrow wings.

Flowers and Fruits : February-March

Occurs occasionally in moist and marshy places. Singh 4340, 7178 (CDRI).

C. hemisphaerica (Roxb.) Wall. ex Clarke, Comp. Ind. 150.1876; FBI 3:316; FUGP 1:434; Raizada, Suppl. FUGP 110.1976.
Artemisia hemisphaerica Roxb. Fl. Ind. 3:422.1832.

Erect, much branched annual herbs. Leaves pinnatisect about 1.5 cm long; segments mucronate. Heads yellow, solitary, peduncled, erect or drooping. Achenes angled, not winged.

Flowers and Fruits : January-March

Occurs occasionally in muddy area of the Park. Singh 7683 (CDRI).

Centipeda Lour.

C. minima (Linn.) A.Br. & Aschers. Ind. Sem. Hort. Berol. App. 6.1867; Srivastava, FG. 173.1976.
Artemisia minima Linn. Sp. Pl. 889.1753.
Centipeda orbicularis Lour. Fl. Cochinch. 2:493.1790; FBI 3:317; FUGP 1:406.

Prostrate, glabrous or sparsely woolly annual herbs. Leaves alternate, subsessile, 0.5-1 cm long, obovate-oblong, crenate. Heads yellow, sessile, solitary axillary, disciform. Achenes 4-angled, minute. Pappus absent.

Local name : Nakchikani
Flowers and Fruits : October-December

Found occasionally in moist shady places. The heads are used in cold and headache. Singh 2895, 7156 (CDRI).

Artemisia Linn.

1. Leaf segments filiform; disc florets sterile ....**A. scoparia**
1. Leaf segments linear-lanceolate; disc florets fertile ....**A. nilagirica**

A. scoparia Waldst. & Kit. Pl. Rar. Hung. 1:66, t.65.1802; FBI 3:323; FUGP 1:435.

Erect, branched, annual or perennial herbs with purple tinged stem. Leaves pinnatisect with filiform segments. Heads minute, greenish-yellow, arranged in panicles. Invol.-bracts oblong, obtuse, shining. Achenes very minute.

Flowers and Fruits : September-October

Occurs rarely in damp area and along river banks of the Park. Singh 7218 (CDRI).

A. nilagirica (Clarke) Pamp. Nuov. Gior. Bot. Ital. 33: 452.1926.
A. vulgaris Linn. var. *nilagirica* Clarke, Comp. Ind. 162.1876.
A. vulgaris auct. pl. non Linn. 1753; FBI 3: 225. 1881. pp.
A. indica Willd. Sp. Pl. ed. 4.3.: 1846. 1804.

Erect hoary pubescent, perennial herbs, 1-2 m tall. Lower leaves petiolate, 5-10 cm long, ovate, 1-2 pinnatisect, white tomentose beneath; upper leaves sessile, linear-lanceolate, 3-fid. Heads brown-yellow, sessile, globose, arranged in racemes. Invol.-bracts woolly. Achenes minute.

Local name : Kirmar
Flowers and Fruits : October-December

Occurs rarely in the forest area in open places. The decoction of the leaves used in fever. Singh 4219, 7640, 8370 (CDRI).

Emilia Cass.

E. sonchifolia (Linn.) DC. in Wight, Contrib. 24.1834; FBI 3:336; FUGP 1:436; Srivastava, FG. 176.1976.
Cachalia sonchifolia Linn. Sp. Pl. 835.1753.

Slender, glaucous, puberulous, annual herbs, rooting near the base. Leaves variable, lower 3-4 cm long, petiolate, lyrate or obovate; upper more or less with amplexicaul or auricled base. Heads purplish, laxly corymbose on slender peduncles. Invol.-bracts linear-oblong, acute. Achenes 5-ribbed, linear-oblong.

Flowers and Fruits : September-November

Occurs frequently in moist shady places and in forests. Singh 7633, 7715 (CDRI).

Cirsium Adans.

C. arvense (Linn.) Scop. Fl. Carn. 2:126.1772; Srivastava, FG. 174.1976.
Serratula arvensis Linn. Sp. Pl. 820.1753.
Cnicus arvensis Hoffm. Deutschl. Fl.1(2):130.1804; FBI 3:362; FUGP 1:438.

Erect, leafy annual herbs. Leaves 5-10 cm, long, sessile, oblong-linear or obovate, woolly beneath, margins tipped with spines. Heads deep pink to purple, dioecious, solitary fascicled or corymbose. Achenes smooth. Pappus pale-brown.

Flowers and Fruits : February-April

Found commonly in moist waste places. Singh 2846, 7184, 8303 (CDRI).

Saussurea DC.

S. heteromalla (D.Don) Hand.-Mazz. Synp. Sin. Pl. 7:1152.1936; inobs; Raizada & Saxena in Indian For. 92(5):322.1966.
Carduus heteromalla D.Don, Prodr. Fl. Nep. 1825.
Saussurea candicans Clarke, Comp. Ind. 232.1876; FBI 3:373; FUGP 1:440; Srivastava, FG. 180.1976.

Erect, cottony, annual herbs. Radical leaves 7-10 x 2-2.5 cm, oblong, sinuate, pinnately lobed near the base, pubescent above and white tomentose beneath; upper leaves few, sessile, lanceolate. Heads pale-red, long stalked, solitary or corymbose. Invol.-bracts rigid, lanceolate with sharp acuminate points. Pappus white.

Flowers and Fruits : March-May

Occurs occasionally in the open areas in waste lands. Singh 2919, 7122, 7274, 8369 (CDRI).

Amberboa Less. *nom. cons.*

A. ramosa (Roxb.) Jafri in Scientist (Karachi) III:29.1959.
Carduus ramosus Roxb. Fl. Ind. 3:407.1832.
Volutarella divaricata Benth. & Hook.f. Gen. Pl. 2:476.1873 in parte; FBI 3:383; FUGP 1:442.
V. ramosa (Roxb.) Santapau, Pl. Saur. 22. 1953; Srivastava, FG. 182.1976. *Oligochaeta ramosa* (Roxb.) Wagenitz in Verroffent, Geobot. Inst. 37: 323. 1962.

Erect or straggling, stiff, dichotomously branched, annual herbs. Leaves variable, sessile, 3-8 cm long, obovate to oblanceolate, usually pinnatifid, lobed. Heads purplish, ovoid, homogamous. Invol.-bracts many seriate, spinescent at tips. Achenes angled, dull brown. Pappus brown, silvery.

Flowers and Fruits : February-May

Found commonly in dry waste places and in cultivated fields in sandy soil. Singh 7611, 8313 (CDRI).

Cichorium Linn.

C. intybus Linn. Sp. Pl. 813.1753; Bailey, Man. Cult. Pl. 983; FBI 3:391; FUGP 1:451; Raizada, Suppl. FUGP 107.1976.

Erect or prostrate herbs with white sap. Leaves 4-8 x 1-1.5 cm, oblanceolate, stem clasping. Heads blue, sessile, homogamous. Achenes glabrous. Pappus minute, scaly.

Local name : Kasani
Flowers and Fruits : April-May

Occurs rarely as an escape, cultivated in the area. Singh 7359 (CDRI).

Youngia Cass.

Y. japonica (Linn.) DC. Prodr. 7: 194. 1838; Kitamura in Act. Phytotax. Geobot. 24: 19. 1969; Srivastava, FG. 183.

1976.

Prenanthes japonica Linn. Mant. 107.1767.

Crepis japonica Benth. Fl. Hongk. 194.1861; FBI 3:395; FUGP 1:445.

Erect, slender, glabrous, annual herbs. Leaves mostly radical and rosette forming, 5-15 x 2-3 cm, sinuately lobed or pinnatifid, membranous. Floral branches slender, erect, corymbosely branched near the top. Heads yellow, homogamous. Involucral-bracts narrowly oblong with hyaline margins. Achenes reddish-brown. Pappus silvery.

Flowers and Fruits : February-April

Occurs occasionally in moist shady places in the Park. Singh 2903, 7112, 7248, 7249 (CDRI); Inayat 22271 (CAL).

Lactuca Linn.

1. Leaf base sagittate, auricled; heads subumbellate ....**L. polycephala**
1. Leaf base cordate or cuneate; heads panicled ....**L. dolichophylla**

L. polycephala Clarke, Comp. Ind. 266.1876; FBI 3:410; FUGP 1:447.

Slender, glabrous annual herbs. Radical leaves petiolate, 8-10 cm long, linear-lanceolate; stem leaves sessile, 4-7 cm, oblong-lanceolate, acuminate, base sagittate with acute auricles. Heads yellow, arranged in subumbellate corymbs on floral shoots. Invol.-bracts in two series, linear-oblong, green. Achenes red brown, ribbed. Pappus silvery.

Flowers and Fruits : September-December

Found occasionally in moist places. Singh 2902, 7170, 7247 (CDRI).

L. dolichophylla Kitamura in Add. Reports 150.1966.

L. longifolia DC. Frodr. 7:135.1838; FBI 3:405.

Erect, glabrous, annual herbs. Leaves sessile, 16-18 cm long, linear-lanceolate, acuminate, base cordate or cuneate, membranous. Heads yellow, long peduncled, arranged in panicles. Invol.-bracts pale-green, linear-oblong. Achenes dark-brown. Pappus white.

Flowers and Fruits : October-December

Found occasionally in moist places and in grasslands. Singh 2915, 7288 (CDRI); Inayat 22270 (CAL).

Launaea Cass.

1. Leaves linear-lanceolate or narrowly obovate .... **L. acaulis**

1. Leaves pinnatifid, or sinuately lobed:

 2. Heads terminal on paniculately branched on floral; shoots.............. .... **L. asplenifolia**

 2. Heads in racemes on floral shoots and its branches ...**L. procumbens**

L. acaulis (Roxb.) Babc. ex Kerr. in Craib, Fl. Siam, Enum. 2:299.1936; Kitamura loc. cit. 143.1969.
Prenanthes acaulis Roxb. Fl. Ind. 3:403.1832.
Crepis acaulis Hook.f. in FBI 3:369.1881; FUGP 1:445.

Glabrous, perennial herbs. Leaves radical, 5-9 cm long, linear-lanceolate, entire or denticulate, acute. Floral shoots 5-11 cm long, dichotomously branched. Heads yellow on bracteolate peduncles. Invol.-bracts in 2-series with hyaline margins. Achenes ellipsoid, ribbed.

Flowers and Fruits : October-January

Occurs occasionally in waste places. Singh 7288, 7624, 8367 (CDRI) Inayat 22265 (CAL).

L. asplenifolia (Willd.) Hook.f. in FBI 3:415.1881; FUGP 1:449; Srivastava, FG. 178.1976.
Prenanthus asplenifolia Willd. Sp. PP. 3:1540.1803.

Perennial suberect, glabrous herbs. Leaves shortly petiolate, 11-13 cm long, sinuately lobed or pinnatifid, entire or denticulate. Flowering shoots 26 cm long, many from the base, bearing yellow paniculate heads. Invol.-bracts with membranous margins. Achenes angled, ribbed.

Local name : Gobhi
Flowers and Fruits : October-April

Found frequently in cultivated fields, wastelands and on rocky places. Singh 2870 (CDRI).

L. procumbens (Roxb.) Ramayya & Rajgopal in Kew Bull. 23(3):463.1969.
Prenanthes procumbens Roxb. Fl. Ind. 3:404.1832.
Launaea fallax (Jaub. & Spach.) O. Kuntze, Rev. Gen. Pl. 351.1891; Srivastava, 179.1976.
L. nudicaulis auct. pl. non. Hook.f. in FBI 3:416.1881; (sensu stricto) FUGP 1:450. non *Microrhynchus nudicaulis* (Linn.) Less.

Prostrate, glabrous herbs with yellow sap. Leaves radical, usually sessile, 5-9 cm long, sinuately lobed or pinnatifid with white teeth. Floral shoots spreading from the radical leaves. Heads yellow, sub-racemose and in terminal clusters, rarely solitary. Achenes columnar, ribbed, shorter than pappus.

Local name : Gobhi
Flowers and Fruits : September-March

Found commonly in waste lands, sandy and rocky places. Singh 7285, 7612 (CDRI).

Sonchus Linn.

S. oleraceus Linn. Sp. Pl. 794.1753; FBI 3:414; FUGP 1:448.

Erect, glabrous or sparsely glandular hairy, annual herbs. Leaves 7-13 cm long, lanceolate, acute, margins spiny, auricles spreading. Heads yellow, arranged in umbellate cymes. Achenes muriculate, 3-ribbed on ech side. Pappus white.

Flowers and Fruits : January-April

Occurs commonly in wastelands and in cultivated fields. Singh 2806, 7609, 8331 (CDRI); Saran 26184 (LWG).

57. CAMPANULACEAE

1. Corolla bilipped; anthers connate ....Lobelia
1. Corolla regular; anthers free:
   2. Flowers blue in lax corymbose panicles ....Wahlenbergia
   2. Flowers grey-purple or white in panicled clusters ....Campanula

Lobelia Linn.

L. alsinoides Lamk. Encycl. 3:588.1795; M.A. Rau in Bull. Bot. Surv. India 2. 10:45.1969.
L. trigona Roxb. Fl. Ind. 1:506.1820; FBI 3:423; FUGP 1:452.

Succulent, branched, glabrous, annual herbs. Branches ascending, rooting near the base, trigonous. Leaves subsessile, 1-2 x 0.3-1 cm, ovate, acute, crenate, glabrous. Flowers blue, axillary, long peduncled, longer than leaves. Capsules ellipsoid. Seeds trigonous, narrowed at both ends.

Flowers and Fruits : September-December

Found commonly in the moist shady places along forest roads. Singh 4184, 8511 (CDRI).

Wahlenbergia Schrad. ex Roth

W. marginata (Thunb.) DC. Mon. Campan. 142. 1830; Srivastava, FG. 183.1976. *Campanula marginata* Thunb. Fl. Jap. 89.1784.
Wahlenbergia gracilis Schrad. Blumenb. 38. 1827; FBI 3:429; FUGP 1:453.

Erect or decumbent, glabrous annual herbs. Leaves 2-7 cm long, linear-oblong, sessile, dentate. Flowers blue arranged in lax corymbose panicles. Capsules erect, 2-3-celled, tapering into a pedicel. Seeds ellipsoid, smooth, yellowish-brown.

Flowers and Fruits : November-March

Found frequently in damp shady places in forests. Singh 7110, 7128, 7638 (CDRI) Saran 26301, 26380 (LWG).

Campanula Linn.

C. benthamii Wall. ex Kitamura Fl. Afghan. 377. 1960; Thaker & Sabinis in JBNHS 16(1):228.1972; Raizada, Suppl. FUGP 138.1976.

C. wallichii C.R. Babu in JBNHS 65(3):808-809.1968 (Dec.)

C. canescens Wall. ex DC. Prodr. 7:473.1838 (non Roth 1827); FBI 3:439; FUGP 1:454.

Erect, hispid annual herbs. Leaves alternate, sessile, 4-5 x 0.3-1 cm, oblong-lanceolate, crenate. Flowers white or grey-purple in panicled clusters. Capsules minute.

Flowers and Fruits : December-March

Found frequently in moist shady places in forests and near river banks. Singh 2854, 7102, 7105, 7109, 8407 (CDRI); Inayat 22216 (a) (DD).

58. SPHENOCLEACEAE

Sphenoclea Gaertn.

S. zeylanica Gaertn. Fruct. 1:113, t. 24.1788; FBI 3:438; FUGP 1:454; Airy Shaw in Fl. Males. 4:27.1948; Srivastava, FG. 184.1976.

Erect, amphibious, annual herbs with fistular stems. Leaves 6-7 x 0.7 x 1 cm, alternate, lanceolate, entire, acute. Flowers greenish-white in dense erect spikes. Capsules wedge-shaped, enclosed by persistent calyx. Seeds minute, brown.

Flowers and Fruits : August-November

Occurs frequently in swampy area in rice fields. Singh 7482, 8488 (CDRI).

59. PLUMBAGINACEAE

Plumbago Linn.

P. zeylanica Linn. Sp. Pl. 151.1753; FBI 3:480; FUGP 1:457; P.C. Kanjilal, For. Fl. UP. 3:393.1966; Srivastava, FG. 184.1976.

Sub-scandent or spreading shrubs with striate stems. Leaves 4-6 x 2-3 cm, ovate, entire, membranous, glabrous. Flowers white in long spiciform racemes. Calyx narrowly tubular, 5-ribbed, sticky, densely glandular, persistent. Capsules oblong, enclosed within calyx.

Local name : Chitrak, chita
Flowers and Fruits : September-March

Found occasionally straggling in bushes along railway lines and waste places. The plant is used in medicine. Singh 7514, 7927 (CDRI); Inayat 22334 (DD).

60. PRIMULACEAE

1. Corolla-lobes imbricate .... **Primula**
1. Corolla-lobes twisted:
   2. Flowers white; capsules opening by valves .... **Lysimachia**
   2. Flowers blue; capsules splitting transversely .... **Anagallis**

Primula Linn.

P. umbellata (Lour.) Bentv. in Fl. Males 6(2):191.1962; Raizada, Suppl. FUGP 142.1976; Srivastava, FG. 185.1976.
Drosera umbellata Lour. Fl. Coch. ed. 1:186.1790.
Androsace saxifragaefolia Bunge, Mem. Ac. St. Petersb. 2:127.1835; FBI 3:496; FUGP 1:460.

Glandular, pubescent, annual herbs. Leaves 2-3 cm long, ovate or suborbicular, cordate, crenate. Flowers white on long peduncles, arranged in umbels. Calyx deeply 5-cleft. Capsules globose, whitish. Seeds minute, angular.

Flowers and Fruits : December-March

Found occasionally in damp waste places in sal forests. Singh 7104, 7636 (CDRI); Inayat 22335 (DD).

Lysimachia Linn.

L. candida Lindl. subsp. **obovata** R. Kunth in Pflazeureich 22:301.1905.
L. obovata Buch.-Ham. ex Hook.f. in FBI 3:502.1882; FUGP 1:460; Srivastava, FG. 185.1976.

Erect or procumbent, glaucous, slender herbs. Leaves 3-12 x 1-2 cm, obovate, spathulate or oblanceolate, acute or obtuse, the lower one long petiolate. Flowers white in terminal racemes, pedicels longer than the branch. Corolla lobes obovate. Capsules 0.3 cm long, bursting irregularly.

Flowers and Fruits : March-October

Found rarely in swampy places of the Park. Singh 7952 (CDRI).

Anagallis Linn.

A. arvensis Linn. Sp. Pl. 148.1753; FBI 3:506; FUGP 1:461; Srivastava, FG. 185.1976.

Erect or procumbent, glabrous, annual herbs with quadrangular branches. Leaves sessile, 1.5-2 x 1.6 cm, cordate, acute, entire. Flowers blue, solitary axillary on slender, recurved peduncles. Capsules many seeded, dehiscing transversely.

Flowers and Fruits : November-March

Occurs commonly as a weed in waste places and cultivated fields. Singh 2841, 7579, (CDRI); Saran 26350 (LWG).

61. MYRSINACEAE

1. Flowers pink; anthers longer than the filaments ....Ardisia

1. Flowers greenish-white; anthers not longer than the filaments ....Embelia

Ardisia Swartz *nom. cons.*

A. solanacea Roxb. Pl. Cor. 1:27, t.27, 1795; FUGP 1:464; P.C. Kanjilal, For. Fl. UP. 3:398.1966; Srivastava, FG.186.1976.
A. humilis A. DC. in TLS 17:118.1837 (non Vahl); FBI 3:529.

Much branched shrubs. Leaves 6-16 x 2-5 cm, obovate or oblanceolate, subacuminate, base acute, entire subcoriaceous. Flowers pink in axillary corymbose racemes. Fruits subglobose, depressed, purple black when ripe.

Flowers : April-May; Fruits : October-January

Found commonly in shady places in sal forests as undergrowth. Singh 2937, 3041, 4166, 7253 (CDRI); Inayat 22326, 22327 (DD).

Embelia Burm.f.

E. villosa Wall. ex Roxb. Fl. Ind. 2:289.1824; FUGP 1:463; P.C. Kanjilal, For. Fl. UP. 3:398.1966.

Large rambling shrubs with densely rusty-villous branches and pale lenticels. Leaves 9-12 cm long, obovate elliptic, shortly acuminate, base cuneate, entire, margin ciliate. Flowers 4-merous, white in long axillary racemes.

Flowers and Fruits : April-October

Found rarely in the area in sal forests. Inayat 22325 (DD).

62. SAPOTACEAE

1. Calyx lobes 6-8; staminodes present .... **Manilkara**

1. Calyx lobes 4-5; staminodes absent .... **Madhuca**

Manilkara Adans.

M. hexandra (Roxb.) Dub. in Ann. Mus. Col. Marseille (Ser. 3) 3:9.1915; P.C. Kanjilal, For. Fl. UP. 401.1966.
Mimusops hexandra Roxb. Pl. Cor. 16, t. 15.1795; FBI 3:549; FUGP 1:467.

Small or medium-sized, evergreen trees with greyish-black or grey-white bark. Leaves 7-12 x 3-4 cm, elliptic or elliptic-obovate, coriaceous, obtuse, glabrous. Flowers dull white, solitary or fascicled. Fruits ellipsoid, 1-seeded; 1 cm long, reddish-yellow when ripe.

Local name : Khirni
Flowers : October-January; Fruits : January-February

Frequently cultivated in forest nursery at Bankati range for its edible fruits. Singh 7207 (CDRI); Inayat 22410 (DD).

Madhuca J.F. Gmel.

M. longifolia (Koenig). Macb. var. **latifolia (Roxb.) Chev.** Contr. Gray Herb. Harv. Uni. NS 53:17.1918; Royen in Blumea 10:53.1960; Saldanha & Nicolson, Fl. Hass. Distt. 193.1976.
Bassia longifolia Koenig in L. Mant. Alt. App. 563: 1771.
Bassia latifolia Roxb. Pl. Cor. 1:2, t.19.1795; FBI 3:544; FUGP 1:45.
Madhuca indica J.F. Gmel. Syst. Nat. 2.799.1791; P.C. Kanjilal, For. Fl. UP. 3:400.1966; Srivastava, FG. 187.1976.

Large or medium-sized, deciduous trees with dark-grey bark. Leaves clustered at the end of branches, 12-17 x 5-7.5 cm, oblong-elliptic, coriaceous, base acute, slightly acuminate, pubescent when young. Flowers cream-coloured on long rusty-tomentose, pedicels, clustered at the end of branches. Fruits globose or ovoid.

Local name : Mahua, Koyina
Flowers : March-April; Fruits : June-July

Occurs frequently in sal forests. The flowers and fruits are edible. Oil is extracted from the seed which is edible and also used in soap making. Singh 4363, 7271 (CDRI).

63. EBENACEAE

Diospyros Linn.

1. Leaves oblong-lanceolate, glabrous ....**D. peregrina**

1. Leaves broad-ovate, tomentose beneath ....**D. exsculpta**

D. peregrina (Gaertn.) Gurke in Engl. & Prantl, Nat. Pfam. 4(1):164.1891; Maheshwari, Fl. Del.; P.C. Kanjilal, For. Fl. UP. 3:407.1966; Srivastava, FG. 188.1976.
Embryopteris peregrina Gaertn. Fruct. 1:145, t.29.f. 1799.
Diospyros embryopteris Pers. Syn. 2:264.1807; FBI 3:556; FUGP 1:472.

I have not collected this plant during my collection trips, but I have included this on the authority of P.C. Kanjilal who reports its occurrence from the South Kheri forest areas. Hence, a brief description is reproduced below.

Medium-sized trees with black bark. Leaves ovate-oblong, rounded at the base, obtuse or acuminate, coriaceous. Flowers cream-coloured, unisexual. Fruits globose.

D. exsculpta Buch.-Ham. in TLS 15:110.1827.
D. tomentosa Roxb. Hort. Beng. 40.1814, nom. nud. et Fl. Ind. 2:532.1824; FBI 3:564; FUGP 1:469; P.C. Kanjilal, For. Fl. UP. 3:408.1966.

Medium-sized, deciduous trees with black bark. Young parts and inflorescence densely grey or rusty tomentose. Leaves 6-16 x 3-8 cm, ovate-oblong, base rounded, acute, coriaceous, tomentose beneath when young, ultimately glabrescent. Flowers white, unisexual; male in cymes; female flowers solitary, subsessile. Fruits globose, smooth, yellow when ripe.

Local name : Tendu
Flowers and Fruits : April-June

Occurs frequently in sal forests. The fruits are used medicinally to cure diarrhoea and dysentery. The leaves are used for making biri (local cigarette). Singh 2858, 3025, 7358, 7740 (CDRI); Inayat 22320 (DD).

64. OLEACEAE

1. Flowers in terminal or axillary cymes; climbing shrubs ....**Jasminum**
1. Flowers in axillary panicles; trees ....**Linociera**

Jasminum Linn.

1. Leaves imparipinnate compound ....**J. grandiflorum**
1. Leaves simple, ovate-lanceolate:
    2. Calyx pubescent:
        3. Flowers sessile or subsessile in compact cymes ...**J. multiflorum**
        3. Flowers pedicelled in lax cymes:
            4. Calyx teeth subclavate ....**J. arborescens**
            4. Calyx teeth linear, triangular ....**J. roxburghianum**
    2. Calyx glabrous ....**J. brevipetiolatum**

J. grandiflorum Linn. Sp. Pl. ed. 2.9.1762; FBI 3:603; FUGP 1:477; P.C. Kanjilal, For. Fl. UP. 3:416.1966; Srivastava, FG. 189.1976.
J. officinale Linn. forma *grandiflorum* (Linn.) Kobuski in JAA 13:161.1932.

I have included this taxon on the basis of the Working Plan of the South Kheri Forest division. I could not collect any plant in wild state. Hence, a brief description is reproduced below.

Large scandent shrubs. Leaves compound with imparipinnate leaflets. Flowers white tinged with pink, fragrant.

Local name : Chameli
Planted in Rest House campus for its fragrant flowers.

J. multiflorum (Burm.f.) Andrews Bot. Repos. 8,t.496.1807; P.C. Kanjilal, For. Fl. UP. 3:413.1966.
Nyctanthes multiflora Burm.f. Fl. Ind. 5, t.3.f.1.1768.
Jasminum pubescens Willd. Sp. Pl. 1:37.1797; FBI 3:592; FUGP 1:475.

Scandent shrubs with young branches densely fulvous-tomentose. Leaves simple, opposite, 9 x 3.5 cm, ovate, base rounded, acute or acuminate. Flowers white in terminal many flowered umbels. Calyx pubescent, teeth 1 cm long.

Local name : Kund
Flowers : November-April

Occurs occasionally in forest areas in scrubs. Singh 7134, 8400 (CDRI).

J. arborescens Roxb. Fl. Ind. 1:95.1820; FBI 3:594 (incl. var. *latifolia* Clarke; FUGP 1:475; P.C. Kanjilal, For. Fl. UP. 3:414.1966.

Suberect or scandent shrubs. Leaves simple, opposite, 3-8 x 1.5-3 cm, ovate-lanceolate, acute, base rounded, glabrous. Flowers white in lax trichotomous terminal cymes. Fruits ellipsoid, black.

Local name : Newadi
Flowers : April-May; Fruits : June-August

Occurs occasionally in open areas. Singh 4117 (CDRI); Saran 26142, 26246 (LWG).

J. roxburghianum Wall. ex DC. Prodr. 8:314.1844; FBI 3:594.

Suberect or scandent pubescent shrubs. Leaves simple, opposite, 4-9 x 2-3.5 cm, ovate-oblong, pubescent, base rounded, shortly acuminate. Flowers white in lax, 3-chotomous, pubescent cymes. Calyx densely fulvous.

Flowers : April-May; Fruits : June-August

Occurs occasionally in open area. Singh 2862 (CDRI).

J. brevipetiolatum Duthie in Brandis Indian Trees. 450. 1921; M.A. Rau in Bull. Bot. Surv. India Suppl. No. 2. 10: 47. 1969.

Shrubs. Leaves simple, opposite, 5-15 cm long, elliptic-lanceolate, acute or shortly acuminate, subcoriaceous, entire or slightly recurved. Flowers white, arranged in lax-terminal corymbs or panicled cymes. Fruits ellipsoid.

Flowers : April-May; Fruits : June-August

Found rarely in the forest areas. Inayat 22151,22151 (a) (DD).

Linociera Swartz

L. intermedia Wight var. **roxburghii** Clarke in Hook.f. FBI 3: 609. 1882; FUGP 1:479.

Small trees. Branches covered with white lenticels. Leaves 10-14 x 2.5-4 cm oblong or oblanceolate, acute or shortly acuminate, base cuneate.

Found rarely in the area. Inayat 22332 (DD). (Specimen vegetative).

65. SALVADORACEAE

Salvadora Linn.

S. oleoides Decne. in Jacq. Voy. Bot. 140, t. 144.1844; FBI 3:620; FUGP 1.481; P.C. Kanjilal, For. Fl. UP. 3:421.1966.

I have included this taxon on the basis of the Working Plan of the South Kheri Forest division. I could not collect the species during my exploration. Hence, a brief description is reproduced below.

Evergreen shrubs or small trees with grey bark. Young leaves ovate. Mature leaves linear-lanceolate, acute, often mucronate, glabrous, coriaceous. Flowers greenish-white in short compact panicled spikes. Drupes globose.

66. APOCYNACEAE

1. Anthers free from the stigma; cells rounded at the base:

   2. Carpels 2, syncarpous; plants armed .... **Carissa**

   2. Carpels 2, united usually by the style; plants unarmed:

      3. Leaves alternate .... **Thevetia**

      3. Leaves opposite or whorled:

         4. Leaves opposite:

            5. Shrubs; calyx glandular .... **Tabernaemontana**

            5. Trees; calyx eglandular .... **Holarrhena**

         4. Leaves whorled:

            6. Flowers pinkish-white; fruits drupe .... **Rauvolfia**

            6. Flowers white; fruits of slender follicle .... **Alstonia**

1. Anthers cohering in a cone round the style and stigma; cells spured at the base:

   7. Trees or large shrubs:

      8. Leaves whorled; anthers included .... **Nerium**

8. Leaves opposite; anthers exserted .... Wrightia

7. Climbers:

9. Flowers white; anthers exserted .... Vallaris

9. Flowers greenish-white; anthers included........................ Ichnocarpus

Carissa Linn. *nom. cons.*

C. opaca Stapf ex Haines in indian For. 47:378.1921; P.C. Kanjilal, For. Fl. UP. 3:424.1966.
C. spinarum auct. non. Linn. 1753.

Large thorny evergreen shrubs with suberect or scandent branches. Leaves 2.5-3.5 x 1.5-2.5 cm, ovate, or elliptic-oblong or suborbicular, acute, glabrous, coriaceous. Flowers white, fragrant, arranged in terminal and axillary corymbose cymes. Fruits ovoid or elliptic, dark purple when ripe.

Local name : Jangali Karonda
Flowers : March-June; Fruits : October-January

Occurs commonly in mixed forest and in open areas. The ripe fruits are edible. Singh 2827, 3095, 4339, 7127, 7872, 8335 (CDRI); Inayat 22147 (CAL); Saran 26249, 26460 (LWG).

Thevetia Linn. *nom. cons.*

T. peruviana (Pers.) Merr. Philip. J. Sci. Bot. 9:130.1914; Shah in JBNHS 59(1):321.1962; P.C. Kanjilal, For. Fl. UP. 3:436.1966; Srivastava, FG. 197.1976.
Cerbera peruviana Pers. Syn. 1:267.1805.
C. thevetia Linn. Sp. Pl. 209.1753.
Thevetia nerifolia Juss. ex Steud. Nom. 2:680.1841.

Evergreen shrubs or small trees. Leaves 7-15 x 0.6 cm, linear, revolute, tapering at both ends, dark green with prominent mid rib. Flowers yellow or orange, fragrant in sub-terminal cymes. Drupes angular, subglobose, green.

Local name : Kanail
Flowers and Fruits : Most part of the year

Commonly planted in Rest House campus for its ornamental flowers. Singh 7251 (CDRI).

Tabernaemontana Linn.

T. divaricata (Linn.) R.Br. in Roem. & Schult. Syst. 4:427.1819; Santapau in RBSI 16(1):167.1953; Srivastava, FG. 196.1976.
Nerium divaricatum Linn. Sp. Pl. 209.1753.
Tabernaemontana coronaria Willd. Enum. Hort. Berol. 275.1809; FBI 3:646.
Ervatamia coronaria Stapf in This. Dyer. Fl. Trop. Africa 4:127.1902; FUGP 1:488.

Evergreen glabrous, dichotomously branched shrubs. Leaves opposite, 10-14 x 2-3 cm, oblong-lanceolate, acuminate or caudate. Stipules cup-like. Flowers white in axillary or terminal few flowered cymes. Corolla tube 2.5 cm long.

Local name : Chandani
Flowers : May-October

Commonly planted in Forest Rest House campus. Singh 7326 (CDRI).

Holarrhena R.Br.

H. antidysenterica (Roth) A.DC. in DC. Prodr. 8:413.1844; FBI 3:644; FUGP 1:487; P.C. Kanjilal, For. Fl. UP. 3:428.1966; Srivastava, FG. 194.1976.
Echites antidysenterica Roth, Nov. Pl. Sp. 138.1821.
Nerium antidysenterica Linn. Sp. Pl. 209.1753 (nomen ambigum).
H. pubescens Wall. ex G.Don.

Medium-sized trees with grey-white bark. Leaves opposite, subsessile, 10-14 x 5-6 cm, ovate to elliptic-oblong, acuminate, base rounded or acute. Flowers creamy-white in terminal cymes. Follicles two, slender, 23 cm long, curved. Seeds 2 cm long, linear.

Local name : Kurchi, Dudhi
Flowers : May-June; Fruits : October-December

Occurs commonly in sal forests. The bark is used in dysentery and fever. Singh 2946, 3097, 4128 (CDRI); Saran 26241 (LWG).

Rauvolfia Linn.

R. **serpentina** (Linn.) Benth. ex Kurz, For. Fl. Burma 2:171.1877; FBI 3:632; FUGP 1.484; Santapau in RBSI 16(1):166.1953; Sulochana in JIBS 38:578.1959; Srivastava, FG. 196.1976.
Ophioxylon serpentinum Linn. Sp. Pl. 1043.1753.

Erect, glabrous undershrubs with pale bark. Leaves in whorls of three, 9-17 x 3-4 cm, oblanceolate or elliptic-lanceolate, acute or shortly acuminate. Flowers pinkish-white in long, peduncled terminal or lateral cymes. Drupes dark purple when ripe.

Local name : Dhawarbarua
Flowers and Fruits : September-February

Occurs occasionally in the sal forests in shady places. The root is used in fever and for treating madness Singh 2947, 7545, 7827, 8457 (CDRI); Inayat 22150 (b) (DD).

Alstonia R.Br. *nom. cons.*

A. **scholaris** (Linn.) R.Br. in Mem. Werb. Soc. 1:75.1810; FBI 3:642; FUGP 1:485; Santapau in RBSI 16(1):166.1953; P.C. Kanjilal, For. Fl. UP. 3:427.1966; Srivastava, FG. 192.1976.
Echites scholaris Linn. Mant. 55.1967.

Large evergreen trees with grey-white bark. Leaves in-whorls of 4-7, coriaceous, 7-14 x 1.5-4 cm, oblong-lanceolate or obovate, obtuse, crowded at the end of branches. Flowers greenish-white in panicled, pubescent cymes. Follicles two, slender, circular, in clusters from the peduncles.

Local name : Chitwan
Flowers : November-January; Fruits : March-June

Occasionally met in the area; probably planted. The bark is used for the treatment of leucoderma. Singh 4267, 7726 (CDRI).

Nerium Linn.

N. indicum Mill. Gard. Dict. n.2.1768; Santapau in RBSI 16(1):170.1953; P.C. Kanjilal, For. Fl. UP. 3:432.1966; Srivastava, FG. 195.1976.
N. odorum Soland. in Ait. Hort. Kew. 1:297.1782; FBI 3:655; FUGP 1:491.

Large, evergreen ornamental shrubs. Leaves usually ternate, linear-lanceolate, 9-16 x 1-2 cm, acute, coriaceous. Flowers pink or white, fragrant, arranged in terminal panicled cymes.

Local name : Kaner
Flowers : March-June

Duthie reports its occurrence as wild in Kheri forests but I could not get it in wild condition. Commonly planted near temples and in Forest Rest House campus. Inayat 22148 (DD); Saran 26378 (LWG).

Wrightia R.Br.

W. tomentosa (Roxb.) Roem. & Schult. Syst. 4:414.1819; FBI 3:653; FUGP 1:490; Nagan in Ann. Miss. Bot. Gard. 52:147.1965; P.C. Kanjilal, For. Fl. UP. 3:430.1966.
Nerium tomentosum Roxb. Fl. Ind. 2:6.1824.

Small deciduous trees with dark-grey or pale-brown bark. Young parts softly tomentose. Leaves 6-10 x 2-4.5 cm, elliptic-oblong, membranous, caudate or acuminate, softly tomentose on both surfaces. Flowers cream coloured, arranged in pubescent corymbose cymes. Follicles 13-15 x 1-2 cm, subcylindric, rough with white lenticular dots, laterally compressed.

Local name : Dudhi
Flowers : April-June; Fruits : October-December

Occurs occasionally in South Kheri Forest division along forest roads. Singh 7435, 7902 (CDRI).

Vallaris Burm.f.

V. solanacea (Roth) O. Kuntze, Rev. Gen. Pl. 2:417.1891; P.C. Kanjilal, For. UP. 3:433.1966; Srivastava, FG.197.1976.
Peltanthera solanacea Roth, Nov. Pl. Sp. 132.1821.
Vallaris heynei Spreng. Syst. 1:635.1824; FBI 3:650; FUGP 1:488.

Large climbing or straggling shrubs with ash-grey lenticular branches. Leaves opposite, 8-9 x 2.5-2.7 cm, ovate-oblong or elliptic, subacuminate. Flowers creamy-white in lax, axillary, pubescent cymes. Corolla subrotate. Fruits 2-connate, tapering follicles.

Local name : Dudhibel, Bakarthana
Flowers : February-April; Fruits : November-December

Found commonly in the open areas climbing on trees. The stem is used for making basket. Singh 2939, 7242, 8399 (CDRI).

Ichnocarpus R.Br. *nom. cons.*

I. frutescens (Linn.) R.Br. in Aiton, Hort. Kew. 2:2.69.1811; FBI 3:669; FUGP 1:492; P.C. Kanjilal, For. Fl. UP. 3:434.1966; Srivastava, FG. 195.1976.
Apocynum frutescens Linn. Sp. Pl. 213.1753.

Large evergreen, climbing shrubs. Young branches and inflorescence rusty-tomentose. Leaves elliptic-oblong or ovate-lanceolate, 5-9 x 1.5 cm, acute, glabrous above, rusty-villous beneath. Flowers greenish-white, in terminal and axillary cymes. Follicles 8 cm long, linear.

Flowers : October-November; Fruits : January-April

Found commonly in open areas and in forests climbing on small trees. Singh 2856, 3077, 4213, 7504, 7869 (CDRI); Inayat 22143 (a) (DD); Saran 2895 (LWG).

67. ASCLEPIADACEAE

1. Filaments usually free; pollinia granular:
    2. Flowers greenish-purple ....Hemidesmus
    2. Flowers pale-yellow ....Cryptolepis
1. Filaments connate into a tube; pollinia waxy:
    3. Anthers with membranous appendages:
        4. Stem erect ....Calotropis
        4. Stem twining:
            5. Leaves linear-lanceolate ....Oxystelma
            5. Leaves ovate-cordate or oblong-lanceolate:
                6. Follicles thick, pointed, ribbed ....Dregia
                6. Follicles slender not ribbed:
                    7. Flowers yellow ....Gongronema
                    7. Flowers greenish-yellow ....Tylophora
    3. Anthers without membranous appendages ....Leptadenia

Hemidesmus R.Br.

H. indicus (Linn.) R.Br. in Aiton f. Hort. Kew. ed. 2.2:75.1811. FBI 4:5; FUGP 1:95; Santapau in Bull. Bot. Surv. India vol. 3(1)17.1961; P.C. Kanjilal, For. Fl. UP. 3:441.1966; Srivastava, FG. 198.1976.
Periploca indica Linn. Sp. Pl. 211.1753.

Slender, twining or prostrate undershrubs with aromatic rootstocks. Leaves 5-7 x 1.2 cm, elliptic-oblong to linear-lanceolate, often variegated above near midrib, acute, glabrous. Flowers greenish-purple in subsessile, clustered, axillary cymes. Calyx long, ovate, acute with ciliate margins. Corolla lobes valvate with 5 coronal scales. Follicles divaricate, slender.

Local name : Kapuri
Flowers and Fruits : August-December

Occurs frequently in moist shady places. The roots are used in medicine. Singh 7415, (CDRI).

Cryptolepis R.Br.

C. buchanani Roem. & Schult. Syst. Veg. 4:409.1819; FBI 4:5; FUGP 1:495; Santapau & Irani in Bot. Mem. Univ. Bombay 4:91.1960; P.C. Kanjilal, For. Fl. UP 3:442.1966.

Large, glabrous, twining shrubs. Leaves elliptic-oblong, 8-13 x 3.5-4 cm, apiculate, coriaceous, smooth above, glaucous and white beneath; secondary nerves transversely united in intramarginal veins. Flowers pale-yellow in short axillary, panicled cymes. Corolla twisted with 5 coronal scales, inserted within the tube. Follicles 5-9 cm long, straight, divaricate, tapering.

Flowers : April-June; Fruits : July-October

Occurs frequently in open places climbing on small trees and shrubs. Singh 4259, 7166 (CDRI); Inayat 22178 (DD).

Calotropis R.Br.

1. Leaves sessile or subsessile:
    2. Flower buds ovoid; corolla lobes spreading .... **C. gigantea**
    2. Flower buds hemispheric; corolla lobes erect .... **C. procera**
1. Leaves petiolate .... **C. acia**

C. gigantea (Linn.) R.Br. in Ait. Hort. Kew. ed. 2, 2:78.1811: FBI 4:17; FUGP 1:497; Santapau & Irani in Bot. Mem. Univ. Bombay 4:17.1960; P.C. Kanjilal, For. Fl. UP.3:444.1966; Srivastava, FG. 199.1976.
Asclepias gigantea Linn. Sp. Pl. 214.1753.

Large, stout shrubs or small trees, covered with appressed, white, cottony tomentum. Leaves sessile, 12-14 x 5.5-6 cm, elliptic-oblong or obovate-oblong, acute, base cordate or amplexicaul. Flowers blue, arranged in umbellate lateral cymes. Corolla lobes spreading, ovate-lanceolate with 5 coronal processes and a recurved spur at the base. Follicles 7-12 cm long, recurved, smooth.

Local name : Akahua
Flowers : December-July; Fruits : March-August

Found commonly in open area in waste places. The stem is used for making fine fibre used for fishing net. Singh 2882, 4312, 7216, 8316, 8317 (CDRI).

C. procera (Ait.) R.Br. in Ait. Hort. Kew. ed. 2, 2:78. 1811; FBI 4:18; FUGP 1:498; Santapau & Irani loc. cit. 19, 1960; P.C. Kanjilal, For. Fl. UP. 3:445.1960; Srivastava, FG. 199.1976.
Asclepias procera Ait. Hort. Kew. 1:305.1789.

Large, erect, hoary-tomentose shrubs. Leaves sessile, ovate-oblong, 13 x 8 cm, cordate, acute. Flowers purplish-red, arranged in axillary and terminal corymbose cymes. Corolla lobes erect, acute. Follicles 6-12 cm long, recurved, smooth. Seeds flat with a silky white coma.

Local name : Madar, Manar
Flowers and Fruits : March-July

Occurs commonly in open waste places and along canal banks. Singh 2878 (CDRI).

C. acia Buch.-Ham. in TLS 14:247.1824; FBI 4:18; FUGP 1:498; P.C. Kanjilal, loc. cit. 446.1966.

Erect, stout herbs; young parts floccose tomentose. Leaves petiolate, 12-17 x 7-10 cm, obovate or oblong, acute or shortly acuminate. Flowers purplish-arranged in branched umbellate cymes. Calyx lobes lanceolate, acuminate. Corolla lobes erect with pink, ciliate, corona lobes. Follicles subglobose, 9 cm long.

Flowers and Fruits : March-July

Occurs rarely in grasslands and open area. Inayat 23723 (DD).

Oxystelma R.Br.

O. secamone (Linn.) K. Schum. in Engl. & Prantl, Nat. Pfam. 4(2):229.1895; Santapau & Irani loc. cit. 67.1960; Srivastava, FG. 200.1976.
Periploca secamone Linn. Mant. 2:216.1771.
Oxystelma esculentum R.Br. in Mem. Wern. Soc. 1:40.1810; FBI 4:17; FUGP 1:500; P.C. Kanjilal, For. Fl. UP. 3:447.1966.
Sarcostemma secamone (L.) Bennet in Fl. Howrah 285. 1979.

Slender, glabrous, much branched, perennial twiners. Leaves linear-lanceolate, 1-12 x 0.4 cm, acute. Flowers white, purple veined inside, drooping in lax cymes. Corolla rotate with an annular villous ring. Staminal corona of 5 erect lobes with the tips projecting above stigma. Follicles 5 cm long, ovoid-lanceolate.

Local name : Dudhi bel
Flowers and Fruits : July-November

Occurs commonly in open areas, climbing on small trees and shrubs along river banks. Singh 4356, 7492 (CDRI); Inayat 22161 (DD).

Dregea E. Mey. *nom. cons.*

D. volubilis (Linn.f.) Benth. ex Hook.f. in FBI 4:46.1883; Santapau & Irani loc. cit. 42.1960; Maheshwari, Fl. Del

219.1963.
Asclepias volubilis Linn.f. Suppl. 170.1781.
Marsdenia volubilis T. Cooke, in Fl. Pres. Bombay 2:166.1904; FUGP 1:505.

Large twining shrubs with lenticellate branches; young parts tomentose. Leaves 6-12 x 4-6.5 cm, ovate-cordate, acuminate, base rounded or cordate, softly pubescent when young. Flowers greenish-yellow in dense axillary, drooping, pubescent umbellate cymes. Follicles paired, divaricate, 13 cm long, pointed.

Flowers : April-September; Fruits : September-November

Found frequently twining on small trees and shrubs in open areas. Singh 4227, 7456, 7826, 8479 (CDRI); Inayat 22178 (DD).

Gongronema Decne.

G. nepalense Decne. in DC. Prodr. 8:624.1844; FBI 4:33; FUGP 1:504; P.C. Kanjilal, For. Fl. UP. 3:454.1966.

Large twining shrubs; young parts pubescent. Leaves ovate, or elliptic-oblong, 9-14 x 4-7 cm, acuminate, base rounded or sub-cordate. Flowers yellow on long peduncled umbellate cymes. Corolla campanulate, with 5 erect fleshy scales of corona. Follicles slender 6-7 cm long.

Flowers : August-September; Fruits : November-December

Occurs rarely in the Park in Dudhwa range. Sis Ram 11478 (DD).

Tylophora R.Br.

1. Leaves ovate-oblong, apiculate; follicles slender ....**T. indica**
1. Leaves elliptic-orbicular or oblong; follicles globose, pointed..**T. rotundifolia**

T. indica (Burm.f.) Merr. in Philip. J. Sci. (Bot.) 19:373.1921; Santapau & Irani loc. cit. 88.1960; P.C. Kanjilal, For. Fl.

UP. 458.1966; Raizada, Suppl. FUGP 156.1976.
Cynanchum indicum Burm.f. Fl. Ind. 70.1768.
Tylophora asthmatica W.&A. in Wight, Contrib. 51.1834; FBI 4:44.

Perennial, twining herbs with many long, fleshy roots. Leaves ovate-oblong, 6 x 2.5 cm, acute or apiculate, pubescent or glabrous, base rounded or cordate. Flowers greenish-yellow in lax, pubescent, umbellate cymes.

Flowers and Fruits : August-February

Occurs rarely in the Park in Dudhwa range. Singh 2941 (CDRI) Kheri-Dudhwa, Henderson S.n. (DD).

T. rotundifolia Buch.-Ham. in Wight, Contrib. 50.1834; FBI 4:43; FUGP 1:510; Santapau & Irani loc. cit. 86; P.C. Kanjilal, For. Fl. UP. 3:458.1966.

Slender, prostrate trailing herbs with woody root stocks. Leaves 9-10 x 5-5.5 cm, ovate-elliptic or suborbicular or oblong, coriaceous, mucronate or acute, base rounded to subcordate. Flowers many in lax, branched umbellate, pubescent, cymes. Follicle single, glabrous, subacute, 7 cm long.

Flowers and Fruits : July-October

Found frequently in the Mailani range of South Kheri forest division in sal forests. Singh 7471 (CDRI); Inayat 23718 (a) (CAL).

Leptadaenia R.Br.

1. Twining, leafy shrubs .... **L. reticulata**
1. Erect, leafless shrubs .... **L. pyrotechnica**

L. reticulata (Retz.) W.&A. in Wight, Contrib. 47.1834; FBI 4:63; FUGP 1:511; Santapau & Irani loc. cit. 60.1960; P.C. Kanjilal, For. Fl. UP. 3:459.1966.
Cynanchum reticulatum Retz. Obs. Bot. 2:15.1781.

Diffuse, much branched, climbing shrubs with corky bark on old stems. Leaves ovate-cordate, 5.5-7.6 x 2.5-3.5 cm, coriaceous, acute, pubescent when young. Flowers pale-yellow, arranged in lateral or subaxillary umbellate, pubescent cymes. Follicles solitary, straight, smooth with a curved beak.

Flowers : July-August; Fruits : October-December

Found rarely in open waste places in South Kheri Forest division in Gola range. Singh 7602, 7728 (CDRI).

L. pyrotechnica (Forsk.) Decne. in Ann. Sc. Nat. (ser 2) 9:269.1838; Santapau & Irani loc.cit. 59.1960; P.C. Kanjilal, For. Fl. UP. 3:460.1966.
Cynanchum pyrotechnicum Forsk. Fl. Aegypt.-Arab. 53.1775.
Leptadaenia spartium Wt. Contrib. 48.1834; FBI 4:64; FUGP 1:511.

Erect, glabrous much branched leafless shrubs. Leaves few, often unequal, present usually on young branches. Flowers yellow, arranged in umbellate, pubescent cymes. Follicles terete, solitary tapering to a slender beak.

Flowers : June-December

Occurs rarely in the area in wastelands. Saran 22172 (LWG); Khalil & party 19804 (LWG).

68. LOGANIACEAE

1. Leaves glabrous; annual herbs ....Cynoctonum
1. Leaves white-tomentose beneath; shrubs ....Buddleja

Cynoctonum Gmel.

G. mitreola (Linn.) Britton in Mem. Torr. Bot. Club. 5.258.1894; Leenhouts in Fl. Males 6:375; t.37.1962; Santapau & Wagh in Bull. Bot. Surv. India 5:106.1963; Srivastava, FG. 201.1976. (Fig. 8).

Ophiorhiza mitreola Linn. Sp. 150.1753.
Mitreola oldenlandioides Wall. ex DC. Prodr. 9:9.1845; FBI 4:79; FUGP 1:515.

Erect, glabrous, annual herbs. Leaves opposite, 6 x 2 cm, membranous, ovate-oblong, acute or acuminate; stipules connate in transverse ridge. Flowers white in terminal and axillary dichotomous cymes; unilateral along the branches. Capsules wedge shaped. Seeds trigonous.

Flowers and Fruits : August-November

Found occasionally in moist shady places in sal forests. Singh 4195, 7852 (CDRI).

Buddleja Linn.

B. asiatica Lour. Fl. Cochinch. 72.1790; Bailey, Man. Cult. Pl. 804; FBI 4:82; FUGP 1:516.

Large, evergreen shrubs; young parts grey or white tomentose. Leaves lanceolate, 6-8 x 1.2 cm, acuminate, serrulate or entire, white-tomentose beneath. Flowers white in slender drooping, axillary and terminal panicled spikes. Capsules ellipsoid.

Flowers and Fruits : February-May

Occurs occasionally near river banks and in moist shady places in Dudhwa range of the Park. Singh 7108, 7244, 8391 (CDRI).

69. GENTIANACEAE

1. Leaves alternate, long petiolate; aquatic herbs....................**Nymphoides**
1. Leaves opposite, sessile; terrestrial herbs:
    2. Ovary 2-celled; flowers blue ....**Exacum**
    2. Ovary 1-celled; flowers pink or white:
        3. Style filiform:
            4. Flowers pink, regular ....**Centaurium**

4. Flowers white, irregular ....**Canscora**

3. Style very short or absent ....**Swertia**

Nymphoides Medic.

N. indicum (Linn.) O. Kuntze, Rev. Gen. Pl. 429.1891; Santapau in Bull. Bot. Surv. India 3:20.1961; Srivastava, FG. 203.1976.
Menyanthes indica Linn. Sp. Pl. 207.1753.
Limnanthemum indicum Thw. Enum. 205.1861; FBI 4:131; FUGP 1:525.

Aquatic floating, glabrous herbs. Floating leaves, petiolate, orbicular-cordate, 6-12 cm in diameter, margins sinuate, glabrous. Flowers white with yellow centre, in clusters at the base of the petiole. Corolla lobes with papillose hairs, not crested down the middle. Capsules subglobose.

Flowers and Fruits : April-October

Found commonly in ponds, jhils etc., Singh 7697, 8340 (CDRI); Saran 26214 (LWG).

Exacum Linn.

E. tetragonum Roxb. Hort. Beng. 10. 1814, nom. nud. et Fl. Ind.: 398. 1820; FBI 4:95; FUGP 1:518.

Erect, glabrous, annual herbs. Leaves sessile, opposite, 5-9 x 1-2.5 cm, ovate-lanceolate, 5-nerved. Flowers blue, arranged in terminal panicles. Calyx ovate, acuminate, keeled. Corolla lobes broadly elliptic, acute. Capsules subglobose.

Flowers and Fruits : October-December

Found rarely in Dudhwa range of the Park in moist shady places. Singh 7790, 7933 (CDRI).

Centaurium Hill.

1. Flowers sessile in lax dichotomous cymes ....**C. ramosissimum**

1. Flowers long pedicellate in branched panicled dichotomous cymes ....**C. centaurioides**

C. ramosissimum (Pers.) Syn. 1:283.1805; FBI 4:101; FUGP 1:520.

Erect, glabrous, annual herbs. Leaves opposite, sessi 2.5-3 cm long, ovate-elliptic or lanceolate, ascending, acute. Flowers rosy, sessile in conspicuously branched axillary and terminal, lax dichotomous cymes. Calyx linear, equalling the corolla tube.

Flowers and Fruits : February-May

Occurs frequently in moist sandy soil and in cultivated fields. Singh 2836 (CDRI).

C. centaurioides (Roxb.) Rolla Rao & Hemadri in JBNHS 67(2):357.1970; Raizada, Suppl. FUGP 157.1976.
Chironia centaurioides Roxb. Hort. Beng. 16.1814, nom. nud. & Fl. Ind. ed. Carey & Wall. 1:565.1832.
Centaurium roxburghii (D.Don) Druce in Rep. Bot. Exch. Club. Brit. Isles, 4:614.1916; Santapau in RBSI 16(1):181.1953; Srivastava, FG. 203.1976.
Erythraea roxburghii D.Don, Syst. 4:206.1837; FBI 4:102; FUGP 1:520.

Erect annual herbs with 4-angled stems. Radical leaves in rosette, 2-3 cm long, obovate or oblong, obtuse; cauline leaves 2 cm long, opposite, sessile, oblong or linear. Flowers pink in dichotomous cymes with a long pedicellate flowers in each fork. Capsules linear.

Flowers and Fruits : February-April

Found commonly in moist and waste places. Singh 7171 (CDRI).

Canscora Lamk.

C. decussata (Roxb.) Roem. & Schult. Syst. Veg. Mant. 1:229.1822; FBI 4:104; FUGP 1:523; Srivastava, FG. 202.1976.
Pladera decussata Roxb. Fl. Ind. 1:418.1820.

Erect, branched annual herbs with 4-winged stems. Leaves opposite, sessile, 2-4 x 1 cm, oblong-lanceolate, 3-nerved, acute. Flowers white in lax cymes. Calyx 4-winged. Capsules oblong, shorter than calyx.

Flowers and Fruits : August-November

Found frequently in moist and muddy places. Singh 4207, 7850 (CDRI).

Swertia Linn.

S. angustifolia Buch.-Ham. ex D.Don Prodr. Fl. Nepal. 127. 1825; Burtt in Notes R. Bot. Gdn. Edinb. 26(3): 274. 1965; FUGP 1:524 (incl. var. *elegans* (Clarke) Burkill).

Dwarf, annual herbs. Stems 4-angled and slightly winged. Leaves sessile, 2 cm long, linear-lanceolate, acute. Flowers bluish-white, axillary or in leafy cymes. Calyx oblong-linear, acute, longer than corolla. Corolla lobes ovate-oblong, short, acute, provided with nectaris at the base.

Local name : Chirata
Flowers and Fruits : September-October

Occurs rarely in moist shady places in the Park. Singh 7670 (CDRI); Inayat 22315 (DD).

70. HYDROPHYLLACEAE

Hydrolea Linn. *nom. cons.*

H. zeylanica (Linn.) Vahl, Symb. Bot. 2:46.1791; FBI 4:133; FUGP 1:526; Santapau in RBSI 16(1):184.1953; Srivastava, FG. 203.1976.
Nama zeylanica Linn. Sp. Pl. 226.1753.

Erect or decumbent, somewhat succulent herbs, often rooting at the nodes. Leaves 3-8 x 1.5 cm, lanceolate, membranous, narrowed at both ends. Flowers blue, in short terminal or lateral racemes. Calyx glandular, pubescent, 3-nerved. Capsules ovoid, enclosed in calyx.

Flowers and Fruits : September-October

Occurs commonly in marshy places and in rice fields. Singh 4270, 7846 (CDRI).

71. BORAGINACEAE

1. Ovaries entire or slightly 4-lobed ....**Heliotropium**
1. Ovaries 4-lobed; nutlets usually 4:
    2. Anthers connivent in a cone; styles subterminal ....**Trichodesma**
    2. Anthers not connivent in a cone; styles gynobasic:
        3. Nutlets glochidiate ....**Cynoglossum**
        3. Nutlets tuberculate:
            4. Flowers white ....**Bothriospermum**
            4. Flowers bright yellow ....**Arnebia**

Heliotropium Linn.

H. strigosum Willd. Sp. Pl. 1:743.1798; FBI 4:151; FUGP 1:537; Srivastava, FG. 206.1976.

Prostrate or procumbent, much branched, strigose, perennial herbs. Leaves 0.7-2 cm long, linear, hirsute, acute. Flowers white. Fruits globose, of 4-nutlets.

Flowers and Fruits : July-March

Found commonly on moist wastelands. Singh 3056, 4116 (CDRI); Inayat 22208 (DD); Saran 26324, 26333 (LWG).

Trichodesma R.Br. *nom. cons.*

T. sedgwickianum Banerjee in Bull. Bot. Soc. Beng 16:10.1962 (April 1964); Srivastava, FG. 207.1976.

T. amplexicaule DC. Prodr. 10:172.1846; Sedgwick in RBSI 6:347.1913 (non Roth).
T. amplexicaule Clarke in FBI 4:153.1883 in part.
T. indicum R.Br. sensu Duthie in FUGP 1:530.1960.

Erect, hispid, branched annual herbs. Leaves opposite, sessile, 6-7.5 x 1.5-2 cm, oblong-lanceolate, acute, base subsagittate or cordate. Flowers funnel-shaped, pale-blue or violet, solitary in leaf opposed or extra axillary leafy racemes. Nutlets smooth.

Flowers and Fruits : August-April

Occurs commonly in wastelands in grasses. Singh 2825, 3033, 4284, 8330 (CDRI).

Cynoglossum Linn.

1. Nutlets with deep margin ....**C. zeylanicum**
1. Nutlets scarcely margined ....**C. lanceolatum**

C. zeylanicum (Vahl ex Hornem.) Thunb. ex Lehm. Neve Schriften Naturf. Ges. Halle 3(2):20.1817; Kazmiin JAA 52:344.1971.
Anchusa zeylanica Vahl ex Hornem. Enum. Hafn. 3.1807.
Cynoglossum denticulatum var. *zeylanica* Clarke in FBI 4:157.1883; FUGP 1:540.

Erect, branched, hispid annual herbs. Leaves 4-6 x 1-2 cm, ovate-lanceolate, acute, denticulate, clothed with stiff bulbous hairs. Flowers distinct in terminal and axillary racemes. Calyx lobes bristly hairy. Corolla lobes rounded, spreading, white with blue tinge. Nutlets glochidiate and margined.

Flowers and Fruits : October-January

Found occasionally in wastelands and amongst grasses. Singh 7294, 8492 (CDRI) Inayat 23735 (a) (DD).

C. lanceolatum Forsk. Fl. Aegypt.-Arab. 41.1775 (non Heyne); FBI 4:156; FUGP 1:540; Srivastava, FG.

205.1976.

C. micranthum Desf. Tabl. Hort. Ecol. Bot. ed. 1:220.1804; FBI 4:156.

Erect, hispid annual herbs. Leaves alternate, 6-8 x 1-1.5 cm, lanceolate, membranous, nerves prominent beneath. Flowers white, tinged with blue in long racemes with curved tips. Nutlets shortly ovate, densely glochidiate, obscurely margined.

Flowers and Fruits : August-October

Found commonly in wastelands and in shady places in sal forests. Singh 2964, 4143, 7931 (CDRI).

Bothriospermum Bunge

B. tenellum (Hornem.) Fisch. & Mey. Ind. Sem. Hort. Petrop. 1:24.1835; FBI 4:167; FUGP 1:540; Srivastava, FG 204.1976.

Anchusa tenella Hornem. Hort. Hafn. 1:176.1813.

Prostrate or diffuse, hispid annual herbs. Leaves alternate, 1-3 cm long, ovate-oblong, softly hairy, entire, subacute. Flowers white, distant or the upper in close, bracteate racemes. Nutlets tuberculate, oblong.

Flowers and Fruits : February-June

Occurs frequently in moist shady places along river and nallah. Singh 2898, 7139, 7334, 7663 (CDRI); Inayat 22186 (b) (DD).

Arnebia Forsk.

A. hispidissima (Lehm.) DC. Prodr. 10:94.1845; FBI 4:176; FUGP 1:541.

Lithospermum hispidissimum Lehm. Icon. Nov. Strip. t.39.1821.

Diffuse or prostrate, hispid annual herbs with red tap root. Branches arising from a woody base, clothed with white, stiff spreading hairs. Leaves sessile, 2-2.5 cm long, lanceolate to linear-lanceolate, densely hispid. Flowers yellow in secund, hirsute spikes. Nutlets small, ovoid, tuberculate.

Flowers and Fruits : October-April

Found occasionally on dry sandy soils in open situation. Singh 7614, 8304 (CDRI).

72. EHRETIACEAE

1. Styles 4-fid .... **Cordia**
1. Styles 2-fid .... **Ehretia**

Cordia Linn.

C. dichotoma Forst. f. Prodr. 18.1786; Santapau in RBSI 16(1):185.1953; P.C. Kanjilal, For. Fl. UP. 3:467.1966; Srivastava, FG. 207.1976.
C. obliqua Willd. Phytogr. 4. t.4.1794; FBI 4:137. Excl. var. *C. myxa* auct. plur. non Linn. 1753.

Medium-sized, deciduous trees with grey or brown bark. Leaves 7-15 x 6.5 cm, ovate, elliptic to suborbicular, coriaceous, cuneate. Flowers white in lax, terminal or axillary cymes. Drupes ovoid, apiculate.

Local name : Lasura, Litora
Flowers : March-April; Fruits : May-July

Occurs occasionally on the edge of the sal forests. Fruits are edible. Singh 2942, 7161, 8403 (CDRI).

Ehretia Linn.

1. Leaves entire .... **E. leavis**
1. Leaves serrate .... **E. acuminata var. serrata**

E. laevis Roxb. Pl. Cor. 1:42, t. 56.1796; FBI 4:141; FUGP 1:531; P.C. Kanjilal, For. Fl. UP. 3:472.1966; Srivastava FG. 208.1976.

Medium-sized, deciduous trees with white, smooth bark. Leaves 8-13 x 5-9 cm, elliptic or obovate, acuminate, entire. Flowers white in unilateral or scorpioid cymes. Drupes orange when ripe, globose.

Local name : Chamror
Flowers and Fruits : February-May

Found occasionally in the mixed forests. Singh 2823, 2894, 7214, 7588, 7654 (CDRI); Inayat 22206 (a) (DD).

E. acuminata R.Br. var. **serrata** (Roxb.) Johns in JAA 32:23.1951.
E. acuminata R.Br. sensu Clarke in FBI 4:141.1882; FUGP 1:531; P.C. Kanjilal, loc. cit. 471.1966.

Moderate-sized, deciduous trees with grey bark. Leaves alternate, 7-14 x 2.5-5 cm, elliptic or ovate, oblong, serrate, acuminate. Flowers white, arranged in dense compound panicles. Drupes ovoid.

Local name : Punnia
Flowers and Fruits : March-May

Found occasionally in open areas in Dudhwa range of the Park. Singh 2911, 7163, 8384 (CDRI).

73. CONVOLVULACEAE

1. Flowers numerous in racemes or panicles ....**Porana**
1. Flowers solitary or in cymes:
    2. Style 2, 2-fid:
        3. Flowers white; capsule 2-valved ....**Volvulopsis**
        3. Flowers blue; capsule 4-valved ....**Evolvulus**

2. Style 1, undivided:

4. Fruits dehiscent:

5. Stigma globose:

6. Capsule 4-valved .... **Merremia**

6. Capsule circumsessile:

7. Stem winged .... **Operculina**

7. Stem not winged .... **Ipomoea**

5. Stigma linear, elongated .... **Convolvulus**

4. Fruits indehiscent .... **Rivea**

Porana Burm.f.

P. paniculata Roxb. Pl. Corom. 3:31,t.235.1835; FBI 4:222; FUGP 1:544; Ooststr. in Fl. Males. 4:404.1953; P.C. Kanjilal, For. Fl. UP. 3:480.1966; Srivastava, FG. 215.1976.

Large, climbing, grey-tomentose shrubs. Leaves 9-12 x 4-5 cm, ovate, cordate, acuminate, entire, hairy on both surfaces. Flowers white, numerous in large axillary and terminal drooping panicles. Capsules membranous, hairy, 1-seeded.

Flowers and Fruits : November-February

Found occasionally climbing on trees in Dudhwa Forest Rest House campus. Singh 4174, 7971 (CDRI).

Volvulopsis Roberty

V. nummularia (Linn.) Roberty in Candollea 14:28.1952; Raizada in Indian For. 94:453.1968 et Suppl. FUGP 166. 1976; Srivastava, FG. 215.1976.
Evolvulus nummularius Linn. Sp. Pl. ed. 2.392.1762.

Prostrate, hairy, perennial herbs, rooting at the nodes. Leaves 2 x 1 cm, ovate, elliptic or orbicular, rounded or subcordate at the base, rounded at the apex. Flowers white, 1-2, axillary. Calyx ovate-oblong, sparsely hairy. Corolla rotate to broadly funnel-shaped. Capsules globular, 2-seeded.

Local name : Musa parni, Musakan.
Flowers and Fruits : August-February

Occurs frequently in moist shady places. The plant is used in medicine. Singh 3052 (CDRI).

Evolvulus Linn.

E. alsinoides Linn. Sp. Pl. ed. 2:392.1762; FBI 4:220; FUGP 1:546; Ooststr. in Fl. Males. 4:395.1953; Stearn. in Taxon 21:649.1972; Srivastava, FG. 210.1976.
Convolvulus alsinoides Linn. Sp. Pl. 157.1753.

Prostrate, diffuse, much branched, hairy, annual herbs. Branches slender, spreading from a woody rootstock. Leaves lanceolate to ovate, 0.5-2 x 0.2-0.5 cm, obtuse, mucronate, pubescent beneath. Flowers blue in 1 to 3-flowered, axillary filiform peduncles. Capsules globose, 4-seeded, hairy.

Local name : Shankhpushpi
Flowers and Fruits : July-March

Occurs commonly in moist places in sal forests and other wastelands. The plant is used in medicine and for brain tonic. Singh 3076, 4323, 7114, 8366 (CDRI).

Merremia Dennst. ex Endlicher *nom. cons.*

1. Flowers yellow .... **M. hederacea**

1. Flowers purplish .... **M. dissecta**

M. hederacea (Burm.f.) Hallier f. in Engl. Bot. Jahrb. 18:118.1894; Ooststr. in Blumea 3:302.f.l.e.f.m.n. 1939; Srivastava, FG. 215.1976.
Evolvulus hederaceus Burm.f. Fl. Ind. 77. t.30.f.2.1768.
Merremia chryseides Hallier f. in Engl. Bot. Jahrb.

16:552.1893; FUGP 1:550.
Ipomoea chryseides Ker-Gawl. in Bot. Reg. t.1. 270.1818; FBI 4:206.

Twining, glabrous, annual herbs. Leaves 4-10 cm long, ovate, often 3-lobed, deeply cordate, apiculate, repand or crenate; petiole 5-6 cm long. Flowers yellow, in dense, axillary cymes. Capsules ovoid or sub-quadrangular, 2-celled. Seeds trigonous, tomentose.

Flowers and Fruits : September-December

Occurs frequently climbing on small trees and shrubs. Singh 4346, 7801 (CDRI).

M. dissecta Hallier f. in Engl. Bot. Jahrb. 16:552.1893; FUGP 1:551; P.C. Kanjilal, For. Fl. UP. 3:488.1966; Raizada, Suppl. FUGP 165.1976.
Ipomoea sinuata Ortega, Decad. 7:84.1800; FBI 4:214.

Twining, hirsute, annual herbs. Leaves palmately 7-partite; lobes 5-6 cm long, lanceolate, acute, crenate, petioles 5 cm long. Flowers purple with strongly lineate bands.

Flowers : November-December

Occurs rarely in the Dudhwa Forest Rest House campus. Singh 4249 (CDRI).

Operculina Silva-Manso

1. Flowers white in few flowered cymes ....**O. turpethum**
1. Flowers yellow in 2-flowered raceme ....**O. petaloidea**

O. turpethum (Linn.) Silva-manso, Enum. Subst. bras. 16 & 49.1836; FUGP 1:552; P.C. Kanjilal, For. Fl. UP. 3:482.1966.
Convolvulus turpethum Linn. Sp. Pl. 155.1753.
Ipomoea turpethum R.Br. Prodr. 485.1810; FBI 4:212.

Climbing, perennial herbs with winged stems. Leaves alternate, 7-8 x 4-5 cm, ovate, subacute, mucronate, base cordate; petiole 3 cm long. Flowers white in 3 to 5-flowered cymes; arranged on stout peduncles. Capsule globose, enclosed in the enlarged bristle calyx lobes.

Flowers and Fruits : October-March

Found rarely in the open areas climbing on small trees and shrubs. Singh 3088, 4234 (CDRI); Inayat 22299 (DD).

O. petaloidea (Choisy) van Ooststr. in Blumea 3:369.1939; P.C. Kanjilal, For. Fl. UP. 3:483.1966.
Ipomoea petaloidea Choisy, Convol. Or. 69.1833; FBI 5:212.

Large scandent herbs. Leaves ovate-lanceolate, 7 x 1.5 cm, acute, base rounded, entire. Flowers yellow in 2 to 3-flowered, axillary racemes. Capsules ovoid. Seeds minutely velvety.

Flowers and Fruits : November-March

Found frequently climbing on small trees and bushes in open areas. Singh 7124 (CDRI); Inayat 22305 (a), 2295 (b) (DD).

Ipomoea Linn.

1. Leaves simple, entire:
    2. Corolla tubular or salver-shaped:
        3. Stem muricate .... **I. muricata**
        3. Stem not muricate:
            4. Flowers scarlet, red:
                5. Leaves ovate-cordate .... **I. hederifolia**
                5. Leaves with filiform segments .... **I. quamoclit**
            4. Flowers purplish-white .... **I. barlerioides**
    2. Corolla campanulate or funnel-shaped:
        6. Scandent shrubs .... **I. carnea var. fistulosa**

6. Herbaceous or woody twiners or trailers:

7. Aquatic or amphibious plants ....I. aquatica

7. Terrestrial plant:

8. Flowers in sessile or sub-sessile head ....I. eriocarpa

8. Flowers in pedunculate, 1 to few flowered cymes:

9. Stem prostrate; leaves obtuse ....I. rumicifolia

9. Stems twining; leaves acute or acuminate:

10. Flowers pinkish-white ....I. sepiaria

10. Flowers creamy-white ....I. obscura

1. Leaves palmately lobed or compound:

11. Flowers white; prostrate hirsute herbs ....I. quinata

11. Flowers violet or pale-blue; twining herbs:

12. Sepals densely hirsute with brown hairs ....I. pentaphylla

12. Sepals not densely hirsute ....I. cairica

I. muricata (Linn.) Jacq. Hort. Schoenbr. 3:40.t.323.1798; FBI 4:197; Santapau in RBSI 16(1):191.1953; Ooststr. in Fl. Males. ser.1.4:481.1953; P.C. Kanjilal, For. Fl. UP. 3:485.1966.
Convolvulus muricatus Linn. Mant. 44.1767.
Calonyction muricatum (Linn.) G.Don, Gen. Syst. 4:264.1838; FUGP 1:559.

Glabrous, slender, twiners with muricate stems. Leaves ovate-cordate, 5-9 x 4-7 cm, shortly acuminate. Flowers violet, in few flowered cymes. Sepals awned, elliptic-lanceolate. Corolla salver-shaped, pubescent outside. Capsules globose, smooth, apiculate.

Flowers and Fruits : September-February

Found occasionally in the open area twining on small trees. Singh 4360 (CDRI).

I. hederifolia Linn. Syst. ed. 10.925.1759; M.A. Rau in Bull. Bot. Surv. India Suppl. No.2.10:35.1969.

I. angulata Lamk. Tabl. Encycl. 1:464.1793; Santapau in RBSI 16(1):193.1953; Ooststr. in Fl. Males. Ser. 1.4.481.1953.
Ouamoclit phoenicea Choisy in Mem. Soc. Phys. Geneve 6:433.1833.
Ipomoea coccinea Clarke in FBI 4:199.1883 (non Linn.)

Slender, glabrescent twiners. Leaves ovate-cordate, entire or lobed, 10 x 5 cm, acuminate. Flowers red in few-flowered, long peduncled cymes. Capsules globose.

Flowers and Fruits : October-March

Probably an escape from the garden, twining on bushes. Singh 2916, 4245 (CDRI).

I. quamoclit Linn. Sp. Pl. 159.1753; FBI 4:199; Santapau in RBSI 16(1):191.1953; Srivastava, FG. 214.1976.
Quamoclit pinnata (Desr.) Bog. Hort. Maurit. 224.1837.

Glabrous, slender, twiners. Leaves pinnately divided into filiform segments. Flowers scarlet in few flowered peduncled cymes. Corolla funnel-shaped.

Flowers : September-December

Grown in garden, escape in the area. Singh 7408 (CDRI).

I. barlerioides (Choisy) Benth. ex Clarke in Hook.f. FBI 4:201.1883; FUGP 1:553; P.C. Kanjilal, For. Fl. UP. 3:485.1966.
Aniseia barlerioides Choisy, Mem. Soc. Phys. Geneve 6:484.1834.

Hispid, slender, trailing herbs. Leaves alternate, ovate-oblong, 5-6 x 2.5 cm, base cordate, acute, often mucronate, hairy on both surfaces. Flowers purple, solitary axillary, funnel-shaped.

Flowers and Fruits : September-November

Occurs occasionally in moist sandy places in sal forests. Singh 7473, 7751, 7860 (CDRI); Inayat 23745 (a) (DD).

I. carnea Jacq. subsp. **fistulosa** (Mart. ex Choisy) Austin in Taxon 26(213):237.1977.

I. fistulosa Mart. ex Choisy in DC. Prodr. 9:349.1845 (Jan.); Ooststr. in Fl. Males (Ser 1) 4:509.1954; Backer & Bakh. f. Fl. Java 2:493.1965; Raizada, Suppl. FUGP 163.1976; Srivastava, FG. 212.1976.

I. crassicaulis (Benth.) B.L. Robinson in Proc. Amer. Acad. 51:530.1916; Ooststr. in Fl. Males. Ser. 1.4(4):485.1953; P.C. Kanjilal, For. Fl. UP. 3:486.1966.

Perennial, suberect or straggling shrubs with milky juice. Stems stout, terete, fistular, greyish-green. Leaves ovate-cordate, 17-22 x 9-15 cm, entire, acuminate. Flowers bluish-purple or pale lilac, funnel-shaped arranged in lax many flowered axillary and terminal cymes. Capsules ovoid, 2.5 cm long.

Local name : Beheya
Flowers and Fruits : August-March

Occurs commonly in open wetland and in marshes. Singh 4369 (CDRI).

I. aquatica Forsk. Fl. Aegypt-Arab. 44.1775; FBI 4:210; Ooststr. in Fl. Males. Ser. 1.4:473.1953; P.C. Kanjilal, loc. cit. 3:1966; Srivastava, FG. 211.1976.

I. reptans Poir. in Lamk. Encycl. Suppl. 3.460.1814, (non *Convolvulus reptans* Linn. 1753); FUGP 1:555.

Aquatic, glabrescent, trailing annual herbs, rooting at the nodes. Stems hollow. Leaves ovate-triangular or ovate-oblong, 4-16 x 2-4 cm, acute-emarginate, base cordate or hastate. Flowers pale-purple, solitary or in 1 to 5-flowered peduncled cymes. Capsules ovoid, glabrous.

Local name : Karemua
Flowers and Fruits : September-February

Occurs commonly in muddy places near ponds. The shoots and leaves are used as vegetable. Singh 2890, 3082, 4109, 4269 (CDRI).

I. eriocarpa R.Br. Prodr. 484.1810; Ooststr. in Fl. Males. Ser. 1.4:462.1953; FBI 4:204; Srivastava, FG. 212.1976.
I. hispida (Vahl) Roem. & Schult. Syst. 4:238.1819 non Zucc. 1809; FUGP 1:554.
Convolvulus hispidus Vahl, Symb. Bot. 3:29.1794.

Slender, hispid twiners. Leaves ovate-oblong, 6 x 1.3 cm, cordate, acuminate, hairy on both surfaces. Flowers pink-purple in subsessile, short, dense axillary heads. Calyx hairy, ovate, acuminate. Corolla campanulate. Capsules globose, hairy, 4-seeded.

Flowers and Fruits : September-November

Occurs frequently twining on shrubs. Singh 7882 (CDRI).

I. rumicifolia Choisy in Mem. Soc. Phys. Geneve 6:447.1834; FBI 4:207; FUGP 1:554.

Prostrate, annual herbs. Leaves long petiolate, 7-11 x 4-10 cm, ovate-oblong, obtuse-emarginate base cordate, glabrous, entire. Flowers in short laxly few flowered peduncled cymes. Calyx lobes ovate, acute. Corolla tubular, glabrous.

Flowers and Fruits : November-February

Found rarely in the area. Singh 7598 (CDRI); Inayat 22297 (DD).

I. sepiaria Roxb. Hort. Beng. 14. 1814, nom. nud. et Fl. Ind. 1:500.1820; FBI 4:209; FUGP 1:555.

Slender, glabrous or more or less hairy, perennial twiners. Leaves long petiolate, 4-6 x 3-4 cm, ovate, acute, base cordate, entire. Flowers pinkish-white in long peduncled subumbellate cymes. Calyx lobes ovate-elliptic, glabrous with membranous margins. Corolla funnel-shaped. Capsules ovoid, glabrous.

Flowers and Fruits : August-December

Occurs frequently in hedges in open areas. Inayat 22306, 22308 (DD).

I. obscura (Linn.) Ker-Gawl. in Bot. Reg. 3.t.239.1817; FBI 4:207; FUGP 1:556; Srivastava, FG. 213.1976.
Convolvulus obscurus Linn. Sp. Pl. ed. 2.220.1762.

Slender, twining, glabrous or slightly hairy, annual herbs. Leaves 4-6 x 3-5 cm ovate-cordate, acuminate, entire. Flowers creamy-white in long peduncled, 1 to 3-flowered cymes. Clayx ovate, obtuse, apiculate. Corolla funnel-shaped.

Flowers and Fruits : October-December

Occurs occasionally in open areas twining on bushes. Singh 4310, 7518 (CDRI).

I. quinata R.Br. Prodr. 486.1810; FBI 4:214. (Fig. 9).

Prostrate, fulvous-hairy, annual herbs. Leaves palmately 5-lobed; leaflets 3-4.5 cm long, elliptic-lanceolate, entire; petioles hairy. Flowers white funnel-shaped.

Flowers and Fruits : September-December

Found occasionally in sal forests in moist places. Singh 7474, 7858 (CDRI).

I. pes-tigridis Linn. Sp. Pl. 162.1753; FBI 4:204; FUGP 1:557; Santapau in RBSI 16(1):193.1953; Srivastava, FG. 213.1976.
I. pes-tigridia var. *hepaticifolia* Clarke in FBI 4:204.1883.

Diffuse or twining hispid, herbs. Stems clothed with long spreading hairs. Leaves deeply palmate, 5-7 lobed, elliptic-lanceolate, acuminate or acute, fulvous-hairy on both surfaces. Flowers whitish-pink on long peduncled heads. Capsules enclosed within the calyx.

Flowers and Fruits : September-December

Occurs commonly in wastelands and in cultivated fields. Singh 3080, 7453, 7725 (CDRI).

I. pentaphylla Jacq. Ic. 2:t.319.1786.1793 (non Cav.); FBI 4:202; Maheshwari, Fl. Del. 238.1963.

Slender, twining, hirsute annual herbs. Leaves digitately 5-foliolate; leaflets 4-6 x 1.5-2 cm, lanceolate, acuminate. Flowers white, funnel-shaped, on long laxly dichotomous cymes. Capsules ovoid.

Flowers and Fruits : September-December

Found rarely in the area. Inayat 22303 (b) (DD).

I. cairica (Linn.) Sweet, Hort. Brit. 287.1827; Santapau in RBSI 16(1):193.1953; P.C. Kanjilal, For. Fl. UP. 3:486.1966; Raizada, Suppl. FUGP 162.1976; Srivastava, FG.211.1976.
Convolvulus cairicus Linn. Syst. 722.1759.
Ipomoea palmata Forsk. Fl. Aegypt.-Arab. 43.1775; FBI 4:214.

Glabrous, extensive, climbing shrubs. Leaves ovate or orbicular, palmately dissected to the base into 5, elliptic-lanceolate lobes; petioles with small pseudostipules. Flowers pale-purple, campanulate, arranged in long peduncled, 1 to 3-flowered cymes.

Local name : Railway creeper
Flowers and Fruits : October-May

Found as an escape, climbing on trees. Singh 7337, 7405 (CDRI).

Convolvulus Linn.

C. microphyllus Sieb. ex Spreng. Syst. 1:611.1824; FBI 4:218; Srivastava F.G. 209. 1976.
C. pluricaulis Choisy, Convol. Or. 95. 1853; FBI 4: 218; FUGP 1:547; P.C. Kanjilal, For. Fl. UP. 3:488.1966.

Prostrate or suberect, hairy, perennial herbs. Leaves in rosette, 2.5-5 cm long, linear or oblanceolate, hairy on both sides, obtuse, mucronate. Flowers whitish-pink, solitary axillary or paired, funnel-shaped. Capsules oblong, globose, pale-brown, Seeds brown.

Local name : Shankhpushpi
Flowers and Fruits : May-December

Occurs commonly on dry sandy places in wastelands. Singh 2891, 7590, 8312 (CDRI).

Rivea Choisy

R. ornata Choisy, in Mem. Soc. Phys. Geneve 6:409.1834; FBI 4:183; FUGP 1:560.

Grey-tomentose, spreading shrubs. Leaves reniform-orbicular, cordate, 12-14 x 8 cm, white tomentose beneath. Flowers white, arranged in 3-flowered peduncled cyme. Fruits globose, indehiscent.

Local name : Rampan
Flowers and Fruits : July-October

Occurs occasionally in sal forests as undergrowth and in grasslands. Singh 3016, 8503 (CDRI).

74. CUSCUTACEAE

Cuscuta Linn.

C. reflexa Roxb. Pl. Corom. 2Z:3.t.104.1799; FBI 4:225; FUGP 1:543; Ooststr. in Fl. Males. 4:393.1953.

Glabrous, twining, leaflets, stem parasite with fleshy yellow branches. Flowers pale-white, arranged in clustered racemes. Calyx divided into lobes, ovate-obtuse. Corolla lobes reflexed. Stigma diverging. Capsules depressed-globose.

Local name : Amarbel, Akashbanwar.
Flowers and Fruits : September-February

Occurs commonly on small trees and shrubs. Singh 2873, 4257 (CDRI).

75. SOLANACEAE

1. Sepals enlarged, enclosing the fruits:
    2. Flowers usually solitary; yellow .... **Physalis**
    2. Flowers in fascicles, greenish-yellow .... **Withania**
1. Sepals not enclosing the fruits:
    3. Corolla tubular; fruits capsular:
        4. Calyx covering the fruit .... **Nicotiana**
        4. Calyx much shorter than the fruit .... **Datura**
    3. Corolla rotate; fruit a berry .... **Solanum**

Physalis Linn.

P. minima Linn. Sp. Pl. 183.1753; FBI 4:238; Santapau in RBSI 16(1):197.1953; Srivastava, FG. 221.1976.

Erect or decumbent, glabrous, annual herbs. Leaves ovate, 4-6 x 3 cm, entire or crenate, rounded at the base, acute. Flowers yellow, solitary on slender pedicels. Fruiting calyx upto 2.5 cm long. Corolla yellow. Berries enclosed within the inflated 5 to 10-ribbed calyx. Seeds reniform or discoid, muriculate.

Local name : Makoy; Choti Rasbhari.
Flowers and Fruits : July-January

Found rarely in moist shady places and wastelands. The fruits are edible. Singh 2802 (CDRI).

Withania Pauq. *nom. cons.*

W. somnifera (Linn.) Dunal in DC. Prodr. 13(1):453.1852; FBI 4:239; FUGP 2:6; Santapau in JBNHS 47(4):657.1948; P.C. Kanjilal, For. Fl. UP. 3:493.1966; Srivastava, FG. 223.1976.
Physalis somnifera Linn. Sp. Pl. 182.1753.

Erect, stellate-tomentose, annual herbs. Leaves ovate, 6-15 x 4-7 cm, subacute, entire, densely hairy beneath. Flowers greenish-yellow, 3-10 together in umbellate cymes. Fruiting calyx 2 cm long, pubescent or glabrous. Berries red, enclosed in the bladder-like inflated clayx.

Local name : Asgandh
Flowers and Fruits : October-May

Found occasionally in shady places. The root is used for strength, vitality and memory improving. Singh 8565 (CDRI).

Nicotiana Linn.

N. plumbaginifolia Viv. Elench. Pl. Hort. Dinegro 26,t.5.1802; Maheshwari, Fl. Del. 242.1963; Raizada, Suppl. FUGP 170.1976; Srivastava, FG. 220.1976.

Erect, sticky, glandular pubescent, annual herbs. Leaves in rosette at the base; upper ones oblong or elliptic-ovate, 13-19 x 6-7 cm, acute, semiamplexicaul, hairy. Flowers pale-pink in lax racemes. Corolla tube linear. Capsules ovoid.

Local name : Jangli tamakhu
Flowers and Fruits : March-October

Occurs frequently along roadside in moist places and in wastelands. Singh 2801 (CDRI).

Datura Linn.

1. Plants glabrescent; corolla 5-toothed .... **D. metel**

1. Plants pubescent; corolla 10-toothed .... **D. innoxia**

D. metel Linn. Sp. Pl. 179.1753; Santapau in JBNHS 47(4):652-662.1948 et in RBSI 16(1):197.1953; P.C. Kanjilal, For. Fl. UP. 3:494.1966; Srivastava, FG. 219.1976.
D. fastuosa Linn. Syst. 2:932.1759; FBI 4:242.
D. fastuosa var. *alba* Clarke in FBI 4:243.1883.

Glabrescent, bluish black undershrubs. Leaves ovate, about 17 x 9 cm, angled or irregularly toothed, unequal sided at the base, glabrous. Flowers white, tinged with reddish-purple below, funnel-shaped, 16 cm long. Calyx base persistent, reflexed. Capsules globose covered with deltoid spines, dehiscing irregularly. Seeds pale-brown.

Local name : Kala Dhatura
Flowers : November-January; Fruits : January-April

Found frequently in wastelands. Singh 7496, 8415 (CDRI); Inayat 22432 (DD).

D. innoxia Mill. Gard. Dict. ed. 8.n.5.1768; Santapau in JBNHS 47(4):659.1948 et in RBSI 16(1):198.1953; P.C. Kanjilal, For. Fl. UP. 3:495.1966; Raizada, Suppl. FUGP.168.1976; Srivastava, FG. 219.1976.
D. metel Sims in Bot. Mag. t. 1440.1812; FBI 4:243 (non Linn. 1753).

Stout undershrubs with grey tomentum. Leaves ovate-oblong, 13-20 x 6-9 cm, entire, or repand-dentate, acute or acuminate, unequal at the base, softly tomentose on both sides. Flowers white, solitary, 16-18 cm long, funnel-shaped. Calyx reflexed in the fruit. Capsules globose, covered with long spines.

Local name : Dhatura
Flowers and Fruits : July-December

Found frequently in wastelands. The leaves are used in rheumatism. Singh 4349, 7365, (CDRI).

Solanum Linn.

1. Plants armed:

   2. Herbs with prickly diffuse branches; flowers purple ...S. surattense

   2. Shrubs or undershrubs:

      3. Flowers white ....S. viarum

      3. Flowers blue:

         4. Berries 1 cm in diameter ....S. indicum

         4. Berries 2.2 cm in diameter ....S. incanum

1. Plants unarmed:

   5. Flowers in extra axillary umbellata cymes ....S. nigrum

   5. Flowers in corymbose cymes ....S. erianthum

S. surattense Burm.f. Fl. Ind. 57.1768; Santapau in RBSI 16(1):154.ed.2.1967; Srivastava, FG. 222.1976.
S. xanthocarpum Schrad. & Wendl. Sert. Hanov. 1:8, t.2.1795; FBI 4:236; FUGP 2:3; P.C. Kanjilal, For. Fl. UP. 3:492.1966.

Diffuse or prostrate, prickly herbs. Prickles large yellow. Leaves ovate-oblong or elliptic, 5-10 x 3-4.5 cm, deeply lobed or subpinnatifid, spiny. Flowers bluish-purple in extra axillary cymes. Berries globose, yellow, smooth.

Local name : Bhatkataiyya
Flowers and Fruits : December-June

Occurs frequently in waste open lands and in cultivated fields. The roots are used in scorpion sting Singh.7149, 8301 (CDRI); Inayat 22440 (DD).

S. viarum Dunal in DC. Prodr. 13:240.1852; Babu in JBNHS 67(3):609-611.1970-71; Raizada, Suppl. FUGP 177.1976. (Fig.10).

S. khasianum Clarke var. *chatterjeeanum* Sen Gupta in Bull. Bot. Surv. India 3:412-415.1961.

Erect, tomentose, prickly undershrubs with two types of prickles. Leaves ovate, 18 cm long, lobed, angled, toothed, densely hirsute; provided with spines on the midribs and veins. Flowers white, arranged in lateral 1 to 4-flowered corymbose cymes. Berries globose, 3-4 cm across, pale-yellow streaked with green, glabrous.

Local name : Kachriya bhanta
Flowers and Fruits : April-November

Occurs frequently in the open areas and in grasslands. The plant is used in medicine. Singh 2826, 3036, 7985 (CDRI).

S. indicum Linn. Sp. Pl. 187.1753; FBI 4:234; FUGP 2:3.
S. violaceum Ortega, Hort. Matr. Dec. 56.1798; Saldanha & Nicolson, Fl. Hass. Distt. 462.1976.

Erect or diffuse, stellately tomentose, undershrubs armed with short hooked spines. Leaves ovate-oblong, sinuate, 5-10 x 2.5-3 cm, base oblique, spiny along the mid-rib. Flowers blue or violet in extra axillary short cymes. Berries globose, 0.5-1 cm across, yellow.

Flowers and Fruits : Most part of the year

Found frequently in open areas in wastelands. The plant is used in medicine. Singh 2940, 4123, 4222 (CDRI).

S. incanum Linn. Sp. Pl. 188.1753; FBI 4:236; FUGP 2:4.

Much branched, prickly shrubs. clothed with soft stellate pale hairs. Leaves 9.5-10 x 4-5 cm, ovate-elliptic, sinuately lobed, stellately tomentose above, densely woolly beneath. Flowers blue on lateral paired peduncles; one bearing a solitary fertile flower, the other a raceme of males, berries 2-2.5 cm in diameter, ovoid or subglobose, yellow.

Flowers and Fruits : August-October

Occurs rarely in the Sonaripur range of the Park in open areas. Singh 8528 (CDRI).

S. nigrum Linn. Sp. Pl. 186.1753; FBI 4:229; FUGP 2:2; Santapau in RBSI 16(1):195.1953; Srivastava, FG. 222.1976.

Erect, glabrous, unarmed, annual herbs. Leaves ovate-oblong, 6-9 x 2 cm, toothed or lobed, sparsely pubescent. Flowers white in drooping umbellate extra axillary cymes. Berries globose, glabrous, 0.7 cm across.

Local name : Makoi
Flowers and Fruits : Most part of the year

Occurs frequently in moist shady places. The fruits are edible. Singh 2871 (CDRI).

S. erianthum D.Don, Prodr. Fl. Nep. 96.1825; Roe in Taxon 17:176.1968 et Britonia 24:251.1972.
S. verbascifolium auct. non Linn. Sp. Pl. 184.1753. FBI 4:230; FUGP 2:2; P.C. Kanjilal, For. Fl. UP. 3:490.1966.

Unarmed, densely stellately hairy shrubs upto 2.5 m high. Leaves ovate-oblong or elliptic, 16-19 x 4.5-6 cm, entire, acute or subacuminate, pubescent on both surfaces. Flowers white, densely arranged in woolly terminal corymbose cymes. Berries globose, stellately tomentose, yellow when ripe.

Flowers and Fruits : February-September

Found commonly in exposed areas in wastelands. Singh 2874, 4122 (CDRI); Inayat 22442 (DD & CAL).

76. SCROPHULARIACEAE

1. All leaves alternate .... **Verbascum**
1. All leaves or atleast the lower opposite:

2. Corolla distinctly 2-lipped:

3. Anthers 1-celled; leaves sessile ....Striga

3. Anthers 2-celled; leaves shortly petiolate or sessile:

4. Flowers yellow ....Lindenbergia

4. Flowers blue, pink or white never yellow:

5. Stamens 4, perfect, all included in tube:

6. Leaves radical, opposite; flowers pale-blue or white..Mazus

6. Leaves whorled; flowers pale-violet ...Limnophila

5. Stamens 4, 2 perfect in the tube; 2 in the throat reduced to staminodes.

7. Calyx tube ribbed or winged; fruits included....... ....Torenia

7. Calyx tube generally smooth; fruits exserted...... ....Lindernia

2. Corolla not 2-lipped:

8. Flowers 5-merous:

9. Flowers peduncled, bluish-purple ....Bacopa

9. Flowers sessile, pink ....Mella

8. Flowers 4-merous:

10. Stamens 2 ....Veronica

10. Stamens 4 ....Scoparia

Verbascum Linn.

1. Woolly-tomentose herbs; stamens 5 ....V. thapsus

1. Pubescent herbs; stamens 4 ....V. chinense

V. thapsus Linn. sp. Pl. 177.1753; FBI 4.520; FUGP 2:14; Pennell in Acad. Nat. Sci. Phil. Monograph 5:10.1943; P.C. Kanjilal, For. Fl. UP. 3:497.1966.

Erect, woolly-tomentose herbs with winged stems and prolonged leaf bases. Leaves radical and cauline 15-30 x 6-9 cm, obovate-lanceolate, acute or acuminate, entire or crenate. Flowers yellow, crowded in terminal, woolly, long spikes. Capsules ovoid, tomentose.

Local name : Gidar tambakhu
Flowers and Fruits : February-May

Occurs rarely in Dudhwa forest of the Park. The root is used as an antidote to snake bite. Singh 7576 (CDRI); Inayat 22415 (DD).

V. chinense (Linn.) Santapau, Fl. Purandhar 90.1957; Maheshwari, Fl. Del. 249.1963; Srivastava, FG. 232.1976.
Scrophularia chinensis Linn. Mant. 250.1771.
Celsia coromandeliana Vahl, Symb. Bot. 3:79.1794; FBI 4:251; FUGP 2:15.

Erect, pubescent, annual herbs. Radical leaves in rosette, lyrate-pinnatifid; cauline sessile, ovate or oblong-ovate, toothed, passing into floral bracts. Flowers yellow in simple or branched, terminal, long racemes. Capsules sub-globose pale-brown.

Flowers and Fruits : February-April

Found occasionally in moist shady places along river banks. Singh 8409 (CDRI).

Striga Lour.

1. Calyx 10-ribbed; flowers yellow .... **S. asiatica**
1. Calyx 15-ribbed; flowers white .... **S. angustifolia**

S. asiatica (Linn.) O. Kuntze, Rev. Gen. Pl. 2:466.1891; Chatterjee & Bhardwaja in Bull. Bot. Soc. Bengal 9(2): 152. 1955; Hepper in Rhodora 76: 46. 1974; Srivastava, FG. 231.1976.
Buchnera asiatica Linn. Sp. Pl. 630.1753.
Striga lutea Lour. Fl. Cochinch. 22.1790; FBI 4:299; FUGP 2:30.

Erect, slender, annual herbs. Leaves sessile, linear, 1-2.5 cm long, scabrous. Flowers yellow or pink in elongated, interrupted spikes. Calyx 10-ribbed; one rib reaching to the tip of each lobe, others ending at sinus between the lobes. Capsules oblong, dark brown.

Flowers and Fruits : September-January

Occurs occasionally in the grasslands. Inayat 2241 (a) (DD).

S. angustifolia (D.Don) Saldanha in Bull. Bot. Surv. India 5:70.1963; Raizada, Suppl. FUGP 191.1976.
Buchnera angustifolia D.Don, Prodr. Fl. Nepal. 91:1825.
Striga euphrasioides Sensu Benth. (excl. basyonym) Comp. Bot. Mag. 1:364.1835; FBI 4:299; FUGP 2:31.

Erect, strigose, annual herbs. Leaves sessile, linear. 2-4 cm long, scabrous. Flowers white, solitary, axillary, subsessile in interrupted terminal spikes. Calyx 15-ribbed; scabrous; lobes linear-subulate. Corolla tube exceeding calyx. Capsules oblong, shorter than calyx.

Flowers and Fruits : September-November

Found occasionally in moist places often mixed with grasses. Singh 7479, 7700 (CDRI).

Lindenbergia Lihm.

1. Plants 10-30 cm high; calyx shorter than the bracts.................**L. indica**
1. Plants 30 cm to 1 m high; calyx longer than the bracts..**L. macrostachya**

L. indica (Linn.) Vatke in Oester. Bot. Zeits. 25:10.1875; Santapau in JBNHS 49:45.1950; Chatterjee & Bhardwaja loc.cit. 139.1955.
Dodartia indica Linn. Sp. Pl. 633.1753.
Lindenbergia urticaefolia Lehm. in Link & Otto. Sem. Hort. Hanb. 5.1829-1830; FBI 4:262; FUGP 2:33.

Erect, more or less glandular hairy, herbs. Leaves 2-5 x 2-3.5 cm, ovate or elliptic, serrate, acute, villous on both surfaces. Flowers yellow, solitary axillary or in pairs, forming axillary or terminal spicate racemes. Capsules ovoid, exserted from calyx, hairy.

Flowers and Fruits : September-April

Found frequently growing on old walls. Singh 7714 (CDRI).

L. macrostachya (Benth.) Benth. Scroph. Ind. 22.1835; FBI 4:262; FUGP 2:33; Pennell in Acad. Nat. Sci. Phil. Monograph 5:25.1943; Chatterjee and Bhardwaja loc. cit. 140.1955.
Stemodia macrostachya Benth. in Wall. Cat. 3295. nom. nud. 1831.

Glabrous or slightly pubescent, branched, 30 cm-1 m tall herbs. Leaves ovate or oblong-lanceolate, 3-6 cm long, acute, serrate-dentate. Flowers yellow, sessile, in elongated spikes. Calyx longer than the bracts. Capsules ovoid, pubescent.

Flowers and Fruits : February-May

Found rarely along river banks on sandy soil. Singh 7101, 7118, 8393 (CDRI); Inayat S.n. (Kheri).

Mazus Lour.

1. Pedicel glabrous, longer than calyx .... **M. pumilus**

1. Pedicel hirsute, shorter than calyx .... **M. delavayi**

M. pumilus (Burm.f.) van Steenis, Nova Guina n.s. 31.1958; Backer & Bakh. f. Fl. Java. 2:505.1965; Raizada in Indian For. 94(6):445.1968; Shah in JBNHS 66(1):233.1969; Srivastava, FG. 230.1976.
Lobelia pumila Burm.f. Fl. Ind. 186.t.60.f.3.1784.
Mazus japonicus (Thunb.) O. Kuntze, Rev. Gen. Pl. 462.1891; Chatterjee & Bhardwaja loc. cit. 146. 1955.
M. rugosus Lour. Fl. Cochinch. 385.1790; FBI 4:259; FUGP 1:19.

Small glabrous or sparsely hairy, tufted, annual herbs. Leaves radical, obovate-spathulate, 3-6 x 2.5 cm, irregularly crenate. Flowers pale-blue or white in terminal lax racemes; pedicel longer than the calyx. Corolla 2-lipped. Capsules subglobose.

Flowers and Fruits : September-February

Occurs frequently in moist shady places and along river banks. Singh 4251 (CDRI); Saran 26542 (LWG).

M. delavayi Bonati in Bull. Herb. Boiss. ser 2.8:530.1908; Pennell loc. cit. 34. 1943; Saxena in Indian For. 93(5):329.1967; Raizada, Suppl. FUGP 188.1976.

Erect or prostrate, hispid, annual herbs. Leaves sessile, obovate to spathulate, 4-5 x 1-1.5 cm, crenate, cuneate. Flowers white in terminal racemes; pedicels shorter than calyx. Corolla 2-liped. Stamens 4.

Flowers and Fruits : September-December

Occurs occasionally in moist shady places and along river banks. Singh 7116, 7637 (CDRI).

Limnophila R.Br. *nom. cons.*

L. indica (Linn.) Druce, Bot. Exch. Club. Soc. Brit. Isles 3: 420. 1914; Pennell, loc.cit. 26.1943; Chatterjee & Bhardwaja loc.cit. 138.1955; Philcox in Kew Bull. 24:115.1970; Srivastava, FG. 226.1976.

Hottonia indica Linn. Syst. Nat. ed. 10:919.1759.

Limnophila gratioloides R.Br. Prodr. 442.1810; FBI 4:271; FUGP 2:21.

Aquatic or marshy, hirsute or glabrous aromatic, annual herbs. Stems much branched from the base, rooting from the lower nodes. Leaves sessile, upper whorled, 3-nerved, 4-5 cm long, oblong, serrate; submerged leaves dissected. Flowers white, tinged with purple, pedicellate, solitary axillary or in terminal racemes. Fruits ellipsoid or globose, enclosed in clayx.

Flowers and Fruits : September-December

Found commonly in permanent pools, ponds and in marshy places. The plant is variable in size and shape of leaves and arrangement of flowers. Singh 4275, 4328, 7856, 7870 (CDRI).

Torenia Linn.

T. cordifolia Roxb. Pl. Cor. 2:52.t.f.161.1789; FBI 4:276; FUGP 2:23; Pennell, loc. cit. 31.1943; Chatterjee & Bhardwaja loc. cit. 141.1955.

Erect, glabrous, annual herbs. Stems and branches quadrangular. Leaves opposite, petiolate, ovate, 2-4 cm long, glabrous, serrate, acute. Flowers bluish-purple, axillary and usually whorled at the tip of the branches. Capsules oblong, acute, enclosed in persistent calyx.

Flowers and Fruits : August-November

Found frequently as an escape in Dudhwa forest nursery in moist shady places. Singh 7756 (CDRI).

Lindernia Allioni

1. Capsules equalling or shorter than calyx:
    2. Plants glabrous; calyx cleft to the middle ....**L. crustacea**
    2. Plants hairy; calyx cleft to the base ....**L. viscosa**
1. Capsules twice the length of calyx:
    3. Leaves parallel-nerved ....**L. parviflora**
    3. Leaves penninerved:
        4. Flowers white or pink; leaves spiny serrate ....**L. ciliata**
        4. Flowers bluish-purple; leaves not as above ....**L. anagallis**

L. crustacea (Linn.) F. Muell. Syst. Cens. Austr. Pl. 97. 1882; Pennell in Acad. Nat. Sci. Phil. Monograph 5:29.1943; Chatterjee & Bhardwaja Bull. Bot. Soc. Bengal 9(2):142.1955; Srivastava, FG. 228.1976. *Capraria crustacea* Linn. Mant. 87. 1667.
Vandellia crustacea Benth. Scroph. Ind. 35. 1835; FBI 4. 279; FUGP 2:23.

Glabrous, diffuse, branched, annual herbs with angular stems. Leaves opposite, shortly petiolate, ovate, 2.5 x 0.9 cm, penninerved, serrate. Flowers purplish-white, axillary or in racemes. Calyx divided upto the middle. Capsules ovoid-oblong.

Flowers and Fruits : September-December

Found occasionally in moist ground and other marshy places. Singh 7463, 8464 (CDRI).

L. viscosa (Hornem.) Boldingh, Zakfl. Lanb. Java 165.1916; Pennell in JAA 24:251.1943; Srivastava, FG. 229.1976; Raizada, Suppl. FUGP 19.1976.
Gratiola viscosa Hornem. Enum. Pl. Hort. Hafn. 19. 1807.
Lindernia hirsuta (Benth.) Wettst. in Nat. Pflanzen.f. 4. 3b.79.280.1884; Chatterjee & Bhardwaja loc. cit. 143.1955.
Vandellia hirsuta Benth. Scroph. Ind. 36.1835; FBI 4:280; Raizada in JIBS 15:164.1936.

Erect succulent, hairy annual herbs. Lower leaves petiolate, 4-5 x 1.5 cm, oblong, serrulate or entire; upper leaves ovate-oblong, serrate. Flowers bluish-white, in lax terminal racemes. Calyx cleft to the base, linear-lanceolate. Capsules globose.

Flowers and Fruits : September-December

Found occasionally in the moist shady places. Singh 7784 (CDRI).

L. parviflora (Roxb.) Haines, Bot. Bihar & Orissa 4:635.1942; Pennell in Acad Nat. Sci. Phil. Monograph 5:29.1943; Chatterjee & Bhardwaja loc. cit. 144.1955.
Gratiola parviflora Roxb. Pl. Cor. 3:3f. 203.1819 & Fl. Ind. 1:140.1820.
Ilysanthes parviflora Benth. in DC. Prodr. 10:419.1846; FBI 4:283; FUGP 2:25.

Erect, glabrous, annual herbs. Leaves opposite, sessile, 1-2 cm long, ovate-lanceolate or elliptic, entire or serrulate. Flowers white, slender in axillary or terminal racemes. Calyx deeply divided, linear-lanceolate. Capsules ovoid-oblong, acute.

Flowers and Fruits : July-November

Found frequently in moist shady places along forest roads. Singh 7490 (CDRI).

L. ciliata (Colsm.) Pennell in JAA 24:252.1943 et in Acad. Nat. Sci. Ehil. Monograph 5:39.1943; Chatterjee & Bhardwaja loc. cit. 145.1955; Srivastava, FG. 227.1976.
Gratiola ciliata Colsm. Prodr. Dec. Grat. 14.1793.
Bonnaya brachiata Link & Otto. Icon. Pl. Select. 25. t.11.1820; FBI 4:284; FUGP 2:26.

Erect, rigid, glabrous, branched annual herbs. Leaves opposite, sessile, 1-2.5 x 0.5 cm, oblong, margins spinose-serrate, acute. Flowers whitish-pink in lax terminal racemes. Capsules linear-oblong.

Flowers and Fruits : September-December

Occurs frequently in moist places along forest roads. Singh 7428 (CDRI).

L. anagallis (Burm.f.) Pennell in JAA 24.252.1943 et in Acad. Sci. Phil. Monograph 5 39.1943; Chatterjee & Bhardwaja loc. cit. 145.1955; Srivastava, FG. 227.1976.
Ruellia anagallis Burm.f. Fl. Ind. 133.1768.

Bonnaya veronicaefolia Spreng. syst. 1:14.1825; FBI 4:285; FUGP 2:26.

Decumbent or creeping, annual herbs. Leaves opposite, sessile, 2.5 x 1 cm, oblong or obovate-lanceolate, serrate. Flowers bluish-pink or white in terminal and axillary racemes. Calyx deeply divided. Capsules linear, twice as long as the calyx.

Flowers and Fruits : September-December

Found frequently in moist places along forest roads. Singh 3038, 4146, 7422 (CDRI).

Bacopa Aublet *nom. cons.*

B. monnieri (Linn.) Wettst. in Proc. Acad. Sci. Phil. 98.94.1946; Chatterjee & Bhardwaja loc. cit. 133.1955; Srivastava, FG. 225.1976.
Lysimachia monnieri Linn. Cent. Pl. 2:9.1756.
Herpestris monnieri Benth. Scroph. Ind. 30.1835; FBI 4:272.
Moniera cuneifolia Michx. Fl. Bor. Amer. 2:22.1803; FUGP 2:17.
Bramia monnieri Pennell in Acad. Nat. SCP. Phil. Monograph 5:24.1943.

Prostrate much branched herbs with ascending branches. Leaves sessile, opposite, 1.5 x 0.5 cm, oblong or obovate, entire or obscurely crenulate, obtuse. Flowers bluish-purple or white, solitary axillary. Capsules ovoid, glabrous.

Local name : Jalneem, Patalnibia.
Flowers and Fruits : August-November

Found occasionally in wet and marshy places. The plant is used in fever. Singh 3071, 7941 (CDRI).

Mella Vand.

M. hamiltoniana (Benth.) Pennell in Acad. Nat. Sci. Phil. Monograph 5:23.1943; Srivastava, FG. 230.1976.

Herpestis hamiltoniana Benth. in Wall. Number List. Spec. Ind. Mus. n. 3898.1828.
Moniera hamiltoniana T. Cooke Fl. Bomb. 2:357.1905; FBI 4:272; FUGP 2:18.

Erect, glabrous, annual herbs. Stem decumbent at the base, branched, Leaves opposite, subsessile, 2-3 x 0.4-0.5 cm, linear-lanceolate, acute, entire. Flowers bluish-pink, sessile, opposite and solitary axillary. Capsules globose, glabrous. Seeds yellowish-brown.

Flowers and Fruits : September-October

Occurs rarely in Sonaripur range of the Park in marshy places. Singh 8502 (CDRI).

Veronica Linn.

V. anagallis-aquatica Linn. Sp. Pl. 12. 1753; FBI 4:293; FUGP 2:154; Chatterjee & Bhardwaja loc. cit. 148.1955; Srivastava, FG. 232.1976.

Erect, succulent, glabrous, annual herbs with hollow stems. Leaves opposite, 6-10 x 1.5-2 cm, oblong to lanceolate, serrate. Flowers white or pale purple arranged in lax axillary 5-9 cm long bracteate racemes. Calyx lobes 4, unequal. Stamens 2. Capsules oblong-orbicular, emarginate.

Flowers and Fruits : February-April

Occurs commonly in marshy places and along river banks. Singh 7107, 7232, 7290, 8324 (CDRI); G. Saran 26149 (LWG).

Scoparia Linn.

S. dulcis Linn. Sp. Pl. 116.1753; FBI 4:289; FUGP 2:27; Pennell in Acad. Nat. Sci. Phil. Monograph 5:32.1943; Chatterjee & Bhardwaja loc. cit. 132.1955; Srivastava, FG. 231.1976.

Erect, much branched, herbs or undershrubs with angled stems. Leaves opposite or whorled, 2.5-4 x 0.5-0.9 cm, elliptic-obovate, serrate. Flowers white, tetramerous, 3-6 in whorls, forming many flowered racemes.

Flowers and Fruits : Most part of the year.

Found frequently in wastelands and in cultivated fields. Singh 2869, 7977 (CDRI); G. Saran 26432, 26580 (LWG).

77. LENTIBULARIACEAE

Utricularia Linn.

U. aurea Lour. Fl. Cochinch. 1:26.1790; Taylor in Kew Bull. 18:197.1964; M.A. Rau in Bull. Bot. Surv. India Suppl. No. 2.10:57.1969.
U. flexuosa Vahl, Enum. 1:198.1804; FBI 4:329; FUGP 2:38.

Rootless, free-floating, aquatic herbs, submerged except the flowers. Leaves whorled, segments filiform, pectinate, possessing subglobose bladder at its base. Flowers yellow, in erect, 3 to 8-flowered racemes. Capsules subglobose.

Flowers and Fruits : September-January

Occurs occasionally in stagnant water of ponds and ditches etc., Singh 7568 (CDRI).

78. BIGNONIACEAE

1. Perfect stamens 5 .... **Oroxylum**
1. Perfect stamens 4:
    2. Flowers scarlet:
        3. Fruits indehiscent; leaflets 7-9 .... **Kigelia**
        3. Fruits dehiscent; leaflets 9-19 .... **Spathodea**
    2. Flowers purple .... **Stereospermum**

Oroxylum Vent.

O. indicum (Linn.) Vent. Dec. Gen. Nov. Paris 8:1808; FBI 4:378; FUGP 2:42; P.C. Kanjilal, For. Fl. UP. 3:499.1966.
Bignonia indica Linn. Sp-Pl. 625.1753.

Small, deciduous, glabrous trees with greyish-brown, corky bark. Leaves collected at the end of the branches; 2-3 pinnate; leaflets 2-4 pairs, ovate-elliptic, 9-10 x 4-4.5 cm, acuminate, entire, glabrous, oblique. Flowers purplish-green. Capsules 30-60 x 5-6 cm, flat, oblong, septicidally 2-valved.

Local name : Dagdagowa, Sona
Flowers : June-August; Fruits : December-March

Occurs occasionally in open areas and along railway lines of the Park. The stem bark is used in dysentery. Singh 4302, 8544 (CDRI).

Kigelia DC.

K. africana (Lam.) Benth. in Hooker, Niger Fl. 463.1849. Van Steenis in Fl. Males. ser. 1.8:183.1977.
Bignonia africana Lam. Encycl. 1:424-1783.
Kigelia pinnata DC. Prodr. 9:247.1845; Benthall, Trees Calc. 346; Bailey, Man. Cult. Pl. 905; FUGP 2:46; P.C. Kanjilal, For. Fl. UP. 3:507.1966; Srivastava, FG. 235.1976.

Medium-sized trees with ash-coloured bark. Leaves ternate, imparipinnate; leaflets 7-9, elliptic-oblong or obovate, 7-9 x 2-4 cm, coriaceous, toothed or entire. Flowers chocolate-red, in long drooping lax racemes. Fruits woody 30-40 x 12 cm, indehiscent, hanging on the branches.

Local name : Balamkhira
Flowers : March-April; Fruits : November-December

Planted in Bankati range in the Forest Rest House campus. The fruits are used in stomach disorder. Singh 7211 (CDRI).

Spathodea Beauv.

S. campanulata Beauv. Fl. D'Owar. 1:47,t.27-28.1805; Bailey, Man. Cult. Pl. 907; P.C. Kanjilal, For. Fl. UP. 3:508. 1966; Srivastava, FG. 237.1976.

Large or medium-sized, trees with grey bark. Leaves opposite, imparipinnate; leaflets 5-10 cm long, ovate-lanceolate or elliptic, acuminate. Flowers scarlet, in many-flowered terminal racemes or panicles. Calyx spathaceous, leathery. Stamens 4, exserted.

Flowers and Fruits : February-May

Planted in Bellrain range office and Dudhwa Forest Rest House campus. Singh 8375 (CDRI).

Stereospermum Cham.

S. suaveolens (Roxb.) DC. Prodr. 9:211.1845; FBI 4:382; FUGP 2:45: P.C. Kanjilal, For. Fl. UP. 3:504.1966.
Bignonia suaveolens Roxb. Fl. Ind. 3:104.1832.
S. chelonoides (L.f.) DC.

Large deciduous, trees with pale-brown bark; young parts viscid pubescent. Leaves 20-30 cm long, imparipinnate, 5 to 9-foliolate; leaflets 9-10 x 4 cm, elliptic, acuminate, base acute or unequal. Flowers yellow-purple in lax terminal panicles. Capsules 30-40 cm long, cylindric, grey, rough with white dots.

Local name : Padar
Flowers : April-June; Fruits : October-December

Found frequently in mixed forests. The fruits and roots are used in medicine. Singh 4176, 7286 (CDRI).

79. PEDALIACEAE

Pedalium Linn.

P. murex Linn. Syst. 1123. 1759; FBI 4:386; Maheshwari, Fl. Del. 262.1963.

Diffuse, much branched, glabrous, annual herbs. Leaves 5-6 x 2.5 cm, ovate, distantly crenate. Two glands present at the base of petiole and pedicel. Flowers yellow. Capsules indehiscent, 4-sided, pyramidal-ovoid with 4 horizontal spines.

Local name : Gokharu
Flowers and Fruits : August-October

Occurs rarely in sandy open areas. Singh 7422, 7734 (CDRI).

80. MARTYNIACEAE

Martynia Linn.

M. annua Linn. Sp. Pl. 618.1753; Maheshwari, Fl. Del. 263.1963.
M. diandra Glox. Obs. Bot. 14, t.1.1785; FBI 4:386; FUGP 2:48.

Erect, rough, densely glandular hairy, annual herbs or undershrubs. Leaves about 15 x 10 cm, long petiolate, ovate, cordate, dentate, sticky, hairy on both surfaces. Flowers pinkish-purple, diandrous, drooping in lax terminal glandular, sticky racemes. Fruits hard, beaked by two strong curved spines.

Local name : Kauwa, Bagh nakh.
Flowers and Fruits : August-November

Occurs frequently in wastelands in exposed situations. Singh 7400 (CDRI).

81. ACANTHACEAE

1. Climbing shrubs ....**Thunbergia**

1. Erect or prostrate herbs or undershrubs:

   2. Seeds not supported on retinacula ....**Nelsonia**

2. Seeds supported on retinacula:

3. Corolla lobes twisted in bud:

4. Ovules more than 4 in each locule:

5. Corolla 2-lipped:

6. Flowers in axillary whorls ....**Hygrophila**

6. Flowers in terminal spikes ....**Hemiadelphis**

5. Corolla subequally 5-lobed:

7. Flowers in capituliform clusters or contracted cymes..**Hemigraphis**

7. Flowers solitary or few together or in cymes..**Dipteracanthus**

4. Ovules 2 in each locule;

8. Stamens 4 ....**Dyschoriste**

8. Stamens 2 ....**Eranthemum**

3. Corolla lobes imbricate in bud:

9. Ovules 3-8 in each locule ....**Andrographis**

9. Ovules 2 or 1 in each locule:

10. Corolla distinctly 2-lipped:

11. Stamens 4 ....**Lepidagathis**

11. Stamens 2:

12. Anthers cell spurred at the base:

13. Inflorescence secund; bracts dimorphic....... ....**Rungia**

13. Inflorescence not secund; bracts homomorphic..**Justicia**

12. Anthers cell not spurred at the base:

14. Flowers in dichotomously branched cymes forming panicles.........................**Peristrophe**

14. Flowers in dense spikes:

15. Bracts flat, orbicular, clasping 2-flowers..................................**Dicliptera**

15. Bracts linear, clasping solitary flowers ....Adhatoda

10. Corolla lobes sub equal 5; calyx lobes unequal ....Barleria

Thunbergia Retz. *nom. cons.*

T. grandiflora (Roxb. ex Rottl.) Roxb. Hort Beng. 45.1814, nom. nud. et Fl. Ind. 3:34.1832; Santapau, Acanthaceae 9: FBI 4:392; FUGP 2:51; P.C. Kanjilal, For. Fl. UP. 3:527.1966; Srivastava, FG. 251.1976.
Flemingia grandiflora Roxb. ex Rotl. in Ges. Naturf. Fr. Neue Schr. 4:202.1803.

Large woody climbers. Leaves 7-18 cm long, orbicular or ovate, deep cordate, toothed or lobed, coriaceous. Flowers blue, showy, axillary, from the uppermost leaves in racemes. Bracts large.

Flowers : September-November

Found only in Dudhwa Forest Rest House campus. Singh 7768, 7968 (CDRI).

Nelsonia R. Br.

N. canescens (Lamk.) Spreng. Syst. Veg. 1:42.1825; Bremek. in Reinwardtia 3(2):248.1955; Srivastava, FG, 248.1976.
Justicia canescens Lamk. Encycl. 1:41.1791.
Nelsonia campestris R. Br. Prodr. Fl. Nov. Holl. 1:481.1810; FBI 4:394; FUGP 2:52.

Diffuse, villous herbs; young parts densely clothed with white hairs. Upper leaves opposite, sessile, elliptic-oblong, 3.5-4 x 1.5-2 cm, obtuse, entire, pubescent; lower leaves upto 13 cm, long petiolate. Flowers purplish in spikes; bracts silky pubescent; bracteoles represented by 2 opposite, tufts of hairs. Calyx 4-partite. Capsules ovoid, conic.

Local name : Prithipal
Flowers and Fruits : November-April

Found frequently in moist situations in sal forests. The roots are used in the preparation of liquor: Singh 2855, 7157, 8368 (CDRI); G. Saran 26430, 26565 (LWG).

Hygrophila R. Br.

1. Plants armed; calyx 4-partite .... H. auriculata
1. Plants unarmed, calyx 5-partite .... H. salicifolia

H. auriculata (K. Schum.) Heine in Kew Bull. 16(2): 172. 1962; Srivastava, FG. 244.1976.
Barleria auriculata K. Schum. in Schumach. & Tonn. Beskr. Guin. Pl. 285.1827.
Asteracantha longifolia Nees in Wall. Pl. As. Rar. 3:90.1832; FUGP 2.55; Santapau, Acanth. 17; P.C. Kanjilal, loc. cit. 525.1966.
Hygrophila spinosa T. Anders. Thw. Enum. 225.1860; FBI 4:408.

Erect, hispid, armed annual herbs. Leaves sessile, whorled, the 2-outer ones 5-6 cm long, lanceolate, hairy with axillary spines. Flowers bluish-purple, in sessile, axillary whorls. Bracteoles linear. Capsules linear, oblong.

Local name : Talmakhana
Flowers and Fruits : October-December

Found commonly in marshy places and along roadsides near ponds and ditches. The seeds are used in medicine for spermatorrhoea. Singh 2892 (CDRI); G. Saran 26726 (LWG).

H. salicifolia (Vahl) Nees in Wall. Pl. As. Rar. 3:81.1832; Santapau in Bot. Mem. Univ. Bombay No. 2.20.1951.
Ruellia salicifolia Vahl, Symb. 3:84.1794.
Hygrophila angustifolia auct. non. R. Br., Sensu FBI 4:407; FUGP 2:56.

Erect, unarmed annual herbs. Leaves sessile or shortly petiolate, obovate or oblong or upper linear-elliptic, 2-4 x 1 cm, acute, entire. Flowers purple in dense, axillary whorls;

bracts and bracteoles oblong, acute, hispid, ciliate. Calyx 5-partite, lanceolate. Capsules subquadrangular, many seeded.

Flowers and Fruits : October-December

Found occasionally in moist and marshy places. Singh 7580 (CDRI).

Hemiadelphis Nees

H. polyspermus Nees in Wall. Pl. As. Rar. 3:80.1832; Bremek. in Dansk. Bot. Ark. 20:66.1961; Srivastava, FG. 243.1976.
Hygrophila polysperma T. Anders. in JLS 9:456.1867; FBI 4:406; FUGP 2:56; Santapau, Acanth 19.

Erect or procumbent, much branched annual herbs, rooting near the base. Leaves opposite, sessile, 1-2 x 0.5 cm, oblong-ovate, entire, acute. Flowers 2-lipped, purple-white in dense, terminal spikes. Fertile stamens 2. Capsules linear-oblong.

Flowers and Fruits : September-March

Occurs commonly in moist and marshy places. Singh 2845, 7681, 7812, 8329 (CDRI); G. Saran 26708 (LWG).

Hemigraphis Nees

H. hirta (Vahl) T. Anders. in JLS 9:462.1867; FBI 4:422; FUGP 2:191; Srivastava, FG. 244.1976.
Ruellia hirta Vahl, Symb. Bot. 3:84, t.67.1794.

Prostrate, densely hairy, much branched herbs. Leaves opposite, 2-3.5 x 0.5-1.5 cm, ovate, crenate, hirsute. Flowers pale-violet or purple-pink in few flowered heads; bracts-elliptic, pubescent. Capsules linear, glabrous.

Flowers and Fruits : April-June

Found frequently in moist wastelands and in orchard. Singh 2887, 7194 (CDRI).

Dipteracanthus Nees emend. Bremek.

1. Flowers white; capsules glabrous .... **D. suffruticosus**

1. Flowers pale-violet; capsules pubescent............................. **D. prostratus**

D. suffruticosus (Roxb.) Voigt, Hort. Suburb. Calc. 483. 1845; Santapau in Indian For. 79:322.1953; P.C. Kanjilal, For. Fl. UP. 3:512.1966.
Ruellia suffruticosa Roxb. Fl. Ind. 3:53.1832; FBI 4:413; FUGP 2:58.

Erect, pubescent, undershrubs with woody rootstocks. Leaves 3-6 x 1-1.5 cm, obovate or oblanceolate, villous, margins ciliate, obtuse. Flowers white, solitary axillary, subsessile, enclosed by two leafy bracts. Capsules glabrous, oblong.

Flowers and Fruits : August-October

Found frequently in sal forests in moist situations. Singh 7388 (CDRI).

D. prostratus (Poir.) Nees in Wall. Pl. As. Rar. 3:81.1832; Bremek. in Verh. K. Nederl. Akad. Wet. II 45(1):16.1948; Santapau, Acanth. 24; Srivastava, FG. 243.1976.
Ruellia prostrata Poir. in Lamk. Encycl. 6:349.1804; FBI 4:411; FUGP 2:57 (incl. var. *dejecta*).

Diffuse or straggling, stout, sparsely hairy herbs. Leaves opposite, 5-7 x 1.5-3 cm, ovate, elliptic, acute, entire, sparsely hairy. Flowers pale-violet, subsessile, solitary or few together. Capsules clavate, pointed, pubescent.

Flowers and Fruits : August-October

Occurs occasionally in open areas. Singh 7464 (CDRI).

Dyschoriste Nees

D. erecta (Burm.f.) O. Kuntze, Rev. Gen. Pl. 1:485.1891; Santapau in Bot. Mem. Univ. Bombay No. 2:21.1951.
Ruellia erecta Burm.f. Fl. Ind. 135,t.41.f.3.1768.
Dyschoriste depressa Nees in Wall. Pl. As. Rar. 3:81.1832; FUGP 2:62.
Calophanes nagchana Nees in DC. Prodr. 11:109.1847; FBI 4:410.

I have included this plant on the authority of Duthie loc. cit. who reports its occurrence from Kheri forest area. Hence, a brief description is reproduced below.

Prostrate, much branched, glabrescent perennial herbs. Leaves elliptic or ovate-oblong, entire, obtuse. Flowers blue, subsessile in axillary clusters. Capsules glabrous pointed.

Eranthemum Linn.

E. purpurascens Wt. ex Nees in Wall. Pl. As. Rar. 3:106. 1832; P.C. Kanjilal, For. Fl. UP. 3:516.1966 (Plate XVII).
Daedalacanthus purpurascens T. Anders in JLS 9:488.1867; FBI 4:420; FUGP 2:63.

Erect annual herbs. Leaves opposite, petiolate, 13 x 4 cm, ovate, acuminate, entire or serrulate, nerves prominent beneath. Flowers bluish-purple or lilac in dense solitary spike; bracts ovate-lanceolate, acuminate, margins ciliate. Corolla tube funnel-shaped, hairy outside.

Flowers and Fruits : September-November

Found rarely in moist shady places in sal forests of South Kheri Forest division. Singh 7502, 8557 (CDRI).

Andrographis Wall. ex Nees

A. paniculata (Burm.f.) Wall. ex Nees in Wall. Pl. As. Rar. 3:116.1832; FBI 4:501; FUGP 2:66; P.C. Kanjilal, For. Fl. UP. 3:526.1966; Srivastava, FG.241.1976.
Justicia paniculata Burm. f. Fl. Ind. 9.1768.

Erect glabrous, annual herbs. Stems quadrangular or slightly winged. Leaves 3-8 x 1.5-2 cm, ovate-lanceolate, glabrous. Flowers white, spotted with rose purple, in solitary, lax axillary and terminal, unilateral branched racemes. Bracts opposite paired. Capsules linear-oblong, tapering at ends, glabrous.

Local name : Kalmegh.
Flowers and Fruits : September-December

Found only in Gola range in South Kheri Forest division growing on temple walls. The plant is used in medicine for liver ailments. Singh 7835 (CDRI); G. Saran 26492 (LWG).

Lepidagathis Willd.

L. incurva D.Don, Prodr. Fl. Nep. 119. 1825; Santapau, Acanth. 74; Srivastava, FG. 247.1976.
L. hyalina Nees in Wall. Pl. As. Rar. 3:95.1832; FBI 4:521; FUGP 2:74.

Sub-erect, much branched, perennial herbs. Leaves variable, 2-2.5 x 0.5-0.7 cm, oblong-ovate or linear, acute, scabrous, hairy. Flowers white with mottled brown on upper lip, arranged in dense axillary and terminal clustered spikes Capsules oblong, 4-seeded.

Flowers and Fruits : December-April

Occurs occasionally in open places and in grasslands. Singh 7321, 7679 (CDRI); G. Saran 26279, 26325 (LWG).

Rungia Nees

R. pectinata (Linn.) Nees in Wall. Pl. As. Rar. 3:110.1832; Santapau, Acanth. 77; Srivastava, FG. 249.1976.
Justicia pectinata Linn. Amoen. Acad. 4:293.1759.
Rungia parviflora var. *pectinata* Clarke in FBI 4: 550. 1885; FUGP 2:80.

Much branched procumbent annual herbs. Leaves subsessile, opposite, 2-5 x 0.5-1 cm, ovate-elliptic, acute at both ends. Flowers blue in one sided, short, axillary spikes. Bracts dimorphic; the fertile one orbicular, hyaline; barren reduced in size not hyaline. Calyx lobes linear-lanceolate, acuminate. Corolla purple. Stamens 2. Capsules ellipsoid, compressed.

Flowers and Fruits : November-April

Found commonly in wastelands and along roadsides. Singh 4199, 4308, 7631, 7810 (CDRI); G. Saran 26505 (LWG).

Justicia Linn.

J. quinqueangularis Koen. ex Roxb. Fl. Ind. 1:133.1820; FBI 4:536; FUGP 2:77: Santapau, Acanth. 89; Srivastava, FG. 246.1976.
Rostellaria quinqueangularis Nees in Wall. Pl. As. Rar. 3:101.1832.

Slender, glabrous, erect or ascending, annual herbs with angular stems. Leaves sessile, opposite, linear or linear-lanceolate, 3-5 cm long, glabrous. Flowers white with pink spots in long terminal spikes. Bracts linear, obtuse, glabrous. Capsules oblong, pointed, glabrous.

Flowers and Fruits : September-February

Occurs frequently in moist shady places in sal forests. Singh 4309, 7306, 7804 (CDRI).

Peristrophe Nees

P. bicalyculata (Retz.) Nees in Wall. Pl. As. Rar. 3:113.1832; FBI 4:554; FUGP 2:79; Santapau, Acanth. 93; P.C. Kanjilal, For. Fl. UP. 526.1966.
Dianthera bicalyculata Retz. in vet. Akad. Handl. 297,t.9.1775.

Erect much branched, hispid herbs or undershrubs with 6-angled branches. Leaves opposite, petiolate, 6-8 x 2-2.5 cm, ovate, acute or acuminate, sparsely hairy. Flowers pink, in trichotomously branched lax panicles. Corolla bilipped. Stamens 2. Capsules oblong, pointed.

Local name : Choti Hadjor
Flowers and Fruits : September-February

Occurs commonly in shady places in open areas on wall and as an undergrowth of orchard. The plant paste is bandaged on bone fracture. Singh 4246 (CDRI); G. Saran 25135 (LWG).

Dicliptera Juss. *nom. cons.*

D. roxburghiana Nees in Wall. Pl. As. Rar. 3:111.1832; FBI 4:553; FUGP 2:81.

Erect glabrescent herbs. Leaves opposite, 3.5-4 x 1-1.5 cm, ovate-elliptic, acute, pubescent. Flowers pink in terminal and axillary clusters. Bracts cuneate-elliptic or obovate, apiculate, margins ciliate. Corolla bilabiate. Stamens 2, usually exserted. Capsules clavate.

Flowers and Fruits : December-March

Occurs rarely in waste places along roadsides. Singh 7185 (CDRI).

Adhatoda Nees

A. zeylanica Medic., Hist. & Commentat. Acad. Elect. Sci. Theod.-Palat. 6;393.1790; Saldanha & Nicolson, Fl. Hass. Distt. 540.1976.

Justicia adhatoda Linn. Sp. Pl. 15.1753; FUGP 2:76.
Adhatoda vasica Nees in Wall. Pl. Rar. 3:103.1832; FBI 4:540; Santapau, Acanth. 92; P.C. Kanjilal, loc. cit. 3:522.1966; Srivastava, FG. 240.1976.

Diffuse branched shrubs; young parts pubescent. Leaves opposite, 7-14 x 1.3 cm, ovate or elliptic-lanceolate, acuminate. Flowers white with purple strips, in dense axillary peduncled spikes. Bracts elliptic, acute, pubescent. Corolla bilipped. Capsules clavate. Seeds suborbicular.

Local name : Arusa
Flowers and Fruits : October-April

Occurs commonly in wastelands. The leaves decoction is used in cough and cold. Singh 2851 (CDRI); G. Saran 26113, 26729, 26670 (LWG).

Barleria Linn.

1. Plants armed; flowers orange-yellow ....**B. prionitis**

1. Plants unarmed; flowers pale-blue or purple:

   2. Outer calyx segment spinescent ....**B. cristata**

   2. Outer calyx segment strigose, margins ciliate ....**B. strigosa**

B. prionitis Linn. Sp. Pl. 636.1753; FBI 4:488; FUGP 2:69; Santapau, Acanth. 59; P.C. Kanjilal, For. Fl. UP. 3:518.1966; Srivastava, FG. 242.1976.

Erect or diffuse, prickly, undershrubs with grey-white stems. Spines white, 3-4 in the axil of the leaves. Leaves petiolate, 10 x 5 cm, ovate or elliptic, acute at both ends. Flowers orange-yellow in terminal spikes. Outer calyx lobes spine-tipped. Capsules pointed, 2-seeded.

Local name : Bansa
Flowers and Fruits : October-May

Found occasionally near villages. G. Saran 26118 (LWG).

B. cristata Linn. Sp. Pl. 636. 1753; FBI 4:483; FUGP 2:70; Santapau, Acanth. 59; P.C. Kanjilal, For. Fl. UP. 3:519.1966; Srivastava, FG. 242.1976.

Erect or diffuse undershrubs; the branches clothed with appressed yellowish hairs. Leaves elliptic-oblong, 5-12 x 1-

1.4 cm, acute, sparsely hairy. Flowers blue-purple in axillary and terminal short spikes. Outer sepals spinous. Corolla tube funnel-shaped in upper side. Capsules ellipsoid, 4-seeded.

Flowers and Fruits : October-March

Found commonly in grasslands and on the edge of the sal forests. Singh 4173, 7836 (CDRI); G. Saran 26312 (LWG).

B. strigosa Willd. Sp. Pl. 3:379.1800; FBI 4:489; FUGP 2:70; P.C. Kanjilal, For. Fl. UP. 3:520.1966.

I have included this taxon on the authority of P.C. Kanjilal loc. cit. who reports its occurrence from the Kheri forest areas. Hence, a brief description is reproduced below.

Unarmed, under shrubs with strigose branches. Leaves long petiolate, ovate-elliptic, acuminate, strigose specially on nerves beneath. Flowers pale-blue in dense, hairy unilateral axillary spikes. Calyx densely strigose, outer sepals ovate or elliptic with denticulate and ciliate margins. Corolla pale-blue; upper portion funnel-shaped. Capsules apiculate, glabrous.

82. VERBENACEAE

1. Leaves digitately 3 to 5-foliolate .... **Vitex**
1. Leaves simple:
    2. Inflorescence spicate, centripetal:
        3. Flowers in short, capitate or cylindric spikes:
            4. Creeping herbs; fruits dry .... **Phyla**
            4. Erect shrubs; fruits fleshy .... **Lantana**
        3. Flowers in elongated, lax spikes or racemes:
            5. Flowers pink, sessile in spikes .... **Verbena**
            5. Flowers violet, pedicellate in lax branched racemes..**Duranta**
    2. Inflorescence cymose; centrifugal:

6. Flowers zygomorphic; stamens didynamous:

7. Drupes with 1 to 4-celled pyrenes:

8. Trees:

9. Flowers brownish-yellow .... **Gmelina**

9. Flowers greenish-white .... **Premna**

8. Herbs.. **Pygmaeopremna**

7. Drupes with 4-pyrenes or drupels .... **Clerodendrum**

6. Flowers actinomorphic; stamens equal:

10. Flowers pink; shrubs .... **Callicarpa**

10. Flowers white; trees .... **Tectona**

Vitex Linn.

V. negundo Linn. Sp. Pl. 638.1753; FBI 4:565; FUGP 2:90; P.C. Kanjilal, For. Fl. UP. 3:541.1966; Srivastava, FG. 253.1976.

Shrubs or small trees. Branches quadrangular, densely white-tomentose. Leaves 3 to 5-foliolate; leaflets stalked, lanceolate, 10-16 x 2.5-4 cm, acuminate, entire or slightly crenate, white-tomentose beneath. Flowers blue in loose clusters, arranged in large terminal panicles. Drupes black.

Local name : Sambhalu
Flowers and Fruits : All seasons

Occasionally planted near village side. The plant is used in fever. Singh 8558, 8567 (CDRI).

Phyla Lour.

P. nodiflora (Linn.) Greene in Pittonia 4:46.1899; Santapau in RBSI 16(1):237.1953; P.C. Kanjilal, For. Fl. UP. 3:549.1966; Srivastava, FG. 257.1976.
Verbena nodiflora Linn. Sp. Pl. 20.1753.

Lippia nodiflora A. Rich. in Michx. Fl. Bom.-Amer. 2:15.1803; FBI 4:563; FUGP 2:84.

Prostrate slender, creeping much branched perennial herbs rooting at the nodes. Leaves obovate, 2.5 x 0.7 cm, cuneate, serrate above. Flowers pale-pink, sessile, numerous in dense globose axillary long peduncled spikes. Pyrenes 2, dry, 1-seeded, plano-convex.

Flowers and Fruits : April-August

Occurs commonly in moist places in grasses and along river banks. Singh 3075, 4115, 7234 (CDRI); G. Saran 26706 (LWG) Hira Lal 7191 (LWG).

Lantana Linn.

L. camara Linn. var. **aculeata** (Linn.) Moldenke in Torreya 34:9.1934; Maheshwari, Fl. Del. 277; P.C. Kanjilal, For. Fl. UP. 3:531.1966; Srivastava, FG. 256.1976.
L. aculeata Linn. Sp. Pl. 627.1753.

Straggling or climbing, aromatic shrubs with minute prickles on the branches. Leaves 5-7 x 2.5-4 cm, ovate or ovate-oblong, crenate-serrate, acute, scabrous on both surfaces. Flowers orange-yellow or pink in axillary long peduncled, capitate spikes. Drupes black when ripe.

Local name : Lal tahani
Flowers and Fruits : Throughout the year

Found frequently in wastelands. The twigs are used for making baskets. Singh 2883 (CDRI); Hira Lal 6871 (LWG).

Verbena Linn.

V. officinalis Linn. Sp. Pl. 20. 1753; FBI 4:565; FUGP 2:85.

Erect or decumbent, branched herbs with quadrangular branches. Leaves 5-10 cm long, variously lobed, scabrous.

Lower petiolate, pinnatifid, coarsely toothed, upper sessile, usually 3-partite, densely hairy on both surfaces. Flowers pink fading to white, in dense bracteate long spikes. Pyrenes 4, subcylindric, smooth.

Flowers and Fruits : April-October

Occurs rarely in the Park on wastelands. Singh 4150, 7217, 8556 (CDRI).

Duranta Linn.

D. repens Linn. Sp. Pl. 637.1753; Bailey, Man. Cult. Pl. 843; P.C. Kanjilal, For. Fl. UP. 3:549.1966; Srivastava, FG. 254.1976.
D. plumieri Jacq. Select. Am. 186.t.176.1763.

Erect evergreen shrubs with or without spines. Leaves 5 x 2 cm, obovate or oblong-lanceolate, serrate above, acute, cuneate. Flowers violet in axillary and terminal racemes. Drupes small, orange. Pyrenes 4.

Flowers and Fruits : Most part of the year

Planted as a hedge plant in Forest Rest House campus. Singh 4372 (CDRI).

Gmelina Linn.

G. arborea Roxb. Hort. Beng. 46.1814, nom. nud. et Pl. Cor. 3:42. t.246.1815; FBI 4:581; FUGP 2:88; P.C. Kanjilal, For. Fl. UP. 3:536.1966; Srivastava, FG. 255.1976.

Medium-sized, deciduous trees with light grey bark. Leaves long petiolate, ovate, 10-15 x 4-6 cm, acuminate, entire, base cuneate. Flowers brownish-yellow in lateral or terminal, panicled cymes. Drupes pyriform or obovoid, smooth.

Local name : Gambhar
Flowers and Fruits : February-May

Found frequently in the mixed forests. The wood is used for making drums. Singh 2897 (CDRI).

Premna Linn.

P. latifolia Roxb. var. **mucronata** Clarke in FBI 4:578.1885; P.C. Kanjilal, For. Fl. UP. 3:539.1966; Srivastava, FG. 257.1976.
P. mucronata Roxb. Fl. Ind. 3:80.1832; FUGP 2:89.

Small or medium-sized, deciduous trees with grey bark. Leaves 9-10 x 3-3.5 cm, ovate, acuminate, entire or undulate, dark green above, pale beneath, hairy on the midrib. Flowers greenish-white in corymbs. Calyx 5-toothed, clothed with spreading hairs. Corolla greenish-white. Drupes globose.

Flowers and Fruits : Most part of the year

Found occasionally on the edge of the sal forests. Singh 2935, 7256 (CDRI); G. Saran 26475 (LWG).

Pygmaeopremna Merr.

P. herbacea (Roxb.) Moldenke in Phytologia 2:54.1941; M.A. Rau in Bull. Bot. Surv. India 2.10:62.1969.
Premna herbacea Roxb. Fl. Ind. 3:80.1832; P.C. Kanjilal, For. Fl. UP. 3:540.1966.

Small almost stemless herbs from a woody rootstock. Leaves sessile, 5-14 x 3-6 cm, upper opposite, obovate, coarsely toothed; lower in rosettes, base cuneate, obtuse. Flowers white or greenish-yellow, in terminal corymbose panicles. Drupes globose, black when ripe.

Flowers : September-December; Fruits : May-July

Occurs frequently in grasslands and in open burnt areas. The root is used in rheumatism and gout. Singh 7399. 7678, 8388 (CDRI) G. Saran 26273 (LWG).

Clerodendrum Linn.

1. Leaves ternately whorled:

    2. Flowers purplish-white; leaves serrate .... C. serratum

    2. Flowers cream-white; leaves not serrate .... C. indicum

1. Leaves opposite; flowers white tinged with pink .... C. viscosum

C. serratum (Linn.) Moon, Cat. 46 No. 382.1824; Spreng. Syst. 2:758.1825; FBI 4:592; FUGP 2:92; P.C. Kanjilal, For. Fl. UP. 3:544.1966.
Volkameria serrata Linn. Mant. 90.1767.

Erect, glabrous, perennial herbs or undershrubs. Stems 4-angled. Leaves sessile, ternately whorled or rarely opposite, ovate-oblong 18-20 x 5-6 cm, sharply serrate, base cuneate. Flowers purplish-white, arranged in elongated, compact, pubescent, panicled cymes. Calyx campanulate, dentate. Corolla purple-white. Drupes broadly obovoid, 1 to 4-lobed, dark-purple when ripe.

Local name : Barang
Flowers and Fruits : April-September

Occurs frequently along forest roads and in grasslands. The plant is used in ayurvedic medicine. Singh 3058, 4266, 7895, 8546 (CDRI).

C. indicum (Linn.) O. Kuntze, Rev. Gen. 586.1891; Bailey, Man. Cult. Pl. 845; P.C. Kanjilal, For. Fl. UP. 3:546.1966; Srivastava, FG. 253.1976.
Siphonanthus indica Linn. Sp. Pl. 109.1753.
Clerodendrum siphonanthus R. Br. in Ait.f. Hort. Kew. 4:65.1812; FBI 4:595; FUGP 2:93.

Tall glabrous perennial shrubs with hollow ridged stems. Leaves ternately whorled, subsessile, 7-13 x 2.5 cm, narrowly lanceolate, entire, acuminate. Flowers creamy-white in lax, axillary and terminal panicled cymes. Bracts reddish when young. Calyx accrescent, enlarged and scarlet in fruits. Drupes of 4-connate drupels in a persistent enlarged calyx.

Flowers and Fruits : September-December

Found frequently in grasslands. The plant is used in ayurvedic medicine. Singh 2948, 3057, 7913 (CDRI); G. Saran 26466 (LWG); Hira Lal 7088 (LWG).

C. viscosum Vent. Jard. Malm. t.25.1803; Raizada in Indian For. Rec. 5(1):12.1958; Santapau in Bull. Bot. Surv. India 3:14.1961; P.C. Kanjilal, For. Fl. UP. 3:545.1966; Srivastava, FG. 254.1976.
C. infortunatum auct. non Linn. 1753.

Softly tomentose, undershrubs or shrubs with 4-angular branches. Leaves opposite, 12-23 x 6-10 cm, ovate-cordate, acuminate, denticulate or entire, softly hairy. Flowers white, tinged with pink in lax, large terminal sub-corymbose panicles. Calyx segments divided to the base. Stamens exserted. Drupes bluish-black when ripe, subglobose.

Local name : Bhant
Flowers and Fruits : December-July

Occurs commonly in sal forests as an undergrowth. Singh 2834, 7147, 7599 (CDRI); G. Saran 26700, 26235, 26502 (LWG).

Callicarpa Linn.

1. Peduncles and cymes branches stout; calyx pubescent.....**C. macrophylla**
1. Peduncles and cymes branches slender; calyx glabrous ....**C. longifolia var. lanceolaria**

C. macrophylla Vahl, Symb. Bot. 3:13,t.53.1794; FBI 4:568; FUGP 2:86; P.C. Kanjilal, For. Fl. UP. 3:533.1966; Srivastava, FG. 253.1976 (Plate XVIII).

Erect branched, stellately tomentose shrubs. Leaves 17-28 x 5-8 cm, ovate or ovate-lanceolate, acuminate, crenate-serrate, base cuneate, stellately tomentose beneath. Flowers pink, arranged on axillary, stout peduncles, in lax dichotomous cymes. Calyx lobes triangular, hairy. Drupes globose, white.

Local name : Daya
Flowers and Fruits : August-December

Occurs frequently in the sal forest and in open areas. The fruits are edible. Singh 2909, 2959, 4167, 7929, 8484 (CDRI).

C. longifolia Lamk. var. **lanceolaria** Clarke in Hook. f. FBI 4:570.1885; P.C. Kanjilal, For. Fl. UP. 3:534.1966; Raizada, Suppl. FUGP 211.1976.

Erect shrubs; young parts densely stellate-pubescent. Leaves 10-14 x 2.5-3 cm, narrow-lanceolate, acuminate, serrulate, glandular-dotted beneath, base cuneate. Flowers rose or purple on spreading slender peduncles in lax cymes. Calyx truncate, glabrous. Drupes white.

Flowers : March-April; Fruits : June-September

Found rarely in Mailani Range of South Kheri Forest division. Sis Ram 22516 (DD).

Tectona Linn.f. *nom. cons.*

T. grandis Linn.f. Suppl. 151.1781; FBI 4:570; FUGP 2:87; P.C. Kanjilal, For. Fl. UP. 3:534.1966.

Large deciduous trees with pale-brown or grey bark. Leaves large, elliptic or obovate, entire, cuneate at the base, acute, stellate-tomentose beneath. Flowers dull-white, numerous in large dichotomous or trichotomous branched, terminal panicles. Drupes globose, enclosed by the inflated calyx.

Local name : Sagon
Flowers : September-October; Fruits : November-Dec.

Commonly planted by the Forest department for its valuable timber. Singh 4336 (CDRI).

83. LAMIACEAE (*nom. alt.* LABIATAE)

1. Flowers in capitate whorls ....**Acrocephalus**

1. Flowers whorl various, not capitate:

    2. Stamens declinate, 4, all perfect:

        3. Lower lip of corolla flat ....**Ocimum**

        3. Lower lip of corolla concave or boat-shaped:

            4. Leaves opposite, long petiolate ....**Hyptis**

            4. Leaves opposite or whorled, subsessile ....**Plectranthus**

    2. Stamens erect, spreading or ascending:

        5. Perfect stamens 2 ....**Salvia**

        5. Perfect stamens 4, didynamous:

            6. Anthers monothecous:

                7. Flowers unisexual; stamens not bearded..**Colebrookea**

                7. Flowers bisexual; stamens bearded:

                    8. Leaves usually whorled, sessile; corolla lobes equal...**Eusteralis**

                    8. Leaves opposite, petiolate; corolla lobes unequal...**Pogostemon**

            6. Anthers dithecous:

                9. Calyx 15-ribbed ....**Nepeta**

                9. Calyx 5-10 ribbed:

                    10. Corolla 5-fid, campanulate ....**Perilla**

                    10. Corolla 2-lipped, not campanulate:

                        11. Herbs:

                            12. Flowers white ....**Leucas**

                            12. Flowers blue ....**Ajuga**

                        11. Shrubs or undershrubs:

                            13. Flowers deep orange in axillary whorls..................................**Leonotis**

                            13. Flowers bluish-red in panicled cymes or axillary whorls...............**Anisomeles**

Acrocephalus Benth.

A. hispidus (L.) Nicolson & Sivadasan in Taxon 29: 324.1980; *Gomphrena hispida* Linn. Sp. Pl. ed. 2: 326. 1762.

A. indicus (Burm.f.) O. Kuntze, Rev. Gen. Pl. 511. 1891; Mukerjee, Lab. Ind. Emp. 29.1940; Santapau in RBSI 16(1):243.1953; Srivastava, FG. 261.1976.

Acrocephalus capitatus Benth. in Wall. Pl. As. Rar. 2:18.1832; FBI 4:611; FUGP 2:102.

Erect annual herbs. Leaves opposite, 3-7 x 0.7-1 cm, narrowly ovate, cuneate, coarsely serrate, sub-acute, membranous. Flowers white, in dense subglobose heads. Nutlets oblong, ellipsoid, compressed.

Flowers and Fruits : September-December

Occurs commonly in moist shady places in sal forests. Singh 4165, 7851 (CDRI).

Ocimum Linn.

1. Calyx glabrous within .... **O. sanctum**

1. Calyx hairy within:

 2. Fruiting calyx 3 mm long .... **O. basilicum**

 2. Fruiting calyx 6 mm long .... **O. americanum**

O. sanctum Linn. Mant. 1:85.1767; FBI 4:609; FUGP 2:98; Mukerjee, Lab. Ind. Emp. 17.1940; P.C. Kanjilal, For. Fl. UP. 3:553.1966; Srivastava, FG. 267.1976.

Much branched, woody herbs. Leaves 2.5-5 x 1.5-2 cm, ovate, elliptic-oblong, entire or serrate, pubescent on both surfaces. Flowers purplish-pink, whorled in racemes, forming panicles. Nutlets broad-ellipsoid, smooth.

Local name : Tulsi
Flowers and Fruits : November-March

Planted in Dudhwa Forest Rest House campus. The plant is sacred and used for improving memory and for skin diseases. Singh 7995 (CDRI).

O. basilicum Linn. Sp. Pl. 597.1753; FBI 4:608; FUGP 2:99; Mukerjee, Lab. Ind. Emp. 18.1940; P.C. Kanjilal, For. Fl. UP. 3:555.1966; Srivastava, FG. 266.1976.

Erect much branched aromatic pubescent herbs, often purple in colour. Leaves ovate-lanceolate, 2.5-6 x 1.5-2 cm, acute, entire or toothed. Flowers whitish-pink or purple in whorls on simple or branched racemes. Nutlets ellipsoid.

Local name : Dawna
Flowers and Fruits : Greater part of the year

Cultivated in the gardens of Forest Rest House campus. Singh 7996 (CDRI).

O. americanum Linn. Sp. Pl. 833.1753 (non Jacq.); Keng, Gdns. Bull. Straits Settlem. 24:126.1969; Santapau in RBSI 16(1):242.1953; P.C. Kanjilal, For. Fl. UP. 3:554.1966; Srivastava, FG. 266.1976.
O. cannum Sims. in Bot. Mag. t. 2452.1824; FBI 4:607; FUGP 2:99; Mukerjee loc. cit. 17.1940.

Erect branched, pubescent, annual herbs. Leaves 2.5-5 x 1.5-2.5 cm, elliptic or elliptic-lanceolate or oblong, acute, entire or sub-crenate. Flowers white-purplish, decurved on short pedicels, arranged in whorls on erect lax inflorescence. Nutlets ellipsoid.

Local name : Jangali-tulsi, Mirmiri
Flowers and Fruits : August-February

Occurs rarely in the open wastelands. The plant is used as insecticide. Singh 8000 (CDRI).

Hyptis Jacq.

H. suaveolens (Linn.) Poit. in Ann. Mus. Natl. Hist. Nat. 7:472,t.29.f.2.1806; FBI 4:630; Mukerjee loc. cit. 63;

Santapau in RBSI 16(1):247.1953; Raizada, Suppl. FUGP 220.1976; Srivastava, FG. 262.1976.
Ballota suaveolens Linn. Syst. Nat. ed. 10.1100.1759.

Tall branched, aromatic, annual herbs. Leaves opposite, long petiolate, ovate; 3-7 x 2-2.5 cm, hairy, rounded at the base. Flowers blue, 2-4 together on an axillary peduncle or racemiform cymes. Calyx throat villous. Corolla 2-lipped. Stamens 4.

Local name: Ban tulsi
Flowers and Fruits : September-February

Found only in open wastelands. The plant is used as insecticide. Singh 7813 (CDRI).

Plectranthus L'Herit. *nom. cons.*

P. ternifolius D.Don, Prodr. Fl. Nepal.117.1825; FBI 4:621; FUGP 2:103; Mukerjee, Lab. Ind. Emp. 40.1940; P.C. Kanjilal, For. Fl. UP. 3:557.1966.

Tall densely tomentose, herbs or undershrubs. Leaves opposite and ternately whorled, subsessile, ovate-lanceolate, 4-9 x 1.5-2.5 cm, acuminate, serrate. Flowers whitish-pink or purple, subsessile in dense villous, panicled cymes. Nutlets trigonous, acuminate.

Flowers and Fruits : September-December

Found frequently in the Dudhwa range of the Park in grasslands and along railway lines in open areas. Singh 4243, 7934 (CDRI).

Salvia Linn.

S. plebeia R.Br. Prodr. 501.1810; FBI 4:655; FUGP 2:118; Mukerjee loc. cit. 111. 1940; Santapau in RBSI 16(1): 246. 1953; Srivastava, FG. 268.1976.

Erect branched, pubescent, annual herbs. Stems 4-angular and grooved. Leaves opposite, 4-9 x 1-2 cm oblong-lanceolate, crenate, subacute. Flowers bluish-white in compact, spicate, pubescent, panicled racemes. Nutlets ovoid, brown.

Flowers and Fruits : September-April

Found frequently in moist places and along river banks. Singh 2804, 2842, 4138, 7158, 7179, 8383 (CDRI).

Colebrookea Smith

C. oppositifolia Sm. Exot. 2:111,t.115.1805 (*Colebrookia);* FBI 4:642; FUGP 2:107; Santapau in RBSI 16(1):245.1953; Mukerjee loc. cit. 84.1940 *(Colebrookia);* P.C. Kanjilal, For. Fl. UP. 3:558.1966; Srivastava, FG. 262.1976.

Large much branched, white-tomentose shrubs. Leaves oblong-lanceolate, 10-23 x 3-8 cm, acuminate, crenate, pubescent. Flowers white in axillary or panicled spikes. Calyx 5-partite, feathery with white hairs. Nutlets oblong-ovoid.

Local name : Daya, Bansa.
Flowers and Fruits : December-April

Occurs frequently in the Park in sal forests and in open areas. Singh 7669, 7926, 8353 (CDRI).

Eusteralis Rafin.

E. stellata (Lour.) Panigrahi in Phytologia 32(6):474.1976.
Mentha stellata Lour. Fl. Cochin. 2:361.1790.
Dysophylla verticillata Benth in Wall. Pl. As. Rar. 1:30.1830; Srivastava, FG. 262.1976.

Erect branched, glabrous, annual herbs. Leaves 4-6 in whorls, sessile, 2.5-7 x 0.7 cm, linear-lanceolate, entire or obscurely toothed, obtuse. Flowers purplish, in terminal and axillary dense cylindrical pubescent spikes. Calyx teeth stellately spreading and conspicuous in fruits. Stamens bearded. Nutlets orbicular, compressed.

Flowers and Fruits : September-November

Found frequently in marshy areas and along forest roads in moist places. Singh 4280, 7809 (CDRI); D.P. Raturi 2557 (DD).

Pogostemon Desf.

P. benghalensis (Burm.f.) O. Kuntze, Rev. Gen. Pl. 529.1891; Santapau in RBSI 16(1):217.1967; Srivastava, FG. 268.1976.

Origanum benghalensis Burm.f. Fl. Ind. 128. t. 38. f. 3. 1768.

Pogostemon plectranthoides Desf. in Ann. Mus. Natl. Hist. Nat. 2:156.t.6.1808; FBI 4:632; FUGP 2:105; P.C. Kanjilal, For. Fl. UP. 3:559.1966.

Large grey-tomentose undershrubs; often branches tinged with purple. Leaves opposite, petiolate, ovate, 10-18 x 4.5 cm, doubly serrate, acuminate, white tomentose beneath. Flowers white-purple, arranged in dense villous, bracteate whorls of panicled spikes. Calyx shorter than the bracts. Stamens bearded. Nutlets ellipsoid.

Flowers and Fruits : September-April

Occurs frequently in open areas and along the edge of the sal forests. The leaves are used as insecticide. Singh 2828, 7332, 7620, 8314 (CDRI).

Nepeta Linn.

N. hindostana (Roth.) Haines, Bot. Bihar & Orissa 744.1922; Mukerjee loc. cit. 133; Srivastava, FG.265.1976.

Glechona hindostana Roth, Nov. Pl. Sp. 528.1821.

Nepeta ruderalis Buch.-Ham. ex Hook.f. in FBI 4:661.1885; FUGP 2:117.

Erect hairy annual herbs. Branches quadrangular, grooved. Leaves opposite, petiolate, 2-4 x 1.2 cm, ovate or suborbicular, coarsely crenate-serrate. Flowers bluish-purple, in stalked, axillary, dense cymes, forming spiciform thyrse in the upper portion. Nutlets oblong.

Flowers and Fruits : June-November

Found frequently in wastelands and near villages. Singh 2803, 4379 (CDRI).

Perilla Linn.

P. frutescens (Linn.) Britt. in Mem. Torrey. Bot. Club 5:277.1894; Hara in Enum. Sperm. Jap. 1:217.1948.
Ocimum frutescens Linn. Sp. Pl. 597.1753.
Perilla ocimoides Linn. Gen. Pl. ed.(6):578.1763; FBI 4:646; FUGP 2:122; Mukerjee loc. cit. 85.

Erect hairy aromatic annual herbs. Leaves opposite, long petiolate, ovate, 10 x 6 cm, coarsely toothed, acute. Flowers white in whorls of 2, forming long, erect, axillary and terminal racemes. Calyx 5-toothed, campanulate, hairy. Corolla 5-fid. Nutlets globose.

Local name : Bhangjira
Flowers and Fruits : October - December

Found occasionally in the Park on the edge of the sal forests. The seed oil is edible and used in rheumatism. Singh 4233, 7924 (CDRI).

Leucas R.Br.

1. Flowers in large, subglobose, terminal whorls only ....**L. cephalotes**
1. Flowers both in axillary and terminal whorls:
    2. Mouth of calyx oblique:
        3. Calyx teeth very short:

4. Throat of calyx glabrous and wide .... **L. hyssopifolia**

4. Throat of calyx hispid above and contracted .... **L. aspera**

2. Mouth of calyx not oblique:

3. Calyx teeth long, slender. .... **L. nutans**

5. Flowers sessile in many flowered whorls:

6. Calyx silky-villous .... **L. mollissima**

6. Calyx hispid .... **L. mollissimia** var. **scaberula**

5. Flowers slightly pedicellate, arranged in 2 to 4-flowered whorls... **L. biflora** var. **procumbens**

L. cephalotes (Roth) Spreng. Syst. 2:743.1825; FBI 4:689; FUGP 2:114; Mukerjee, Lab. Ind. Emp. 168.1940; Srivastava, FG. 264.1976.
Phlomis cephalotes Roth, Nov. Sp. Pl. 262.1821.

Erect pubescent annual herbs. Stems and branches angular, clothed with spreading hairs. Leaves ovate or lanceolate or elliptic, 5-8 x 1.5 cm, crenate-serrate. Flowers white, in dense, globose, terminal whorls. Bracts many, foliar. Nutlets smooth, brown.

Local name : Guma
Flowers and Fruits : September-March

Occurs as a weed in cultivated fields and in forest clearing. The plant is used for antifertility and snake bite. Singh 4250, 7698 (CDRI).

L. hyssopifolia Benth. in Wall. Pl. As. Rar. 1:60.1830; FBI 4:690; FUGP 2:112; P.C. Kanjilal, For. Fl. UP. 3:562.1966.

Erect hispid perennial herbs. Leaves sessile, 2.5-5 cm long, linear, obtuse, margin recurved, entire. Flowers in small axillary and terminal whorls. Bracts slender, subulate. Calyx mouth oblique, wide, glabrous.

Flowers and Fruits : September-December

Occurs occasionally in the area. Inayat 22389 (CAL).

L. **aspera** (Willd.) Spreng. Syst. 2:743.1825; FBI 4:690; FUGP 2:113; Mukerjee loc. cit. 166; Srivastava, FG. 263.1976.

Phlomis aspera Willd. in Link, Enum. Pl. Hort. Berol. 2:621.1822.

Erect or diffuse, much branched, hairy annual herbs. Leaves elliptic-linear, 5-8 x 1.5-2 cm, entire or crenate. Flowers white in terminal and axillary whorls. Bracts linear, filiform, ciliate. Calyx teeth curved, mouth oblique. Nutlets oblong, smooth, brown.

Local name : Guma
Flowers and Fruits : August-February

Occurs as a weed in cultivated fields. The plant decoction is used in fever. Singh 8310 (CDRI); G. Saran 26662 (LWG).

L. **nutans** Spreng. Syst. 2:743.1825; FBI 4:688; FUGP 2:114.

Dwarf diffusely branched, hispid annual herbs. Leaves opposite, 2.5-4 x 0.6 cm, lanceolate or linear-oblong, serrate, obtuse. Flowers white, sessile in dense axillary and terminal whorls. Calyx teeth long, slender, decurved, hairy. Nutlets oblong-elliptic.

Flowers and Fruits : September-December

Found occasionally in wastelands. Singh 4204 (CDRI).

L. **mollissima** Wall. Pl. As. Rar. 1:62.1830; FBI 4:682; FUGP 2:115; Mukerjee loc. cit. 183; P.C. Kanjilal, For. Fl. UP. 3:560.1966; Srivastava, FG. 265.1976.

Straggling, much branched, hoary tomentose herbs. Stems 4-angled. Leaves opposite, 3-5 x 1.5-2.5 cm, ovate or ovate-lanceolate, pubescent above, tomentose beneath, serrate, acute. Flowers white in whorls. Calyx silky-villous. Nutlets oblong-obovoid.

Flowers and Fruits : November-April

Found frequently in moist places along forest roads. Singh 2992, 4316, 7923 (CDRI).

L. mollissimia Wall. var. **scaberula** Hook.f. in FBI 4:682.1885; FUGP 2:115; Mukerjee loc. cit. 184.

Hispid hairy, annual herbs. Leaves opposite, 1.5-3 x 0.7-1 cm, ovate or elliptic-lanceolate, hairy, serrate, acute. Flowers white in many flowered whorls. Calyx hispid.

Flowers and Fruits : October-March

Found occasionally in sal forests in moist shady places. Singh 2910 (CDRI); Inayat 22390 (CAL).

L. biflora R.Br. var. **procumbens** (Desf.) Gamble, Fl. Madras 1151.1924; Srivastava, FG. 264.1976.
Phlomis biflora Vahl, Symb. 3:77.1794.
Leucas procumbens Desf. in Mem. Mus. Par. 11:7. t.3, f.2. FBI 4:683; FUGP 2:115; P.C. Kanjilal, For. Fl. UP. 3:561.1966.

Procumbent, perennial herbs. Leaves petiolate 4-5.6 x 2.5-3 cm, ovate-elliptic, serrate, acute. Flowers white, pedicellate in 1 to 4-flowered whorls only. Calyx 10-nerved, ciliate. Nutlets obovoid, truncate at the apex.

Flowers and Fruits : September-December

Found frequently in moist wastelands. Singh 3050 (CDRI).

Ajuga Linn.

A. bracteosa Wall. ex. Benth. in Wall. Pl. As. Rar. 1:59.1830; FBI 4:703; FUGP 2:121; Mukerjee, Lab. Ind. Emp. 225; Srivastava, FG. 261.1976.

Erect decumbent, softly hairy annual herbs. Stem obtusely 4-sided. Leaves petiolate, 2.5-4 cm long, obovate-spathulate, sinuate-toothed, lower surface tinged with purple. Flowers pale-blue in axillary whorls, arranged in spikes. Calyx villous. Stamens exserted. Nutlets 4, deeply rugose.

Flowers and Fruits : November-April

Found rarely in moist places along forest roads in the Park. The plant is used in skin disease. Singh 7658 (CDRI).

Anisomeles R.Br.

A. indica (Linn.) O. Kuntze, Rev. Gen. Pl. 2:502.1891; Mukerjee, loc. cit. 152; Backer & Bakh.f. Fl. Java 2:624.1965; Raizada, Suppl. FUGP 217.1976; Srivastava, FG. 261.1976.
Nepeta indica Linn. Sp. Pl. 571.1753.
Anisomeles ovata R.Br. in Ait.f. Hort. Kew. 3:364.1811; FBI 4:672; FUGP 2:109.

Erect, softly pubescent undershrubs. Stems acutely tetragonous, grey-tomentose. Leaves 6-12 x 3-5.6 cm, ovate-cordate, coarsely crenate, acuminate, pubescent on both surfaces. Flowers bluish-red, in axillary whorls forming a dense interrupted, spicate inflorescence. Stamens 4, exserted. Nutlets black, smooth, polished.

Local name : Gopali

Flowers and Fruits : September-March

Occurs frequently in open grasslands. Singh 3078, 4190 (CDRI).

Leonotis R.Br.

L. nepetifolia (Linn.) R.Br. Prodr. 503.1810; FBI 4:691; FUGP 2:116; Mukerjee, loc. cit. 185 (nepetaefolia); P.C. Kanjilal, For. Fl. UP. 3:562.1966; Srivastava, FG. 265.1976.
Phlomis nepetifolia Linn. Sp. Pl. 586.1753 (nepetaefolia).

Tall erect puberulous undershrubs. Stem 4-angled, grooved. Leaves ovate, 7-10 x 4 cm, crenate-serrate, acute, puberulous on both surfaces. Flowers orange in axillary whorls forming verticillaster. Bracts linear, spine pointed. Calyx teeth spinescent, unequal. Corolla 2 cm long, orange, 2-lipped.

Flowers and Fruits : October-March

Occurs frequently in wastelands in open situations. Singh 3096, 7829 (CDRI).

84. NYCTAGINACEAE

1. Armed climbing shrubs; bracts coloured ....**Bougainvillea**
1. Unarmed straggling herbs; bracts green ....**Boerhavia**

Bougainvillea Comm. ex Juss. corr. Spach nom. cons.

B. spectabilis Willd. Sp. Pl. 2: 348. 1799; Bailey, Man. Cult. Pl. 358; P.C. Kanjilal, For. Fl. UP. 3:564.1966; Srivastava, FG. 270.1976.

Woody armed, climbing shrubs; young branches softly pubescent. Leaves 5-6 x 2.5-3 cm, ovate to orbicular, acute, entire, pubescent when young. Flowers yellow, clustered at the end of the branches. Bracts 3, pink, showy, exceeding the flowers.

Flowers : January-July

Planted in forest nurseries and in compounds. Singh 7304 (CDRI).

Boerhavia Linn.

B. diffusa Linn. Sp. Pl. 3.1753; FUGP 2:127; P.C. Kanjilal, For. Fl. UP. 3:564.1966; Srivastava, FG. 269.1976.

Spreading much branched herbs; stems swollen at the nodes. Leaves in unequal pairs, 3.5 x 3 cm, ovate-oblong or suborbicular, margins undulate. Flowers pink or white, minute, bracteate, in axillary and terminal panicled umbels. Fruits clavate, 5-ribbed.

Local name : Gadapurna
Flowers and Fruits : Major part of the year

Occurs commonly in wastelands and on old walls. The roots are used in jaundice. Singh 2835, 7433, 7613, 7883, 8311 (CDRI).

85. AMARANTHACEAE

1. Leaves opposite:

    2. Flowers arranged in long spikes:

        3. Sepals subequal; terrestrial herbs .... **Achyranthes**

        3. Sepals unequal; aquatic herbs .... **Centrostachys**

    2. Flowers arranged in dense globose or cylindrical heads:

        4. Anthers monothecous:

            5. Staminal filaments fused below middle; stigma capitate .... **Alternanthera**

            5. Staminal filaments completely united forming a tube; stigma bifid .... **Gomphrena**

        4. Anthers dithecous .... **Psilotrichum**

1. Leaves alternate:

    6. Staminodes present .... **Aerva**

    6. Staminodes absent:

        7. Utricle single-seeded:

            8. Flowers unisexual or polygamous .... **Amaranthus**

            8. Flowers bisexual .... **Digera**

        7. Utricle many-seeded:

            9. Erect herbs; fruits capsular .... **Celosia**

            9. Scandent undershrubs; fruits berry .... **Deeringia**

Achyranthes Linn.

1. Leaves obovate or suborbicular, rounded at apex .... **A. aspera**

1. Leaves ovate or elliptic, acute or acuminate.. **A. aspera var. porphyristachya**

A. aspera Linn. Sp. Pl. 204.1753; FBI 4:730; FUGP 2:140; P.C. Kanjilal, For. Fl. UP. 3:569.1966; Srivastava, FG. 271.1976.

var. **aspera**

Erect or subscandent, pubescent, annual herbs. Leaves 3-10 cm long, obovate or suborbicular,rounded at apex, pubescent. Flowers greenish-white pink in long terminal spikes; bracts ovate, acuminate, reflexed; bracteoles ending in a spine. Stamens 3. Utricle oblong, enclosed in the perianth.

Local name : Chitchita, Latjeera
Flowers and Fruits : October-March

Occurs frequently in wastelands and along railway lines. The plant is highly medicinal and is used in snake bite., anti-fertility, eye disases etc. Singh 2831, 4293, 7699 (CDRI); G. Saran 26727, 26493 (LWG).

A. aspera Linn. var. **porphyristachya** Hook.f. in FBI 4:730.1885; Santapau in RBSI 16(1): 251.1953.

Erect or subscandent herbs or undershrubs. Leaves 9-16 x 5-7 cm, ovate-elliptic, membranous, acute or acuminate, glabrous. Flowers greenish-white, deflexed, in terminal, slender spikes. Bracts and bracteoles persistent, ending in a spine. Utricle oblong.

Local name : Chirchiri
Flowers and Fruits : October-June

Occurs frequently along the edges of the forests and in shady places. The plant is used in headache, boils and blisters and abortion. Singh 4309, 7417, 7823 (CDRI).

Centrostachys Wall.

C. aquatica (R.Br.) Wall. ex Moq. in DC. Prodr. 13(2):321.1849; Backer in Fl. Males. 4:87.1949 (Fig. 12).

Achyranthes aquatica R.Br. Prodr. Fl. Nov. Holl. 417.1810; FBI 4:730.

Erect, tall, stout, pubescent, aquatic herbs. Leaves opposite, petiolate, elliptic-lanceolate, 10-12 x 1.5 cm, acuminate. Flowers arranged in long pubescent spikes; bracts membranous, concave; bracteoles embracing the thickened perianth. Perianth rigid, glabrous, outer longer, pungent.

Flowers and Fruits : September-December

Found rarely in ponds of Bellrain range of the Park. Singh 7515 (CDRI).

Alternanthera Forsk.

1. Heads prickly; tepals unequal ....**A. pungens**
1. Heads not prickly; tepals all equal:
   2. Leaves spathulate ....**A. paronychioides**
   2. Leaves elliptic-lanceolate or linear-oblong ....**A. sessilis**

A. pungens H.B. & K. Nov. Gen. & Sp. 2:206.1818; Raizada in Indian For. 76:495.1950; Melville in Kew Bull. 174.1958; Backer & Bakh. f. Fl. Java 1:238.1963; Srivastava, FG. 272.1976.

A. repens Link, Enum. Fl. Hort. Berol. 1:154.1821 (non J.F. Gmel. 1791); van Steenis, Fl. Males. 1:5:554-55.1958.

Prostrate, branched herbs, often rooting at the nodes. Stems clothed with glochidiate hairs. Leaves opposite, 1.2-3.5 x 0.3 cm elliptic-obovate or more or less orbicular, obtuse, entire, tapering at the base. Flowers in axillary spinescent heads. Tepals spinescent in Fruits.

Flowers and Fruits : August-December

Occurs occasionally in wastelands and along railway lines. Singh 7692, 8380 (CDRI).

A. paronychioides St. Hil. Voy. Bres. 2:439.1833; Raizada in Indian For. Rec. 1:233; Santapau in JBNHS 54:476. 1956-57; Srivastava, FG. 272.1976.

Prostrate, creeping annual herbs, rooting at the nodes. Leaves 1-2.5 x 0.5 cm, spathulate or obovate, acute, entire. Flowers white, in dense, sessile, globose axillary heads. Bracts and bracteoles small, scarious, white. Tepals 5, lanceolate, acute. Stamens 3-5, alternating with staminodes. Ovary unilocular, 1-ovuled.

Flowers and Fruits : All seasons

Occurs commonly on moist ground and other wastelands. Singh 7259, 7466, 7604 (CDRI); G. Saran 26660 (LWG).

A. sessilis (Linn.) DC. Cat. Pl. Hort. Bot. Monsp. 4:77. 1813; FBI 4:731; FUGP 2:142; Melville in Kew Bull. 172. 1958; Srivastava, FG. 273.1976.
Gomphrena sessilis Linn. Sp. Pl. 225.1753.

Prostrate, branched, glabrous herbs, often rooting at the nodes. Leaves elliptic or linear-oblong, 2-4.5 x 0.4 cm, glabrous, cuneate at the base. Flowers white, in small, axillary globose or cylindrical heads. Tepals lanceolate, uni-nerved. Stamens 3-5, alternating with staminodes. Utricle obcordate.

Flowers and Fruits : Major part of the year.

Occurs commonly in moist, marshy places and along river banks. The plant is used in eye diseases. Singh 3066, 7155 (CDRI); G. Saran 26552 (LWG).

Gomphrena Linn.

G. celosioides Mart. Bietr. Amar. 93.1825 et in Nov. Acad. Nat. Cur. 13:301.1826; Sandwith in Kew Bull. 29.1946; Santapau in RBSI 16(1): 226.1967; Backer in Fl. Males. ser. 1.4(2):96; Srivastava, FG. 276.1976.

Erect or decumbent, branched, annual herbs, clothed with white shaggy hairs. Leaves 2.5-6 x 1-2 cm, obovate or elliptic, mucronate, cuneate, hairy beneath. Flowers white, arranged in dense, cylindrical or ovoid, terminal, 3-4 cm long spikes, subtended by two small leaves. Perianth lobes 5, pilose on the back. Fruits enclosed by the hardened perianth.

Flowers and Fruits : August-November

Found commonly in open, wastelands and along roadsides. Singh 2985 (CDRI).

Psilotrichum Blume

P. ferrugineum (Roxb.) Moq. in DC. Prodr. 13:(2):279.1849; FBI 4:725; Backer in Fl. Males. loc. cit. 97.1949 (Fig. 13).
Archyranthes ferruginea Roxb. Fl. Ind. 1:675.1820.

Prostrate, glabrous, annual herbs. Leaves opposite, 4-5 x 1-1.5 cm, obovate-oblong, entire. Flowers white, in terminal and in forked, dense clusters of spikes. Tepals 5. Stamens 2, unequal, connate below. Utricle globose.

Flowers and Fruits : September-December

Found rarely on moist ground in sal forests. Singh 4299 (CDRI).

Aerva Forsk. *nom. cons.*

1. Flowers in axillary spikes only ....**A. lanata**
1. Flowers in axillary and terminal spikes ....**A. sanguinolenta**

A. lanata (Linn.) Juss. in Ann. Mus. Par. 11:131.1808; FBI 4:728; FUGP 2:139; P.C. Kanjilal, For. Fl. UP. 3:568.1966; Srivastava, FG. 271.1976.
Achyranthes lanata Linn. Sp. Pl. 204. 1753.

Suberect or prostrate, diffuse, woolly-tomentose herbs. Leaves alternate, 2-3.5 x 1-2 cm, obovate or elliptic-rounded, entire, obtuse, white-woolly beneath. Flowers white, in small, dense, axillary spikes. Utricle ovoid, acute.

Flowers and Fruits : October-March

Found occasionally in cultivated fields and other wastelands. G. Saran & Party 26252 (LWG).

A. sanguinolenta (Linn.) Blume, Bijdr. 547.1825; Backer in Fl. Males. 1.4(2):85; P.C. Kanjilal, For. Fl. UP. 3:567.1966; Srivastava, FG. 272.1976.
Achyranthes sanguinolenta Linn. Sp. Pl. 294.1753.
Aerva scandens Wall. ex Moq. in DC. Prodr. 13:(2):302.1849; FBI 4:729; FUGP 2:138.

Erect or straggling herbs with long, hoary-tomentose branches. Leaves alternate and opposite, 4-6 x 2-2.5 cm, ovate or elliptic-lanceolate, acute-mucronate, softly hairy. Flowers silvery-white, in ovoid or cylindrical, axillary and terminal spikes or lax terminal panicles. utricle broadly ovate, acute.

Flowers and Fruits : October-April

Occurs commonly in wastelands and often straggling among bushes. Singh 2843, 2929, 7564, 7880 (CDRI).

Amaranthus Linn.

1. Plants with spines; bracteoles awned ....**A. spinosus**
1. Plants without spines; bracteoles acute ....**A. viridis**

A. spinosus Linn. Sp. Pl. 991.1753; FBI 4:718; FUGP 2:133; Srivastava, FG. 274.1976.

Erect or diffuse, branched, spinous herbs. Spines axillary, straight, divaricate. Leaves alternate, long-petiolate, 4-10 x 1-2.5 cm, ovate-elliptic, acute, base cuneate, entire. Flowers green, unisexual, in dense axillary clusters and in terminal and axillary interrupted spikes. Male flowers with 5 stamens. Tepals 5. Capsules ovoid.

Local name : Katili choulai.
Flowers and Fruits : Throughout the year

Occurs commonly in open wastelands. The leaves are cooked as vegetable. Singh 4355, 7333 (CDRI); G. Saran 26225 (LWG).

A. viridis Linn. Sp. Pl. ed. 2.1405.1763; FBI 4:720; FUGP 2:136; Srivastava, FG. 275.1976.

Erect, unarmed, annual herbs. Leaves alternate, 2.5-8 x 2-3 cm, ovate or deltoid-ovate, entire, cuneate at base, emarginate at tip. Flowers unisexual, pale-green, in small axillary clusters and in axillary and terminal, panicled spike-like racemes. Tepals 3. Stamens 3. Fruits acute, compressed, suborbicular.

Local name : Chaulai
Flowers and Fruits : July-March

Occurs commonly in wastelands. The leaves are used as vegetable by local people. Singh 4357 (CDRI); G. Saran 26373 (LWG).

Digera Forsk.

D. muricata (Linn.) Mart. Beitr. Amar. 77. 1825 et in Nov. Act. Nat. Cur. 13:265.1826.
Achyranthes muricata Linn. Sp. Pl. ed. 2:295.1763.
Digera alternifolia (Linn.) Aschers. in Schwfth. Fl. Aethiop. 180.1867.
Achyranthes alternifolia Linn. Mant. 50.1767.
Digera arvensis Forsk. Fl. Aegypt.-Arab.65.1775; FBI 4:717; FUGP 2:132.

Slender, erect or prostrate, annual herbs. Leaves variable, 5-7 x 2-3 cm, ovate-elliptic or lanceolate, entire, acute. Flowers pink, in lax, axillary spikes. Fruits compressed, muriculate.

Local name : Lahesua
Flowers and Fruits : August-November

Found frequently as a weed near cultivated fields and in nurseries. The leaves are cooked as vegetable. Singh 2985 (CDRI).

Celosia Linn.

C. argentea Linn. Sp. Pl. 205.1753; FBI 4:714; FUGP 2:131; P.C. Kanjilal, For. Fl. UP. 3:570.1966; Srivastava, FG. 275.1976.

Erect, glabrous, annual herbs with angular stems. Leaves alternate, 3-1 x 2.5-4 cm, linear or ovate-lanceolate, entire, acute. Flowers pinkish-white arranged in terminal dense spikes. Capsules ellipsoid.

Flowers and Fruits : September-February

Occurs commonly as a weed in cultivated fields and other wastelands. Singh 4389, 7738 (CDRI); G. Saran 25379 (LWG).

Deeringia R.Br.

D. amaranthoides (Lamk.) Merr. Interpr. Rumph. Herb. 211.1917; Backer in Fl. Males. 4:71.1949; P.C. Kanjilal, For. Fl. UP. 3:566.1966.
Achyranthes amaranthoides Lamk. Encycl. 1:548.1785.
Deeringia celosioides R.Br. Prodr. Fl. Nov. Holl. 413.1810; FBI 4:714; FUGP 2:130.

Scandent shrubs with striate branches. Leaves alternate, 2.5-5 x 1-3 cm, ovate or ovate-lanceolate, entire, acuminate, base rounded, membranous. Flowers yellowish-green, shortly pedicellate arranged in long racemes or in panicles. Fruits globose, berry.

Flowers and Fruits : September-December

Occurs occasionally in open areas in the bushes along railway lines. Singh 7516, 7765, 7968 (CDRI).

86. CHENOPODIACEAE

Chenopodium Linn.

1. Stigmas 5; aromatic herbs .... **C. ambrosioides**

1. Stigmas 2; non-aromatic herbs:

    2. Seeds smooth .... **C. album**

    2. Seeds rugose .... **C. murale**

C. ambrosioides Linn. Pl. 219.1753; FBI 5:4; Backer in Fl. Males. (Ser.1) 4:101.1949; Srivastava, FG. 278.1976.

Erect, aromatic, profusely branched, annual herbs with angular-ribbed stems. Leaves alternate, 4-9 x 1.5-3 cm, oblong-ovate, sinuate-dentate. Flowers greenish, sessile, clustered in the axil, forming interrupted terminal and panicled spikes. Perianth 5-lobed. Utricle enclosed in the perianth, 1-seeded.

Local name : Kirmari
Flowers and Fruits : March-September

Occurs frequently in moist places and along river banks. The plant is used in piles. Singh 3048, 4169, 7329 (CDRI).

C. album Linn. Sp. Pl. 219.1753; FBI 5:3; FUGP 2:143; Srivastava, FG. 278.1976.

Erect, annual herbs. Stems angular, green, often tinged with purple and coated with white granular pubescence. Leaves variable in size, oblong-lanceolate or rhombic-lanceolate, entire, toothed or lobed. Flowers in short axillary spikes and often forming a terminal thyrse. Seeds orbicular, smooth.

Local name : Bathua
Flowers and Fruits : November-February

Occurs as a weed of winter season crops and also in wastelands. The leaves are used as vegetable. Singh 7186, 7583 (CDRI).

C. murale Linn. Sp. Pl. 219.1753; FBI 5:4; FUGP 2:144; Srivastava, FG. 279.1976.

Erect, glabrous herbs. Leaves 2.5-9 x 3-4.5 cm, rhombic or deltoid-ovate, lobed or toothed, entire, at the base, 3-nerved. Flowers green, in clusters of lax and dense, axillary, divaricate cymes. Seeds rugose, orbicular, black.

Local name : Chamarbathua
Flowers and Fruits : December-March

Occurs occasionally in moist wastelands and in cultivated fields. G. Saran 26365 (LWG).

87. BASELLACEAE

Basella Linn.

B. alba Linn. Sp. Pl. 272.1753; Burm.f. Fl. Ind. 76.1768; FUGP 2:147; van Steenis, Fl. Males. 1.5:300.1957; Srivastava, FG. 279.1976.
B. rubra Linn. Sp. Pl. 272.1753; FBI 5:20; FUGP 2:147.

Glabrous, twining, succulent shrubs with pinkish stems. Leaves ovate-cordate, 11 x 4.5 cm, entire, acuminate; petioles up to 3 cm long. Flowers pink or pale-purple, in lax, axillary pedunculate spikes. Fruits ovoid, black.

Local name : Poy
Flowers and Fruits : August-October

Found frequently twining on shrubs. The leaves and stems are used as vegetable. Singh 7820 (CDRI).

88. POLYGONACEAE

1. **Perianth 4 to 5-merous; stigma capitellate ....Polygonum**

1. **Perianth 3-merous; stigma fimbriate ....Rumex**

Polygonum Linn.

1. Flowers in axillary clusters ....**P. plebeium**

1. Flowers in terminal spikes or panicles:

   2. Nutlets biconvex:

      3. Bracts glabrous ....**P. glabrum**

      3. Bracts hairy:

         4. Plants densely white-woolly ....**P. lanigerum**

         4. Plants pubescent, not woolly ....**P. limbatum**

   2. Nutlets trigonous:

      5. Racemes dense, stout; bracts imbricate; perianth eglandular..**P. barbatum** subsp. **gracile**

      5. Racemes slender filiform; bracts more or less distant, perianth glandular..**P. hydropiper** subsp. **microcarpum** var. **lenticularis**

P. plebeium R.Br. Prodr. 420.1810 (plebejum); FBI 5:27; FUGP 2:149; Srivastava, FG. 283.1976.

Prostrate, branched glabrous herbs. Leaves subsessile, 0.4-2 x 0.2 cm, oblong or linear, entire; stipules hyaline. Flowers pink, solitary axillary or 2-3 together. Perianth pink, free nearly to the base. Nutlets trigonous, black.

Flowers and Fruits : August-April

A very variable plant. Found commonly throughout the area in moist places and other swampy areas. Singh 2809, 2814, 7169, 7221, 7319 (CDRI); Inayat 22735 (CAL).

P. glabrum Willd. Sp. Pl. 2:447.1791; FBI 5:34, pro parte FUGP 2:153; Srivastava, FG. 282.1976.

Erect, glabrous annual herbs, tinged with red. Leaves 5-20 x 2-3 cm, lanceolate, entire, acuminate, glabrous, glandular dotted; stipules eciliate. Flowers pink in erect terminal panicled racemes. Perianth eglandular. Nutlets orbicular, biconvex.

Flowers and Fruits : September-March

Found frequently in marshy places and along river banks. Singh 7828 (CDRI).

P. lanigerum R. Br. Prodr. 419.1810; FBI 5:35; FUGP 2:149.

Erect, robust perennial white-woolly herbs. Stems clothed with white tomentum and swollen above the nodes. Leaves 6-20 x 1.5-2 cm, linear-lanceolate, entire, acuminate, white-woolly beneath; stipules striate, sparsely ciliate. Flowers pink or white, crowded in slender racemes or panicles. Nutlets orbicular.

Flowers and Fruits : September-December

Found commonly in marshy places. Singh 7503, 7884 (CDRI).

P. limbatum Meissn. in DC. Prodr. 15:123.1856; FBI 5:30; FUGP 2:152; Srivastava, FG. 283.1976.

Erect, pubescent annual herbs. Leaves 10-15 x 1.7 cm, lanceolate, acuminate, scabrous on both surfaces; stipules cylindric, strigose, truncate with spreading or reflexed, scabrid ciliate limbs. Flowers cream-coloured, arranged in short erect panicled racemes; bracts hispid. Nutlets orbicular, compressed, convex.

Flowers and Fruits : September-December

Found occasionally in marshy places and in ditches along roadsides. Singh 7661, 7976 (CDRI).

P. barbatum Linn. subsp. **gracile** Danser in Bull. Jard. Bot. Buitenz. (Ser. 3) 8:146.1927; Srivastava, FG. 282.1976.
P. barbatum Hook.f. in FBI 5:37.1886. (non Linn. 1753).
P. serrulatum Hook.f. in FBI 5:38.1886.

Erect or ascending, stout, glabrous annual herbs. Leaves 5-18 x 1.5 cm, linear or elliptic-oblong, acute or acuminate, margins ciliate. Stipules tubular, coarsely ciliate. Flowers white in short erect spiciform panicles. Bracts glabrous. Perianth eglandular. Nutlets trigonous.

Flowers and Fruits : September-March

Found commonly in marshy places and in ditches. Singh 2978, 4226, 4303, 7223 (CDRI).

P. hydropiper Linn. subsp. **microcarpum** Danser var. **lenticularis** Danser in Bull. Jard. Bot. Buitenz. (Ser. 3) 8:189.1927; Maheshwari, Fl. Del. 306.1963.
P. hydropiper Meissn. in DC. Prodr. 14:109.1856, in part; FBI 5:39; FUGP 2:155.

Erect or decumbent, glabrous annual herbs. Leaves 8.15 x 1.5-2 cm, lanceolate, or oblong-lanceolate, tapering at ends, entire; stipules glandular, shortly ciliate. Flowers white, in slender filiform, flexuosus racemes. Nutlets trigonous.

Flowers and Fruits : September-March

Found occasionally in marshy areas and in ditches. Singh 7559, 7920, 8404, 8471 (CDRI).

Rumex Linn.

R. dentatus Linn. Mant. 2:226.1771; FBI 5:59; FUGP 2:158; Srivastava, FG. 284.1976.

Erect, glabrous, annual herbs. Stems grooved, tinged with red. Radical leaves longer than cauline ones, oblong, obtuse, rounded or cordate at the base. Flowers trimerous, green, bisexual, in leafy or leafless whorls. Perianth in 2 whorls. Nutlets acutely trigonous, brown.

Flowers and Fruits : January-June

Found commonly in marshy places and along river banks. Singh 2888, 7138, 7593, 8325 (CDRI).

89. PIPERACEAE

1. Plants succulent, monoecious ....**Peperomia**

1. Plants not succulent, dioecious ....**Piper**

Peperomia Ruiz & Pavon

P. pellucida (Linn.) H.B. & K. Nov. Gen. 1:64.1815; FUGP 2:163; Santapau in RBSI 16(1):257.1953; Srivastava, FG. 285.1976.
Piper pellucidum Linn. Sp. Pl. 30.1753.

Slender much branched glabrous succulent annual herbs. Leaves broadly ovate, 1-3 x 1.5-2 cm, 7-nerved, entire. Flowers small, unisexual, distant in the peduncle with minute, peltate bracts. Fruits globose.

Flowers and Fruits : October-February

Occurs occasionally in moist shady places and on the trunk of trees. Singh 4248 (CDRI).

Piper Linn.

P. longum Linn. Sp. Pl. 29.1753; FBI 5:83; P.C. Kanjilal, For. Fl. UP. 3:576.1966. (Fig. 14).

Creeping or rambling, perennial herbs, rooting below. Leaves alternate, ovate or suborbicular, 6-11 x 5-8 cm, petiolate, base cordate, entire, acuminate. Flowers unisexual, greenish or yellowish, in cylindrical spikes. Perianth absént.

Local name : Peeper, Piparamul
Flowers : August-October; Fruits : Cold season

Occurs frequently in moist shady places in sal forests. The plant is used in medicine for cough and cold. Singh 3042, 7963, 8478 (CDRI).

90. LAURACEAE

Litsea Lamk. *nom. cons.*

1. Perianth lobes incomplete or absent; fruits globose ....**L. glutinosa**

1. Perianth lobes complete, 5 to 6-partite; fruits ovoid...........**L. monopetala**

L. glutinosa (Lour.) C.B. Robins. in Philipp. J. Sc. Bot. 321.1911; P.C. Kanjilal, For. Fl. UP. 3:582.1966.
Sebifera glutinosa Lour. Fl. Cochin. 2:783.1791.
Litsea chinensis Lamk. Encycl. 3:574.1792.

Medium-sized, evergreen trees with grey, corky bark; young parts tomentose. Leaves alternate, 12-17 x 6-9 cm, ovate-oblong or elliptic, entire, acute, base cordate, pubescent on the nerves. Flowers yellow, globose, arranged in axillary, pedunculate umbels. Perianth irregular, segments absent.

Local name : Meda Lakari
Flowers : April-July; Fruits : October-November

Occurs occasionally in the area along the edge of the sal forests. The stem bark is used for bone fracture and gout. Singh 7347 (CDRI).

L. monopetala (Roxb.) Pers. Syn. Pl.2(1):4.1806; P.C. Kanjilal, For. Fl. UP. 3:583.1966.
Tetranthera monopetala Roxb. Pl. Corom. 2: 26. t. 148. 1802.
Litsea polyantha Juss. in Ann. Mus. His. Nat. 6: 211.1805; FBI 5:162; FUGP 2:168.

Medium-sized, evergreen trees with brown bark; young parts and inflorescence rusty-tomentose. Leaves alternate, 11-17 x 5-7.5 cm, elliptic-oblong, ovate or obovate, entire, apex acute or rounded, glabrescent above, glaucous and rusty tomentose beneath. Flowers greenish - yellow, axillary in umbellate heads. Perianth 5-6 partite, oblong-spathulate. Fruits ovoid.

Flowers : March-May; Fruits : October-November

Found frequently in the sal forests. Singh 2925, 7205, 8396 (CDRI).

91. PROTEACEAE

Grevillea R.Br. ex Knight *nom. cons.*

G. robusta A. Cunn. in R.Br. Prodr. Nov. 24.1830; P.C. Kanjilal, For. Fl. UP. 3:585.1966; Srivastava, FG. 285.1976.

Large, evergreen trees with rough bark; young parts and branches rusty-tomentose. Leaves alternate, bipinnate or sometimes tripinnate, pinnae lanceolate, acute, margin recurved, dark green above, grey-silky beneath. Flowers orange, in secund, one-sided racemes. Perianth petaloid; segments 4, reflexed. Stamens 4. Ovaries superior, style long, lateral.

Local name : Silver oak.
Flowers : March-April

Planted in Sonaripur range of the Park in Forest Rest House campus. Singh 8385 (CDRI).

92. LORANTHACEAE

Dendrophthoe Mart.

D. falcata (Linn.f.) Ettings. in Denkschr. Akad. Wissen. Math. Naturw. 32:52,53,58,t.13,f.14.1872; Santapau in RBSI 16(1):262.1953; P.C. Kanjilal, For. Fl. UP. 3:588.1966; Srivastava, FG. 286.1976.
Loranthus falcatus Linn.f. Suppl. 211.1781.
L. longiflorus Desr. in Lamk. Encycl. 3:598.1792; FBI 5:214; FUGP 2:175.

Much branched, glabrous, partial parasites. Leaves opposite, 8-12 x 3-6 cm, ovate-elliptic, obtuse, coriaceous. Flowers deep-orange, crowded in short, unilateral racemes on leafless nodes. Stamens exserted. Stigma knob-shaped.

Local name : Banda
Flowers : November-January

Found occasionally on trees in the forest areas. Singh 4323 (CDRI).

93. EUPHORBIACEAE

1. Flowers naked; arranged in cyathia .... **Euphorbia**

1. Flowers with perianth, not in cyathia:

    2. Leaves 3-foliolate, compound .... **Bischofia**

    2. Leaves simple:

        3. Leaves opposite .... **Trewia**

        3. Leaves alternate:

            4. Leaves peltate .... **Ricinus**

            4. Leaves not peltate:

                5. Leaves palmately 3 to 5-nerved:

                    6. Petals present in one or both sexes; or if absent; calyx petaloid:

                        7. Petals showy in both the sexes ..... **Jatropha**

                        7. Petals smaller than calyx; often absent in female:

                            8. Leaves lobed; seeds not strophiolate .... **Chrozophora**

                            8. Leaves not lobed; seeds with a white strophiole.. **Croton**

                    6. Petals absent in both sexes:

                        9. Herbs or undershrubs:

                            10. Plants with stinging hairs .... **Tragia**

                            10. Plants without stinging hairs .... **Baliospermum**

                        9. Trees .... **Mallotus**

5. Leaves penninerved:

11. Male flowers in axillary fascicles or solitary axillary:

12. Petals present ....**Bridelia**

12. Petals absent :

13. Styles connate in column ....**Glochidion**

13. Styles free or slightly connate at the base:

14. Styles arm much slender:

15. Male flowers with 2 to 3-fids pistillode. ....**Securinega**

15. Male flowers without pistillode:

16. Disc present in the female and often in the male flowers:

17. Fruits dry capsule ....**Phyllanthus**

17. Fruits berry:

18. Shrubs ....**Kirganelia**

18. Trees ....**Emblica**

16. Disc absent in both sexes; combined with calyx lobes ....**Breynia**

14. Styles arm much dilated ....**Drypetes**

11. Male-flowers in axillary panicled spikes or racemes ....**Antidesma**

Euphorbia Linn.

1. Leaves radical; stems underground ....**E. acaulis**

1. Leaves cauline; stems above ground:

2. Leaves opposite and lower alternate:

3. Rootstocks with proliferous shoots; seeds globose, without a strophiole..**E. prolifera**

3. Rootstocks not proliferous; seeds strophiolate.**E. dracunculoides**

2. Leaves all opposite:

4. Involucral glands with a prominent rounded limb..**E. hypericifolia**

4. Involucral glands with very small or obsolete limb:

5. Prostrate herbs:

  6. Capsules hairy only along the keel ....E. chamaesyce

  6. Capsules pubescent or hirsute throughout...................E. thymifolia

5. Procumbent herbs ....E. hirta

E. acaulis Roxb. Hort. Beng. 36.1814, nom. nud. et Fl. Ind. 2: 472. 1824; FUGP 2:182.

E. fusiformis Buch.-Ham. ex D.Don, Prodr. Fl. Nepal. 62. 1825; FBI 5:257; P.C. Kanjilal, For. Fl. UP. 3:599.1966.

Dwarf, glabrous herbs, with stout, underground cylindric rootstock. Leaves radical, subsessile, 9-15 x 2.5-5 cm; oblanceolate, base cuneate, margins curled. Cyathia yellow, arranged in cymes on long peduncle, arising from the root stocks. Involucres subsessile, greenish; lobes spathulate.

Local name : Banmuli
Flowers : April-May; (When plant is leafless)

Occurs frequently in Bankati and Sonaripur ranges of the Park along Indo-Nepal border. The root stocks are used in rheumatism & gout. Singh 8401 (CDRI).

E. prolifera Buch.-Ham. ex D.Don, Prodr. Fl. Nepal. 62.1825; FBI 5:264; FUGP 2:186; Airy Shaw in Kew Bull. 26(2):266.1972.

Erect, glabrous, perennial herbs. Stems arising from a stout woody rootstock. Leaves cauline, alternate 4-5 x 0.3-0.7 cm, sessile, linear-oblong or linear, entire, obtuse, coriaceous. Cyathia arranged in umbels. Involucres with 4-5 glands, yellow, crescent shaped.

Flowers and Fruits : March-May

Found occasionally in sal forests. Singh 7470 (CDRI).

E. dracunculoides Lamk. Encycl. 2:428.1788; FBI 5:262; FUGP 2:185; Srivastava, FG. 293.1976.

Erect, glabrous, dichotomously branched, annual herbs. Leaves alternate and opposite, 2-6 x 0.3-0.5 cm, linear to linear-lanceolate, sessile. Cyathia pale-yellow. Involucre subsessile, solitary, glabrous; lobes ciliate; gland semilunate. Capsules glabrous. Seeds ellipsoid, strophiolate.

Local name : Titali
Flowers and Fruits : March-May

Occurs commonly in cultivated fields on sandy soil. Singh 7623, 8309 (CDRI).

E. hypericifolia Linn. Sp. Pl. 454.1753; FUGP 2:187; Airy shaw in Kew Bull. 26(2):265.1972.
E. parviflora Linn. Syst. (ed.10). 2:1047.1759; Santapau in RBSI 16(1):273.1953; Srivastava, FG. 295.1976.

Erect or decumbent, branched, glabrous, annual herbs. Leaves all opposite, variable in shape, linear-oblong to elliptic-ovate, 0.7-1 x 0.5 cm, serrulate, base obliquely rounded. Cyathia arranged in axillary cymes. Involucres minute, cup shaped, green, provided with conspicuous white-pink limbs. Capsules pubescent. Seeds ellipsoid, 4-angled.

Flowers and Fruits : May-October

Occurs occasionally along river banks and other wastelands in moist situation mixed with grasses. Singh 7510 (CDRI).

E. chamaesyce Linn. Sp. Pl. 455.1753; Shah & Yogi in JBNHS 69(2):448.1972; Raizada, Suppl. FUGP 247.1976.
E. prostrata Ait. Hort. Kew. (ed. 1) 2:139.1879; FBI 5:266; Maheshwari, Fl. Del. 313.1963.

Prostrate, slender annual herbs. Branches minutely hairy, often one side purple. Leaves opposite, 0.5-0.8 x 0.3 cm, obliquely oblong, minutely serrate towards apex, 3-nerved. Involucre campanulate with 4-glands and lobes. Capsules usually bent, prominently keeled and provided with stiff, spreading hairs on the keels.

Flowers and Fruits : Throughout the year

Found frequently in the open area and in lawns. Singh 7316 (CDRI).

E. thymifolia Linn. Sp. Pl. 454.1753; FBI 5:252; FUGP 2:188; Airy Shaw in Kew Bull. 26(2):267.1972; Srivastava, FG. 295.1976.
Chamaesyce thymifolia (Linn.) Millsp. Publ. Field. Mus. Nat. Hist. Bot. ser. 2:412.1909.

Prostrate, pubescent herbs with coppery tinge. Leaves opposite, 0.6 x 0.3 cm, obliquely oblong, rounded, mucronate, serrulate, stipules fimbriate, sparsely hairy. Cyathia usually solitary axillary; gland minute. Capsules erect pubescent, obtusely keeled.

Local name : Laldudhi
Flowers and Fruits : April-October

Found commonly in waste grounds. The plant is used in dysentery. Singh 7281, 8341 (CDRI).

E. hirta Linn. Sp. Pl. 454.1753; FUGP 2:187; Airy Shaw loc. cit. 264; Srivastava, FG. 294.1976.
E. pilulifera auct (non Linn. 1753).
Chamaesyce hirta (Linn.) Millsp. loc.cit. 303.1909.

Procumbent, hispid herbs. Leaves opposite, 2-3.5 x 0.6-1 cm, elliptic or ovate-oblong, acute, oblique, serrulate, hairy. Cyathia axillary and terminal, in capitate cymes. Involucre stalked, cup-shaped. Capsules hairy.

Local name : Dudhia
Flowers and Fruits : All seasons

Occurs commonly throughout the areas in waste ground, lawns etc., Singh 7280 (CDRI).

Bischofia Blume

B. javanica Blume, Bijdr. 1168.1827; FBI 5:345; FUGP 2:190; P.C. Kanjilal, For. Fl. UP. 3:622.1966; Srivastava, FG. 289.1976.

Medium-sized deciduous trees with grey or brownish bark. Leaves alternate, long petiolate, 3-foliolate; leaflets elliptic-oblong to obovate, 5-12 x 4-5.5 cm, crenate-serrate, acuminate, terminal largest. Flowers greenish-yellow, unisexual in lateral and axillary panicled racemes. Fruits globose, reddish-brown.

Local name : Paniyala, Marcha.
Flowers : March-April; Fruits : September-December

Occurs occasionally in Dudhwa range of the Park in moist and marshy shady places. Singh 7911 (CDRI).

Trewia Linn.

T. nudiflora Linn. Sp. Pl. 1193.1753; FBI 5:423; FUGP 2:210; P.C. Kanjilal, For. Fl. UP. 3:624.1966; Airy Shaw in Kew Bull. 26(2):343.1972.

Medium-sized deciduous trees, with smooth brown or grey bark. Leaves opposite, 12-20 x 5-7.5 cm, ovate, base cordate, acuminate, glabrous, entire, 3 to 5-nerved. Flowers pale green; male in pendulous racemes; female solitary or 2-3 together on long peduncle. Fruits globose, greenish-brown.

Local name : Gutel
Flowers : February-April; Fruits : October-December

Occurs frequently in damp and marshy places. Singh 2905, 7188, 7655 (CDRI).

Ricinus Linn.

R. communis Linn. Sp. Pl. 1007.1753; FBI 5:457; FUGP 2:212; P.C. Kanjilal, For. Fl. UP. 3:631.1966; Srivastava, FG. 299.1976.

Evergreen large shrubs or small trees, with grey bark. Leaves alternate, long petiolate, peltate, palmately 7-fid, serrate. Flowers unisexual, cream coloured, in erect terminal racemes. Male flowers occupying the upper portion; stamens many in branched fascicles. Capsules 3-lobed, echinate. Seeds 3, oblong, spotted, arillate.

Local name : Rendi
Flowers and Fruits : Major part of the year

Commonly cultivated near village side for its oil. Singh 7276, 8394 (CDRI).

Jatropha Linn.

1. Flowers yellowish-green .... **J. curcas**
1. Flowers purplish-red .... **J. gossypifolia**

J. curcas Linn. Sp. Pl. 1006.1753; FBI 5:383; FUGP 2:215; Santapau in RBSI 16(1):279.1953; P.C. Kanjilal, For. Fl. UP. 3:632.1966; Srivastava, FG. 296.1976.

Much branched shrubs. Leaves long petiolate, slightly 3 to 5-lobed, ovate, acute, cordate. Flowers unisexual; yellowish-green in panicled cymes; female flowers in fork of cymes. Capsules woody, subglobose.

Local name : Bakrend
Flowers and Fruits : May-November

Planted occasionally near Tharu village in Chandan Chowki. The seed is used for oil. Singh 7959 (CDRI).

J. gossypifolia Linn. Sp. Pl. 1006.1753; FBI 5:383; Maheshwari, Fl. Del. 314.1963; P.C. Kanjilal, For. Fl. UP. 3:633.1966; Srivastava, FG. 296.1976.

Much branched shrubs; young parts pale or deep purple. Leaves alternate, long petiolate, 3 to 5-lobed; lobes elliptic-acute; petioles, stipules and leaf margins covered with glandular, bristle hairs. Flowers purplish-red, unisexual in terminal cymes. Capsules 3-lobed, oblong, brownish.

Local name : Bakrendi
Flowers and Fruits : July-November

Occurs commonly in wastelands and along roadsides. Singh 3070 (CDRI).

Chrozophora Juss. Corr. Benth. & Hook.f. *nom. cons.*

1. Leaves pale-green, eglandular; sepals of male flowers triangular..**C. rottleri**
1. Leaves dark-purple, glandular; sepals of male flowers linear...**C. parvifolia**

C. rottleri (Geis.) Juss. ex Spreng. Syst. 3:850.1826; FUGP 2:206; Srivastava, FG. 291.1976.
Croton rottleri Geis. in Crot. Monogr. 57.1807.
Chrozophora plicata Hook.f. in FBI 5:409.1887.

Erect, stellately hairy, annual herbs. Leaves alternate, long petiolate, blade 3-8 x 1-4.5 cm, ovate to orbicular, rounded, acute, subcordate, palmately nerved, sinuate-toothed, scabrous. Flowers unisexual, yellow, in racemes. Stigma red. Capsules densely stellately tomentose.

Flowers and Fruits : Major part of the year

Found frequently in open sandy places. Singh 7340 (CDRI).

C. parvifolia Klotz. ex Schwfth. Pl. Nilot. 11. 1862; Maheshwari, Fl. Del. 315.1963.

Prostrate, stellately woolly herbs. Branches many often tinged with red. Leaves 2.5-5 x 2-4 cm, dark-purple, 2-glands at the base. Flowers in short axillary racemes. Male flowers with yellow anthers. Capsules stellately woolly without silvery scales.

Flowers and Fruits : April-June

Occurs rarely in the area. Inayat 22494 (DD).

Croton Linn.

1. Herbs or undershrubs; flowers pale-white .... C. bonplandianum

1. Trees; flowers yellowish-green .... C. roxburghii

C. bonplandianum Baill. in Adans. 4:339.1864; P.C. Kanjilal, For. Fl. UP. 3:624.1966; Srivastava, FG. 291.1976; Raizada, Suppl. FUGP 247.1976.

C. sparciflorus Morong. Ann. N.Y. Acad. Sci. 7: 221. 1893.

Erect, much branched, hairy herbs or undershrubs. Leaves ovate-lanceolate, 3-7 x 1.5 cm, serrate, 3-nerved, 2-glands at the base, crowded at the end of the branches. Flowers unisexual, pale-white in terminal, erect, androgynous spikes. Female flowers with 2-extra floral glands. Petals absent in female flowers but present in male. Capsules 3-angled, stellately hairy.

Flowers and Fruits : Major part of the year

Occurs occasionally in waste places near roadsides. Singh 7458 (CDRI).

C. roxburghii Balak in Bull. Bot. Surv. India 3: 39. 1962.

C. oblongifolius Roxb. Hort. Beng. 69.1814, nom. nud et Fl. Ind. 3:685.1832; FBI 5:386; FUGP 2:205; P.C. Kanjilal, For. Fl. UP. 3:623.1966; Airy Shaw in Kew Bull. 26(2):249.1972; Srivastava, FG. 292.1976.

I have included this taxon on the authority of P.C. Kanjilal loc. cit. who reports its occurrence from Kheri forests. Hence, a brief description is reproduced below.

Medium-sized, deciduous tree. Leaves oblong, lanceolate, crenate-serrate, acute, glabrous. Flowers yellowish-green, solitary or in terminal and axillary racemes. Capsules sub-globose.

Tragia Linn.

T. involucrata Linn. Sp. Pl. 980.1753; FBI 5:465; FUGP 2:213; P.C. Kanjilal, For. Fl. UP. 3:629.1966; Srivastava, FG. 300.1976.

Twining or trailing herbs with hispid, stinging hairs. Leaves oblong-lanceolate, 6.5-22 x 1.5-6 cm, acuminate, crenate-serrate, base cordate, hairy. Flowers unisexual, yellowish in racemes; male flowers with 3 perianth and 3 stamens; female flowers with 6-perianth. Pistillode minute. Capsules 3-lobed, hairy, with spreading calyx.

Flowers and Fruits : October-April

Found frequently in open areas and in grasslands twining on bushes. Singh 3060, 4237, 7320 (CDRI).

Baliospermum Blume

B. montanum (Willd.) Muell.-Arg. in DC. Prodr. 15(2): 1125.1866; Shah in JBNHS 60(1):296-298.1963; P.C. Kanjilal, For. Fl. UP. 3:634.1966; Airy Shaw in Kew Bull. Vol. 26(2):222.1972; Srivastava, FG. 289.1976.

Jatropha montana Willd. Sp. Pl. 4:563.1805.

Baliospermum axillare Blume, Bijdr. 604.1825; FBI 5:461; FUGP 2:208.

Erect shrubs. Leaves alternate, variable in size and shape, 5-25 x 3-9 cm, palmately 3 to 4-lobed; upper lanceolate, sinuate-toothed, coriaceous, base rounded or cuneate. Flowers yellowish-green, apetalous, arranged in panicles. Male flowers globose with 10-20 stamens. Capsules obovoid, hairy.

Local Name : Danti
Flowers and Fruits : September-December

Found frequently on the edge of the sal forests. Singh 2849, 7173, 8372 (CDRI); G. Saran 26143 (LWG); Hira Lal 7093 (LWG).

Mallotus Lour.

M. philippensis (Lamk.) Muell.-Arg. in Linnaea 34:196.1865; FBI 5:442; FUGP 2:210 (`philippinensis'); P.C. Kanjilal, For. Fl. UP. 3:626.1966; Airy Shaw loc. cit. 300.1972; Srivastava, FG. 297.1976.
Croton philippense Lamk. Encycl. 2:206.1786.

Small, much branched, evergreen trees with dark grey bark; young parts and inflorescence rusty pubescent. Leaves alternate, 12-19 x 4-10 cm, ovate-oblong or lanceolate, acute-acuminate, base 3-nerved, cuneate, entire, glandular beneath. Flowers unisexual in axillary racemes or spikes. Stamens many. Capsules globose, 3-lobed, red.

Local name : Rohini, Lali
Flowers and Fruits : March-November

Found commonly mixed with *Shorea robusta* and *Terminalia alata* in the forest area. The fruit hairs are used in ayurvedic medicince. Singh 8322 (CDRI).

Bridelia Willd.

1. Trees, plant dioecious ....**B. squamosa**
1. Scandent shrubs; plant monoecious ....**B. stipularis**

B. squamosa Gehrm. in Bot. Jahrb. 41. Beibl. 95: 30. 1908; Santapau in RBSI 16(1): 273. 1953; Srivastava, FG. 290. 1976.
B. retusa Hook.f. in FBI 5:268.1887, ambo pro parte, (non Spreng.) P.C. Kanjilal, For. Fl. UP. 3:601.1966.

Small or medium-sized, deciduous trees with grey bark. Leaves elliptic-oblong or ovate or obovate, 12-14 x 4.5 cm, acute or obtuse at the apex, base rounded, entire, coriaceous. Flowers greenish-yellow, unisexual, in axillary clusters or in long axillary or terminal panicled spikes. Drupes subglobose, fleshy.

Local name : Khaja
Flowers : August-October; Fruits : November-January

Found frequently in sal forests. Singh 2950, 4103, 7898 (CDRI).

B. stipularis (Linn.) Blume, Bijdr. 597.1825; FBI 5:270; FUGP 2:192; P.C. Kanjilal, loc. cit. 603.1966; Airy Shaw in Kew Bull. 26(2):230.1972; Srivastava, FG. 290.1976.
Cluytia stipularis Linn. Mant. 1:127.1767.
Bridelia scandens (Roxb.) Willd. Sp. Pl. 4: 979. 1805 p.p.

Large scandent woody climbers, often with deflexed spines. Leaves 11-14 x 4-5.3 cm, elliptic-oblong or obovate, obtuse or subacute, base rounded or subcordate. Flowers unisexual, greenish-yellow, in dense, axillary clusters or spikes. Drupes oblong, obtuse.

Local name : Khaji
Flowers : May-October; Fruits : December-February

Found frequently in forest areas climbing on trees. Singh 4212,. 4307, 7264 (CDRI).

Glochidion Forst. *nom. cons.*

1. Pedicels of male flowers filiform, upto 2 cm long:
   2. Stamens 3 .... G. assamicum
   2. Stamens 4-6 .... G. lanceolarium
1. Pedicels of male flowers stout up to 0.4 cm long........... G. multiloculare

G. assamicum (Muell.-Arg.) Hook.f. in FBI 5:319.1887; P.C. Kanjilal, For. Fl. UP. 3:618.1966; Airy Shaw in Kew Bull. 26(2): 274. 1972; Raizada, Suppl. FUGP 254.1976.
Phyllanthus assamicus Muell.-Arg. in Flora 48:378.1865; in DC. Prodr. 15(2):297.1866.

Small evergreen, glabrous trees with reddish-brown bark. Leaves 9-13 x 3.2-6.5 cm, elliptic or elliptic-oblong, shortly acuminate, glabrous. Flowers pale-green in axillary fascicles; male long pedicellate; female subsessile. Capsules globose, indistinctly lobed. Seeds with orange aril.

Flowers : March-May; Fruits : October-December

Occurs occasionally in moist shady places in sal forests in Bankati range. Singh 7144, 7196 (CDRI).

G. lanceolarium Voigt, Hort. Suburb. Calc. 153.1845; FUGP 2:194; P.C. Kanjilal, For. Fl. UP. 3:618.1966; Airy Shaw, loc. cit. 277.1972.

Small, evergreen, glabrous trees with grey bark. Leaves elliptic-oblong or lanceolate, 7.5-9 x 1-1.5 cm, coriaceous, acute, glabrous, pale-green beneath. Flowers pale-green, in fascicles; male flowers with filiform pedicels, female flowers sessile in upper axils. Capsules orbicular, globose, 6 to 8-lobed.

Flowers : March-April; Fruits : October-December

Occurs frequently in marshy places and in grasslands. Singh 3059, 7964 (CDRI).

G. multiloculare Voigt. Hort. Suburb. Calc. 152.1845; FBI 5:307; FUGP 2:194; P.C. Kanjilal, loc. cit. 617.1966.

Evergreen shrubs with flexuous, compressed branches. Leaves 3-5 x 1-2.5 cm, oblong-lanceolate or obovate, obtuse or subacute, glaucous, purplish beneath. Flowers pale-green; male flowers on stout pedicels; female flowers on stout pedicels. Capsules orbicular. Seeds red with an orange aril.

Flowers : March-May; Fruits : October-December

Occurs occasionally in marshy places and in grasslands. Singh 4304, 8497 (CDRI).

Securinega Juss. *nom. cons.*

1. Plants armed .... **S. leucopyrus**
1. Plants unarmed .... **S. virosa**

S. leucopyrus (Willd.) Muell.-Arg. in DC. Prodr. 15(2):451.1866; Santapau in RBSI 16(1):275.1953; Maheshwari, Fl. Del. 318.1963; Airy Shaw in Kew Bull. 26(2):340.1972; P.C. Kanjilal, For. Fl. UP. 3:611.1966. *Flueggea leucopyrus* Willd. Sp. Pl. 4:757. 1805.

I have included this taxon on the authority of Duthie loc. cit. who reports its occurrence from Kheri forests. Hence, a brief description is reproduced below.

Large, deciduous shrubs. Branchlets stout, spine like, bearing flowers and leaves. Leaves elliptic to obovate, emarginate, glaucous. Male flowers greenish-yellow, slender, pedicellate. Fruits white, globose.

S. virosa (Roxb. ex Willd.) Baill. Adansonia 6:334.1865-66; Santapau in RBSI 16(1): 275. 1953; Maheshwari, Fl. Del. 318; P.C. Kanjilal, loc. cit. 610. 1966; Airy Shaw loc. cit. 340.1972; Saldanha & Nicolson Fl. Hass. Distt. 384. 1976.
Phyllanthus virosus Roxb. ex Willd. Sp. Pl. 4:578.1805.
Flueggea microcarpa Bl. Bijdr. 580.1826; FBI 5:328; FUGP 2:196.

Much branched, unarmed shrubs. Leaves 3-6 x 1.5-2 cm, obovate or elliptic, base cuneate, rounded at the apex, glaucous beneath. Flowers minute, greenish-yellow, in axillary clusters; male numerous on filiform pedicels; female 1-5 or more. Fruits fleshy, globose.

Flowers and Fruits : March-September

Occurs occasionally in grasslands in open areas. Singh 2906, 4221, 7318 (CDRI).

Phyllanthus Linn. emend. Gaertn.

1. Capsules verrucose .... **P. urinaria**
1. Capsules smooth:
    2. Seeds tubercled .... **P. virgatus**
    2. Seeds with parallel ribs on the back .... **P. fraternus**

P. urinaria Linn. Sp. Pl. 982.1753; FBI 5:293; FUGP 2:200; P.C. Kanjilal, For. Fl. UP. 3:607; Airy Shaw in Kew Bull. 26(2): 325. 1972; Srivastava, FG. 298.1976.

Slender, erect, annual herbs, often tinged with purple. Leaves sessile, 0.9-1.5 x 0.4-0.5 cm, oblong or oblong-elliptic, distichously imbricate, mucronate, rounded. Flowers unisexual, axillary, in a row under the branchlets. Perianth lobes 5. Capsules globose, lobed, verrucose. Seeds brown, transversely ridged.

Flowers and Fruits : July-November

Occurs commonly on moist shady places in sal forests and other wastelands. Singh 4200, 7379 (CDRI).

P. virgatus Forst.f. Fl. Ins. Austr. Prodr. 65.1786; Backer & Bakh.f. Fl. Java 1:469.1963; Airy Shaw in Kew Bull. 26(2):325.1972; Saldanha & Nicolson, Fl. Hass. Distt. 347.1976.
P. simplex Retz. Obs. 5:29.1788; FBI 5:295; FUGP 2:202; P.C. Kanjilal, For. Fl. UP. 3:606.1966.

Erect or ascending, perennial herbs. Leaves sessile, 1.6-2.5 x 0.4-0.5 cm, linear-oblong, distichous, mucronate, glaucous. Flowers white, from an axillary clusters of minute bracts; female flowers solitary, long pedicellate. Perianth lobes 6. Stamens free. Capsules globose, long stalked, smooth. Seeds tubercled.

Flowers and Fruits : August-December

Occurs frequently in waste moist places and in sal forests. Singh 2975, 3054, 4205, 7430 (CDRI).

P. fraternus Webster in Contr. Gray. Herb. 176:53.1955 et in JAA 38:309.1957; Maheshwari, Fl. Del. 320.1963.
P. asperulatus Hutch. in Kew Bull. 1920:27-28.1920; Srivastava, FG. 298.1987.
P. niruri auct. non Linn. 1753; Hook.f. in FBI 5: 298.1887 pro parte; FUGP 2:201.

Erect, glabrous, annual herbs. Leaves subsessile, distichous, 0.6-0.9 x 0.4 cm, oblong-elliptic, rounded, subcordate; stipules lanceolate, acuminate. Flowers yellowish, axillary; male 1-3; female solitary, 5-merous. Stamens fused. Capsules depressed, globose, smooth. Seeds trigonous, pale brown, with parallel ribs.

Local name : Bhumi amla.
Flowers and Fruits : July-November

Occurs commonly as a weed in cultivated fields, forest nurseries and in lawns. The plant is used in jaundice. Singh 4387 (CDRI).

Kirganelia Juss.

K. reticulata (Poir.) Baill. Etud. Gen. Euph. 614.1858; Santapau in RBSI 16(1):277.1953; Maheshwari, Fl. Del. 320; P.C. Kanjilal, For. Fl. UP. 3:607.1966; Srivastava, FG.297.1976.
Phyllanthus reticulatus Poir. in Lamk. Encycl. 5:298.1804; FBI 5:288; FUGP 2:199.

Large, diffuse, glabrous shrubs; branchlets drooping. Leaves 1.5-3 x 0.7-1.5 cm, oblong or elliptic-orbicular, glabrous, obtuse or acute. Flowers axillary, green or purplish; male in fascicles; the female solitary. Fruits globose berry, smooth, dark purple when ripe. Seeds trigonous.

Flowers and Fruits : April-July

Occurs frequently in the open areas in wastelands and near water streams. Singh 2848, 7130, 7265 (CDRI).

Emblica Gaertn.

E. officinalis Gaertn. Fruct. 2:122.1790; Santapau in RBSI 16(1):277.1953; P.C. Kanjilal, For. Fl. UP. 3: 608.1966; Srivastava, FG. 292.1976.
Phyllanthus emblica Linn. Sp. Pl. 982.1753; FBI 5:289; FUGP 2:199; Webster in JAA 38:76.1957.

Small or moderate-sized, deciduous trees with greenish-grey bark. Leaves subsessile, simple, distichous, 1.5 x 0.2 cm, elliptic or linear-oblong, mucronate, pale green. Flowers greenish-yellow in axillary fascicles on lower leaves. Fruits globose, fleshy, pale-yellow.

Local name : Amala
Flowers : February-May; Fruits : November-December

Found occasionally along the edges of the sal forest. The fruits are used in ayurvedic medicine. Singh 3022, 7360 (CDRI).

Breynia J. & G. Forst. *nom. cons.*

B. rhamnoides (Retz.) Muell.-Arg. in DC. Prodr. 15:440.1866; FBI 5:330; FUGP 2:197.
Phyllanthus rhamnoides Retz. Obs. 5:30.1788.
Melanthesa rhamnoides (Retz.) Bl. Bijdr. 591.1826; Maheshwari, Fl. Del. 319.1963; P.C. Kanjilal, For. Fl. UP. 615.1966.

Large, glabrous, shrubs. Leaves distichous, 2-3 x 1-1.5 cm, elliptic or ovate-oblong, obtuse, glaucous beneath; stipules subulate. Flowers unisexual, greenish-yellow; male flowers usually in short pedicels with 3 stamens; female flowers campanulate. solitary. Fruits globose, smooth, fleshy, dull red or purple.

Flowers and Fruits : August-November

Occurs frequently in open areas in moist places. Singh 2961, 7642, 7770, 7825, 8435 (CDRI).

Drypetes Vahl

D. roxburghii (Wall.) Hurusawa in J. Fac. Sci. Univ. Tokyo, sect. III Bot. 6:337.1954; Airy Shaw in Kew Bull. 26(2):257.1972.
Putranjiva roxburghii Wall. Tent. Fl. Nep. 61.1826; FBI

5:336; FUGP 2:202; P.C. Kanjilal, For. Fl. UP. 3:621.1966; Srivastava, FG. 299.1976.

Moderate-sized, evergreen trees with grey-bark. Branches drooping. Leave alternate, 5-10 x 2.5-3 cm, obliquely elliptic-oblong or ovate, acute or shortly acuminate, obscurely serrulate, glabrous, coriaceous. Flowers unisexual, yellowish, male flowers subsessile, densely clustered; female flowers 1-3, long pedicellate. Drupes ellipsoid, tomentose.

Local name : Pinpinia
Flowers : March-May; Fruits : January-March

Found occasionally in Fort range of the Park in damp places. Singh 7237 (CDRI).

Antidesma Linn.

1. Leaves glabrous, acute or acuminate; stamens 2 ....**A. acidum**
1. Leaves pubescent, rounded at both ends; stamens 4-7....**A. ghaesembilla**

A. acidum Retz. Obs. Bot. 5: 30. 1789; Airy Shaw in Kew. Bull. 26: 352. 1972.
A. diandrum (Roxb.) Roth, Nov. Pl. Sp. 369.1821; FBI 5:361; FUGP 2:204; P.C. Kanjilal, For. Fl. UP. 3:612.1966.
Stilago diandra Roxb. Pl. Corom. 2.t.166.1802.

Small deciduous trees or shrubs with smooth grey bark. Leaves alternate, 6-9 x 3-3.5 cm, oblong-lanceolate or oblong-elliptic, acute or acuminate, base cuneate, glabrous. Flowers greenish-white, pedicellate in long slender racemes. Drupes globose, smooth.

Local name : Ban musari, Amlola.
Flowers and Fruits : May-November

Occurs occasionally along the edges of sal forest and in open situation. The ripe fruits are edible. Singh 2960, 4125, 7560 (CDRI).

A. ghaesembilla Gaertn. Fruct. 1:189.t.39.1788; FBI 5:375; FUGP 2:203; Airy Shaw in Kew Bull. 26(2):353.1972; Srivastava, FG. 288.1976.

Large shrubs or small deciduous trees; young branches rusty-tomentose. Leaves 5-11 x 3.5-6 cm, ovate or elliptic-suborbicular, rounded or obtuse at the apex, base subcordate, pubescent. Flowers greenish-yellow in densely tomentose, panicled racemes. Drupes subglobose, red when ripe.

Local name : Banmusariya
Flowers and Fruits : May-November

Occurs occasionally in grasslands. The ripe fruits are edible. Singh 2956, 7143, 8346 (CDRI).

94. ULMACEAE

Holoptelea Planch.

H. integrifolia (Roxb.) Planch. in Ann. Sc. Nat. (ser. 3) 10: 259. 1848; FBI 5:481; FUGP 2;217.
Ulmus integrifolia Roxb. Pl. Cor. 1:56, t. 78.1798.

Medium-sized, deciduous trees with smooth grey bark. Leaves 6-13 x 5-7 cm, elliptic or ovate, acuminate, entire, subcoriaceous. Flowers green in numerous fascicles on the leafless branches. Samaras broadly elliptic, notched at the apex, wings reticulately veined.

Local name : Chilbil
Flowers and Fruits : February-May

Planted in avenues along roadsides. The seeds are edible. Singh 7689 (CDRI).

95. CANNABACEAE

Cannabis Linn.

C. sativa Linn. Sp. Pl. 1027.1753; FBI 5:487; FUGP 2:220; Santapau in RBSI 16(1):295.1953.

Erect, branched dioecious annual herbs. Leaves long petiolate, 4 to 8-foliolate, lobes lanceolate, serrate. Flowers unisexual, greenish-white; males in axillary panicled cymes with 5 stamens. Female flowers crowded with leafy bract in axillary racemes. Nuts crustaceous.

Local name : Bhang, Ganja
Flowers and Fruits : October-December

Found commonly in wastelands and along roadsides. The plant is used in piles and as narcotic for hallucination. Singh 2811 (CDRI); G. Saran 26198 (LWG).

96. URTICACEAE

1. Leaves white pubescent beneath ....**Debregeasia**
1. Leaves not as above:
   2. Leaves ovate-orbicular; female perianth 3 to 5-partite ....**Pilea**
   2. Leaves ovate-lanceolate; female perianth tubular...............**Pouzolzia**

Debregeasia Gaud.

D. longifolia (Burm.f.) Wedd. in DC. Prodr. 16(1):235.1869; Backer & Bakh.f. Fl. Java 2:49.1965; P.C. Kanjilal, For. Fl. UP. 3:644.1966.

Urtica longifolia Burm.f. Fl. Ind. 297.1768.

Debregeasia velutina Gaud. Bot. Bomte Voy. t. 90:1844-46; FBI 5:590; FUGP 2:228.

I have included this taxon on the authority of P.C. Kanjilal, loc. cit. who reports its occurrence from Bankati range of North Kheri Forest division. hence, a brief description is reproduced below.

Large shrubs or small trees with dark brown bark. Leaves oblong-lanceolate, acuminate, serrate, white pubescent beneath. Flowers in heads, arranged in lateral dichotomous cymes. Achenes with fleshy perianth arranged in clusters.

Pilea Lindl.

P. peploides Hook. & Arn. Bot. Beech, Voy. 2:96.1832; FBI 5:554; FUGP 2:222.

Small, succulent, glabrous annual herbs. Leaves opposite, long petiolate, orbicular-ovate, rounded at apex, base cuneate, 3-nerved, crenate. Flowers minute, in dense, axillary short peduncled dichotomous cymes. Fruits ovoid, compressed achenes.

Flowers and Fruits : October-December

Found occasionally in moist shady places. Singh 7132, 8364 (CDRI); Inayat 22483, 35476 (DD).

Pouzolzia Gaud.

1. Male perianth 4-lobed, convex on the back .... **P. zeylanica**
1. Male perianth 5-lobed; concave on the top:
    2. Flowers in terminal bracteate spikes .... **P. pentandra**
    2. Flowers in axillary clusters .... **P. hirta**

P. zeylanica (Linn.) Benn. Pl. Jav. Rar. 67. 1838; Santapau in RBSI 16(1):255.1967.
Parietaria zeylanica Linn. Sp. Pl. 1052.1753.
Pouzolzia indica Gaud. Bot. Freyc. Voy. 503.1826; FBI 5:582; FUGP 2:226.

Erect or trailing perennial herbs. Leaves alternate and opposite, 2.5-5 x 1-1.5 cm, ovate-lanceolate, acute, entire, appressed hairy beneath. Flowers axillary strigose. Perianth 4, Stamens 4. Achenes broadly ovoid, shining.

Flowers and Fruits : September-January

Found frequently in moist wastelands and along river banks. Singh 2989, 3043, 4170 (CDRI); Inayat 22763 (CAL).

P. pentandra (Roxb.) Benn. Pl. Jav. Rar 64.t.14.1838; FBI 5:583; FUGP 2:227.

Urtica pentandra Roxb. Fl. Ind. 3:583.1832.
Gonostegia pentandra Miq. Ann. Mus. Bot. Lugd.-Bot. 4:302.1869.

Tall, erect, glabrous, perennial herbs. Leaves sessile or shortly petiolate, lower opposite, oblong-lanceolate, 4-7 cm long, acute or acuminate; upper floral leaves much smaller. Flowers clustered in the floral leaves forming terminal bracteate spikes. Male flowers pedicellate. Perianth 5-partite. Stamens 5. Perianth in fruits, 2 or 3, broad laterally winged.

Flowers and Fruits : September-January

Found occasionally in moist places along streams. Singh 7885 (CDRI).

P. hirta (Blume) Hassk. Cat. Hort. Borger 80.1844; FBI 5:586; Raizada, Suppl. FUGP 259.1976.
Urtica hirta Blume, Bijdr. 495.1825.
Gonostegia hirta Miq. Ann. Mus. Bot. Lud. Bat. 4:303.1869.

Slender, decumbent or suberect herbs. Leaves opposite, subsessile, 3.5-7.5 x 1-1.4 cm, ovate-lanceolate, entire, membranous, 3-nerved; stipules broad, shortly acuminate. Flowers unisexual in small axillary clusters. Male perianth 5-partite. Stamens 5.

Flowers and Fruits : September-December

Occurs rarely in Dudhwa range of the Park in moist shady places. Singh 7965 (CDRI).

97. MORACEAE

1. Flowers arranged in globose heads or inside fleshy receptacle:
    2. Flowers arranged on the outside of the receptacle ....**Artocarpus**
    2. Flowers arranged on the inside of the closed receptacle ....**Ficus**
1. Flowers arranged in spikes or heads:

3. Male flowers spicate; females in subcapitate or globose heads; perianth fleshy in fruits:
    4. Styles undivided .... **Broussonetia**
    4. Styles 2-fid .... **Morus**
3. Male flowers in heads or spikes; females solitary or 2-3; perianth not fleshy in fruits .... **Streblus**

Artocarpus J.& G. Forst. *non. cons.*

1. Fruits subglobose, smooth .... **A. lakoocha**
1. Fruits cylindric, tubercled .... **A. heterophyllus**

A. lakoocha Roxb. Fl. Ind. 3:524.1832; FBI 5:543; FUGP 2:234.

Large deciduous trees with reddish brown bark. Leaves 14-17 x 5.5-9 cm, elliptic-ovate, or oblong, scabrous, denticulate, acute or shortly acuminate. Flowers in axillary globose, shortly pedunculate heads. Male receptacle orange yellow. Fruits subglobose, nearly smooth, yellow when ripe.

Local name : Barhal
Flowers : March-June; Fruits : October-November

Planted in Gola Forest Rest House campus in South Kheri Forest division. The fruits are edible. Singh 7206 (CDRI).

A. heterophyllus Lamk. Encycl. 3:210.1780; Santapau in JBNHS 57:447,T.1.1960; Corner in Fl. Ceylon 1(2):113.1977.
A. integrifolia auct. (non Linn.f. 1781), FBI. 5: 541. 1888.
A. integra auct. (non Merr. 1917).

Moderate-sized, evergreen, glabrous trees. Leaves elliptic or obovate, 9-20 x 9-12 cm, coriaceous; stipules large caducous. Fruits tubercled, hanging on the old branches.

Local name : Kathal
Flowers and Fruits : March-July

Planted in the campus of the Dudhwa Forest Range Office for its edible fruits: Singh 8348 (CDRI).

Ficus Linn.

1. Leaves opposite ....**F. hispida**
1. Leaves alternate:
    2. Receptacles sessile:
        3. Leaves pubescent beneath ....**F. benghalensis**
        3. Leaves glabrous:
            4. Apex of leaves gradually acuminate ....**F. rumphii**
            4. Apex of leaves caudate-cuspidate ....**F. religiosa**
    2. Receptacles distinctly stalked:
        5. Stamen 1:
            6. Creeping or climbing shrubs; receptacle solitary axillary ....**F. heterophylla**
            6. Trees; receptacle in pairs or clusters:
                7. Receptacle in clusters on leafless branches; leaves scabrous, base unequal ....**F. semicordata**
                7. Receptacle axillary in pair, leaves glabrous, base rounded ....**F. virens var. virens**
        5. Stamens 2-3:
            8. Scandent creeping shrubs, rooting from the nodes ....**F. pumila**
            8. Trees:
                9. Receptacle 4-6 grooved, globose.............**F. oligodon**
                9. Receptacle smooth, globose or pyriform:
                    10. Leaves scabrous, crenate-serrate.......**F. palmata**
                    10. Leaves glabrous, entire..................**F. racemosa**

F. hispida Linn.f. Suppl. 442.1781; King in Ann. Roy. Bot. Gdns. Cal. 1:166.t.154.1887-88; Corner in Gdns. Bull. Sing. 21:89.1965; FBI 5:522; FUGP 2:244; P.C. Kanjilal,

For. Fl. UP. 3:666.1966; Srivastava, FG. 306.1976; Corner in Fl. Ceylon 1(2):154.1977.

Small trees with grey bark. Branches hispid-pubescent. Leaves opposite, long petiolate, 10-18 x 4.5-6.5 cm, ovate-oblong or subovate, crenulate, acuminate, scabrous above, hispid beneath. Receptacles clustered on the old stem and separate leafless branches, forming racemes, peduncled, obovoid, pubescent.

Local name : Kathgular
Receptacle : Most part of the year

Found commonly in open wastelands. Singh 2853, 4120 (CDRI); G. Saran 26117 (LWG).

F. benghalensis Linn. Sp. Pl. 1059.1753; FBI 5:499; FUGP 2:238; King, loc. cit. 1:18,t.13.1887-88; Corner in Gdns. Bull. Sing. 17:381.1960 et Gdns. Bull. Sing. 21:14.1965; P.C. Kanjilal, loc. cit. 657.1966; Srivastava, FG. 305.1976; Corner in Fl. Ceylon 1(2):136.1977.

Large, evergeen trees with grey bark and numerous aerial roots. Leaves alternate, petiolate, 15-20 x 8.5-10 cm, ovate or orbicular-ovate or elliptic, coriaceous, entire, obtuse. Receptacles axillary, sessile, paired, globular, silky pubescent, red.

Local name : Bargad
Receptacle : June-March

Found occasionally in forest areas in open situations. The plant is pollution tolerant. Singh 7356 (CDRI); G. Saran 26117 (LWG).

F. rumphii Blume, Bijdr. 437.1825; FBI 5:512; FUGP 2:241, Corner in Gdns. Bull. Singh. 17:37.1960; P.C. Kanjilal, loc. cit. 660.1966; Srivastava, FG. 307.1976.

Moderate-sized deciduous trees with yellowish-grey, smooth bark. Leaves alternate, petiole 6 cm long; blade 12 x

4 cm, ovate, glabrous, acuminate. Receptacles sessile, paired, axillary, globular, smooth, whitish.

Local name : Gajhar
Receptacle : April-July

Found occasionally in the areas. Singh 4321, 7220, 7668 (CDRI).

F. religiosa Linn. Sp. Pl. 1059.1753; FBI 5:513; FUGP 2:241; Corner in Gdns. Bull. Sing. 17:37.1960 et Gdns. Bull. Sing. 17:37.1960 et Gdns. Bull. Sing. 21:6.1965; P.C. Kanjilal, For. Fl. UP. 3: 661.1966; Srivastava, FG. 307.1976; Corner in Fl. Ceylon 1(2):127.1977.

Large, glabrous trees. Leaves alternate, petiole 9 cm long; blade 13-14 x 8-9 cm, ovate-round, coriaceous, apex caudate, margins repand. Receptacles sessile in pairs, smooth, depressed-globose; basal bracts distinct.

Local name : Pipal
Receptacle : April-September

Found frequently in the area. The plant is pollution tolerant. Singh 7204 (CDRI).

F. heterophylla Linn.f. Suppl. 442.1781; FBI 5:518; FUGP 2:243; Corner in Gdns. Bull. Sing. 21:73.1965; P.C. Kanjilal, For. Fl. UP. 3:664.1966; Srivastava, FG. 306.1976; Corner in Fl. Ceylon 1(2):150.1977.

Straggling, creeping or climbing shrubs; branchlets brown, rough-pubescent. Leaves polymorphic, 6-12 x 2-3.4 cm, ovate or orbicular-ovate to lanceolate, sometimes deeply lobed, scabrous on both surfaces; margins coarsely toothed or repand-dentate. Receptacles stalked, solitary, axillary, globose to pyriform; bracts minute, triangular.

Receptacle : October-May

Found commonly in moist shady and marshy places. This species is quite variable as regards the habit, leaf and receptacle. Singh 7258, 7328, 7667, 7954 (CDRI).

F. semicordata Buch.-Ham. ex J.E. Smith in Rees, Encycl. 14 no. 17.1810; Corner in Gdns. Bull. Sing. 17:449.1960.
F. cunia Buch.-Ham. ex Roxb. Fl. Ind. 3:561.1832; FBI 5:523; FUGP 2:246; P.C. Kanjilal, For. Fl. UP. 3:665.1966; Srivastava, FG. 306.1976.

Small or medium-sized trees with greyish-brown rough bark. Leaves 12-19 x 5-9 cm, oblong-lanceolate or elliptic, serrate or entire, acute or shortly acuminate, base unequal sided, scabrous. Receptacles globose or pyriform, in pairs or clusters, on drooping scaly branches, hispid.

Local name : Khurkhur
Receptacle : September-November

Found frequently in moist area in sal forests. The ripe fruits are edible. Singh 4177, 7199, 7239 (CDRI).

F. virens Ait. Hort. Kew. 3:451.1789; Corner in Gdns. Bull. Sing. 21:9.1965; et Fl. Ceylon 1(2):128.1977.
F. infectoria Roxb. Fl. Ind. 3:351.1832; FBI 5:506; FUGP 2:243.
F. lacor auct. non Vera
F. infectoria var. *lambertiana* King 1887; FBI 5:516.

Var. **virens**

Large or medium-sized, spreading, deciduous trees. Leaves 10-13 x 5-5.6 cm, ovate or ovate-oblong, abruptly acuminate, glabrous, undulate, rounded at the base. Receptacles in axillary pairs, shortly peduncled, globose.

Local name : Pakadi
Receptacle : September-December

Planted along roadsides and near railway stations. The plant is pollution tolerant. Singh 7212 (CDRI).

F. pumila Linn. Sp. Pl. 1060.1753; P.C. Kanjilal, For. Fl. UP. 3:668.1966; Corner in Gdns. Bull. Sing. 21: 52.1965 et Fl. Ceylon 1(2):148.1977 (Fig.15).

Scandent or creeping shrubs, rooting at the nodes. Leaves dimorphic, variable in size, 3-7 x 2.3-6 cm, ovate-elliptic, subcute, base subcordate and slightly oblique. Receptacle solitary axillary, 6-7 x 2-3 cm, pyriform, hirsute, purple when ripe.

Receptacle : September-November

Planted in Dudhwa Forest Rest House campus clinging on mango trees. Singh 790 (CDRI).

F. oligodon Miq. in Ann. Mus. Ludg. Bot. 3:234, 297.1867; Raizada, Suppl. FUGP 261.1976.
F. pomifera Wall. ex King in Ann. Roy. Bot. Gdns. Cal. 1:171.t.215.1887; FBI 5:535.

I have included this taxon on the authority of M.B. Raizada loc. cit. who reports its occurrence from swampy areas in Gola range of South Kheri Forest division.

Medium-sized trees with white bark. Leaves 12.5-22 x 7.5-13 cm, coarsely, base rounded or subcuneate; petiole 3.5-8.6 cm long. Receptacles on leafless branchlets, long peduncled, globose, 4-6 grooved, pubescent, red when ripe.

F. palmata Forsk. Fl. Aegypt.-Arab. 179.1775; FBI 5:530; FUGP 2:247; P.C. Kanjilal, For. Fl. UP. 3:666.1966; Maheshwari, Fl. Del. 327.1963.

Erect, woody shrubs or small trees with dull grey bark. Young parts tomentose. Leaves 9-11 x 6.5-8 cm; orbicular-ovate, acute, dentate or serrate, scabrous above. Receptacles axillary, solitary, 2.5 x 1.2 cm; subglobose or pyriform.

Local name : **Anjiri**
Receptacle : June-October

Found commonly in open areas in waste places etc. The ripe fruits are edible. Singh 4124, 7691 (CDRI); G. Saran 26424, 26675 (LWG).

F. racemosa Linn. Sp. Pl. 1060.1753; Corner in Gdns. Bull. Sing. 21:34.1965; Srivastava, FG. 307.1976.
F. glomerata Roxb. Pl. Cor. t.123.1798; King, Ann. Roy. Bot. Gdns. Cal. 1:173.t.218A. 1887-1888; FBI 5:535; FUGP 2:48.

Large or medium-sized deciduous trees with smooth pale or reddish-brown bark. Leaves 12-17 x 3.5-6 cm; ovate-oblong to elliptic-lanceolate, glabrous, acute. Receptacles clustered, on short leafless branchlets, peduncled, obovoid.

Local name : Gular
Receptacle : April-July

Found occasionally in the forest areas near nallahs etc. The fruits are used as vegetable. Singh 7200, 7257 (CDRI).

Broussonetia L'Herit ex Vent.

B. papyrifera Vent. Tabl. Reg. Veg. 3:547.1794; FBI 5:490; FUGP 2:250; P.C. Kanjilal, For. Fl. UP. 3: 650.1966.

Small or medium-sized trees with smooth, grey bark. Leaves 10-20 x 6.5-10 cm; ovate, acuminate, base oblique, crenate, scabrous above, pubescent beneath. Flowers dioecious; male greenish yellow in cylindric, pubescent spikes; female flowers in compact heads. Fruits fleshy, red when ripe.

Flowers : March-April; Fruits : September-October

Planted in Mailani forest nursery in South Kheri Forest division. Singh 2893 (CDRI).

Morus Linn.

M. alba Linn. Sp. Pl. 986.1753; Maheshwari, Fl. Del. 328.1963; P.C. Kanjilal, For. Fl. UP. 3:649.1966.

Small or medium-sized, deciduous trees. Leaves alternate, long petiolate, blade 6-7 x 4.5-5 cm; ovate-cordate, crenate-serrate, acute. Flowers unisexual. Male spike, elongate, catkin. Female spikes short ovoid. Fruits dark purple.

Local name : Shahtut
Flowers and Fruits : February-June

Planted in Forest Rest houses. The fruits are edible. Singh 2930, 7587 (CDRI).

Streblus Lour. emend. Corner

S. asper Lour. Fl. Cochin. 2:615.1790; FBI 4:489; FUGP 2:232; P.C. Kanjilal, For. Fl. UP. 3:646.1966; Corner in Gdns. Bull. Sing. 19:228.1962; Srivastava, FG. 308.1976; Corner in Fl. Ceylon 1(2):154.1977.

Small, evergreen trees with smooth grey bark. Leaves alternate, 3-5 x 2-2.5 cm; rhomboid-elliptic or obovate, acute, rough, margins subcrenate. Flowers unisexual; male flowers in globose heads, yellowish-green; female flowers solitary or 2-4 together.

Local name : Sihora
Flowers : January-March; Fruits : May-July

Found frequently on the edge of the sal forests. Singh 2857, 7266 (CDRI); G. Saran 26684, 26105 (LWG); Hira Lal 6848 (LWG).

98. SALICACEAE

Salix Linn.

S. tetrasperma Roxb. Cor. Pl. 1:66. f.97.1795; FBI 5:627; FUGP 2:253; P.C. Kanjilal, For. Fl. UP. 3:670.1966; Maheshwari, Fl. Del. 330.1963.

Medium-sized, deciduous trees with grey or brown bark. Leaves 5-8 x 1-2 cm; ovate-lanceolate or lanceolate, serrulate, acuminate, glaucous beneath. Flowers unisexual. Male flowers yellow, sessile in 10-12 cm long catkins; stamens 5-10. Female flowers pedicellate in catkins.

Local name : Sukulbent
Male flowers : October-March

Found frequently along river beds and in moist shady places. Singh 2880, 4217, 7986 (CDRI).

99. CERATOPHYLLACEAE

Ceratophyllum Linn.

C. demersum Linn. Sp. Pl. 992.1753; FBI 5:636; FUGP 2:555; Srivastava, FG. 309.1976.

Slender, submerged, aquatic herbs. Leaves whorled, dissected into filiform, serrate, segments. Flowers unisexual, solitary axillary; male flowers with 10-20 extrose stamens; female flowers with filiform style. Nutlets ovoid or ellipsoid, with a basal spines on either side and apical persistent base of the style.

Flowers and Fruits : September-November

Occurs commonly in lakes, ponds and ditches. Singh 8559 (CDRI); Inayat 22945 (DD).

18. SYSTEMATIC TREATMENT OF MONOCOTYLEDONS

100. HYDROCHARITACEAE

1. Leaves whorled, cauline ....Hydrilla
1. Leaves radical:
    2. Leaves ovate-cordate; spath winged ....Ottelia
    2. Leaves linear, ribbon like; spath not winged ....Vallisneria

Hydrilla Rich.

H. verticillata (Linn.f.) Royle, Ill. 376.1839; FBI 5:659; FUGP 3: 173; Subramanyam, Aq. Ang. 55; Srivastava, FG.

310. 1976.
Serpicula verticillata Linn.f. Suppl. 416.1781.

Slender, leafy, submerged, green aquatic herbs. Leaves 3-5 in whorls, linear-oblong, 1-5 x 0.2 cm. Flowers minute, solitary, unisexual; male flowers pedicellate; female flowers sessile.

Flowers : September-November

Found frequently in tanks and slowly running water of ponds, and ditches. Singh 8354, 8419 (CDRI).

Ottelia Pers.

O. alismoides (Linn.) Pers. Syn. Pl. 1:400.1805; FBI 5:662; FUGP 2:265; Subramanyam, Aq. Ang. 61.1962.
Stratiotes alismoides Linn. Sp. Pl. 335.1753.

Succulent, submerged, aquatic herbs. Leaves radical, long petiolate, blade ovate-cordate, 12 x 8 cm, 7 to 11-nerved, membranous. Flowers white, solitary, bisexual, hypanthium enclosed in a green winged spath. Petals 3. Fruits ellipsoid.

Flowers and Fruits : September-December

Occurs commonly in tanks, ponds and slow running water. Singh 7556, 7708 (CDRI).

Vallisneria Linn.

V. spiralis Linn. Sp. Pl. 1015.1753; FBI 5:660; FUGP 2:263; Subramanyam, Aq. Ang. 57.1962.

Submerged, tufted stemless herbs. Leaves radical, linear, ribbon-like, very long, obtuse or acute, entire or serrulate at the tips, 5 to 9-nerved. Flowers unisexual; male flowers clustered on a short peduncle; female flowers solitary on long spirally coiled, filiform peduncles.

Flowers and Fruits : October-March

Found frequently at the bottom of ponds and ditches. Singh 8564 (CDRI).

101. ORCHIDACEAE

1. Plants epiphytic ....Vanda

1. Plants terrestrial:

    2. Inflorescence terminal on same shoot as leaves:

        3. Leaves linear, papery on the hyaline sheath................Zeuxine

        3. Leaves ovate-lanceolate, with non hyaline sheath......Habenaria

    2. Inflorescence lateral, on separate branches from leaves.......Eulophia

Vanda R.Br.

V. tessellata (Roxb.) Hook. ex G. Don in Loud. Hort. Brit. 372.1830; Santapau & Kapadia Orch. Bombay 220.1966.

Epidendrum tessellatum Roxb. Pl. Cor. t. 42.1795.

Vanda roxburghii R.Br. in Bot. Reg. t. 506.1820; FBI 6:52; FUGP 2:290.

Epiphytes with leafy stems. Leaves 17 x 1.5 cm, thick, coriaceous, flat, recurved, obtusely keeled. Flowers 6-10, bluish, arranged in racemes. Sepals and petals brown, tessellate on the inside. Lips with a conical, pointed spur, bluish dotted with purple, side lobes ovate-acute, mid lobe white with 2-fleshy longitudinal ridges.

Local name : Banda
Flowers : September-October

Occurs rarely in South Kheri Forest division climbing on the branches of *Madhuca longifolia* and *Shorea robusta*. Singh 7501 (CDRI).

Zeuxine Lindl. Corr. Roep.

Z. strateumatica (Linn.) Schltr. Orch. Detschr. N. Guin. 77.1911; Holttum in Fl. Males. 1:131; Santapau & Kapadia in JBNHS 59(1):161.1962; Srivastava, FG. 311.1976.

Orchis strateumatica Linn. Sp. Pl. 943.1753.

Zeuxine sulcata Lindl. Gen. & Sp. Orch. 485.1810; FBI 6:106; FUGP 2:298.

Terrestrial, glabrous herbs, 5-12 cm tall. Leaves *ca* 2-4.5 cm long, linear, sessile on the hyaline truncate sheath, acuminate. Flowers cream-coloured, bracteate, sessile in terminal spikes; bracts ovate-lanceolate, acuminate. Sepals subequal, dorsal adnate with petals forming a hood. Capsules ellipsoid.

Flowers and Fruits : February-March

Occurs rarely on sandy soil in moist places in South Kheri Forest division. Singh 7647 (CDRI).

Habenaria Willd.

H. commelinifolia Wall. ex Lindl. Gen. & Sp. Orch. 325.1835; FBI 6:143; FUGP 2:301.

Terrestrial, leafy herbs. Stems loosely sheathed at the base. Leaves 10-12 x 1 cm, oblong-lanceolate, acute, sheathing at the base. Flowers white in spikes. Sepals scaberulous, dorsal sepal orbicular with short curved beak, lateral pair spreading, hatched shaped, beaked with many arched nerves. Petals obliquely ovate-oblong, obtuse. Anthers tube long, straight, erect.

Flowers : September

Found frequently in sal forests in moist shady places. Singh 7753, 7894 (CDRI).

Eulophia R.Br. *nom. cons.*

1. Flowers and leaves appearing simultaneously; column not produced into a foot:
    2. Leaves grass-like .... **E. graminea**
    2. Leaves elliptic-lanceolate .... **E. explanata**
1. Flowers appearing before leaves; column produced into a distinct foot:
    3. Lateral sepals inserted on the spur of the lip:
        4. Flowers whitish-pink .... **E. hormusjii**
        4. Flowers green-purple .... **E. nuda**
    3. Lateral sepals inserted on the foot of the column, free from the spur of lip:
        5. Flowers purplish-brown .... **E. bicarinata**
        5. Flowers green-yellow .... **E. flava**

E. graminea Lindl. Gen. & Sp. Orch. 182.1833; FBI 6:2; C.L. Malhotra in Bull. Bot. Surv. India 8:77.1965; Raizada, Suppl. FUGP 263.1976.

Terrestrial herbs with fibrous roots. Pseudobulbs thick, obpyriform, epigeal, green, 8-10 cm long, crowned at the top with older leaf sheaths. Leaves grass-like, narrow. Scape lateral, branched, 50-60 cm long, glabrous. Flowers pedicellate, bracteate in racemes. Sepals dull-green, elliptic-lanceolate, acute with reddish-brown veins. Lips 3-lobed, white, cylindrically spurred, with slightly swollen tips. Fruits elliptic-oblong, 2 cm long, yellowish-green, drooping.

Flowers and Fruits : April-May

Occurs rarely in grasslands near Chandan Chowki area of the Park. C.L. Malhotra 31544 (DD).

E. explanata Lindl. Gen. & Sp. Orch. 180.1933; FBI 6:3; FUGP 2:280; M.A. Rau in Bull. Bot. Surv. India Suppl. No.2 10:76.1969.

Terrestrial herbs. Pseudobulbs ovoid, annulated. Leaves at maturity 25 x 7.5 cm, elliptic-lanceolate, acuminate, 7-nerved. Flowers yellow, streaked with reddish-purple, arranged in lax racemes on the scape. Sepals spathulate, acute. Petals oblong, rounded. Lips 3-lobed, adnate by its base to the lateral sepals; basal lobes rounded, erect, forming a short conical sac at the base.

Flowers and Fruits : April-May

Occurs occasionally in the sal forests of the Park. Inayat 23860, 23860(a) (DD).

E. hormusjii Duthie in Ann. R. Bot. Gdn. Calc. 9(2):125; FUGP 2:281; M.A. Rau loc. cit. 76.1969.

Terrestrial herbs with depressed oval tubers. Leaves 2, elliptic, acuminate, developing after flowers. Scape 20-28 cm long, arising from tuber. Flowers whitish-pink, bracteate, arranged in lax racemes. Sepals and petals subequal in length, oblanceolate, acuminate. Lips 3-lobed. Spur cylindric, curved, grooved.

Flowers : April-May

Occurs occasionally in sal forests of the Park. Inayat 22797 (DD).

E. nuda Lindl. in Wall. Cat. 7371.1828 et Gen. & Sp. Orch. 180.1833; FBI 6:5; FUGP 2:282; Santapau & Kapadia Orch. Bombay 115.1966.

Terrestrial herbs with large, spherical, smooth tubers. Leaves 2-3 arrising from the side of the tuber, forming a pseudostem by sheath, 25-30 x 14 cm, linear or elliptic-lanceolate, acute. Flowers bracteate, green-purple, arranged in lax racemes on the scape. Sepals linear-oblong, acute, lateral sepals inserted on the spur of the lip. Lips 3-lobed, longer than the sepals, side lobes suborbicular, erect. Capsules fusiform, ribbed.

Flowers and Fruits : May-September

Found frequently in swampy areas in sal forests. Singh 7544 (CDRI); Inayat 24143, 24143(a), 24414, 24414(a) (DD).

E. bicarinata Hook.f. in FBI 6:6.1890; FUGP 2:282; M.A. Rau in Bull. Bot. Surv. India Supp. no.2. 10:76.1969.

Terrestrial herbs with compressed, irregularly lobed tubers. Leaves 18-30 cm long, linear, acuminate, narrowed into a long petiole. Flowers yellowish-green, tinged with purple-brown, arranged in racemes on long scape. Lateral sepals inserted on the foot of the column, free from the spur of the lip. Lips 3-lobed, spurred.

Flowers : March-April

Occurs occasionally in the grasslands in Dudhwa range of the Park. Inayat 22798, 22798(a) (DD).

E. flava Hook.f. in FBI 6:7.1890; FUGP 2:283; M.A. Rau in Bull. Bot. Surv. India 2:10:76.1969.

Tall robust herbs with tuberous rhizomes. Leaves 3, forming a pseudostem by their sheathing bases, when mature about 1 m long. Flowers lemon-yellow, large, drooping, arranged in long, stout, unbranched scape. Sepals 2.5-3 cm long, oblong-lanceolate. Lips 3-lobed, spurred, lateral lobes large, rounded, mid lobe oblong, yellow. Disc with two small pyriform basal calli and three crenate lamellae.

Flowers : June-July

Found rarely in the forest area of the Park. Inayat 24242(a) (DD).

102. ZINGIBERACEAE

1. Ovaries 1-celled; placentation parietal..**Globba**

1. Ovaries 3-celled; placentation axile:

2. Lateral staminodes broad and well developed ....**Curcuma**

2. Lateral staminode absent or very small:

3. Stigmas subglobose ....**Zingiber**

3. Stigmas with a semilunar pit, ciliate ....**Costus**

Globba Linn.

G. sessiliflora Sims in Bot. Mag. t. 1428; FBI 6:203.1892; Raizada, Suppl. FUGP 271.1976.

Erect perennial herbs, with creeping rhizomes and fleshy fibrous roots. Bulbils present. Leaves subsessile, 10-12 x 2-2.5 cm, elliptic-oblong or lanceolate, base acute, apex acuminate-caudate, glabrous. Flowers yellow arranged in terminal erect panicles. Calyx 1-1.4 cm long. Lips yellow, oblanceolate-spathulate. Capsules globose, verrucose.

Flowers and Fruits : August-November

Occurs occasionally on slopes in damp shady places in sal forests. Singh 7783, 8543 (CDRI).

Note: This species resembles closely with *Globba orixensis* from which it can be easily recognised in having yellow coloured lip and presence of bulbils.

Curcuma Roxb. *nom. cons.* prop.

1. Spikes distinct and develop before the leaves appear ....**C. angustifolia**

1. Spikes rising from the centre of the tufts of leaves.................**C. amada**

C. angustifolia Roxb. in Asiat. Research 11:338. t.3.1810; FBI 6:210; FUGP 2:308 (Plate XXII).

Stemless herbs with tuberous rootstocks. Leaves 30-40 x 5-6 cm, lanceolate, acute. Floral spikes 18-26 cm long, crowded by several empty pink bracts. Flowers yellow, 3-4 together in the axil of each bracts. Calyx 3-lobed. Lateral staminodes oblong, united to the filaments.

Local name : Tikhur
Flowers : April-June

Found commonly in grasslands in open areas. The corm are used for making Tikhur a starch product. Singh 7296, 8527 (CDRI).

C. amada Roxb. in Asiat. Res. 11:341.1810; FBI 6:213.

Herbs with large, sessile, cylindrical, tuberous rootstocks. Leaves long-petiolate in tufts; blade 19-21 x 4-6 cm, oblong-lanceolate, acuminate, narrowed to the base, glabrous. Flowers pale-yellow, bracteate, arranged in spikes.

Local name : Amahaldi
Flowers : July-August

Occurs rarely in sal forests in moist shady places of the Park. The rhizomes are used for sprain. Singh 4278, 8512 (CDRI).

Note: B.L. Burtt & R.M. Smith in Taxon 21:709.1972 have suggested conservation of the generic name *Curcuma* Roxb. against *Curcuma* Linn.

Zingiber Boehm. *nom. cons.*

1. Spikes laterally produced from the rootstock ....**Z. roseum**
1. Spikes terminally produced by leafy stem ....**Z. capitatum**

Z. roseum (Roxb.) Rosc. in TLS 8:348.1807; FBI 6:244; Raizada, Suppl. FUGP 274.1976.
Ammomum roseum Roxb. Pl. Cor. f.126.1800.

Erect, perennial herbs. Leaves 30-35 x 5-6 cm, ovate-oblong, acuminate, entire, pubescent beneath. Spikes lateral dense, oblong on short peduncles. Bracts pale-red. Corolla segments pale-red, 2.5 cm, lips oblong-cuneate, whitish, unspotted. Ovaries villous; styles 5-7 cm long. Fruits ovoid-ellipsoid, 3-gonous, red.

Flowers and Fruits : August-November

Found frequently in sal forests in moist shady places. Singh 7779, 7958 (CDRI).

Z. capitatum Roxb. in Asiat. Res. 11:348.1810; FBI 6:248; FUGP 2:309.

Slender leafy herbs with oval, tuberous rootstocks. Leaves 30-40 x 1-2 cm, linear, acuminate, ascending, pubescent beneath. Spikes terminal at the end of leafy stems, dense. Bracts ovate, closely imbricate, unspotted. Capsules bright red.

Flowers and Fruits : August-November

Occurs commonly in open grasslands and on the edge of the sal forests. Singh 2993, 3027, 4285, 7526, 7757, 7896 (CDRI).

Costus Linn.

C. speciosus (Koenig) Smith in TLS 1:249.1800; FBI 6:250; FUGP 2:310; P.C. Kanjilal, For. Fl. UP. 3:672.1966; Srivastava, FG. 312.1976 (Plate XXI).
Banksea speciosa Koenig in Retz. Obs. 3:75.1783.

Tall, erect, perennial herbs with creeping tuberous rootstocks. Leaves subsessile, arranged spirally, 18-30 x 4-6 cm, elliptic or obovate, acute or acuminate, glabrous above, white pubescent beneath. Flowers bisexual, in terminal dense spikes. Bracts scarlet, 2 cm long. Petals elliptic, bluish-white, acute.

Local name : Kewa
Flowers : July-September

Occurs frequently in the moist shady places. The flowers and leaves are cooked as vegetable. Singh 4252, 7517, 8450 (CDRI).

103. AMARYLLIDACEAE

Curculigo Gaertn.

C. **orchioides** Gaertn. Fruct. 1:63, t.16.1788; FBI 6:279; FUGP 2:319; Srivastava, FG. 313.1976.

Slender herbs with tuberous rootstocks. Leaves radical, subsessile or petiolate, 17-24 x 1-3 cm, oblong-lanceolate, entire, acuminate. Flowers yellow, 1-2 on a short scape. Perianth segments yellow, hairy. Ovary 6 to 8-ovuled. Style filiform. Stigmas 3. Capsules sessile in the axil of bracts.

Local name : Kali musali
Flowers and Fruits : April-September

Found commonly in moist shady places in sal forests. The roots are used as tonic. Singh 2966, 3018, 4114 (CDRI).

104. DIOSCOREACEAE

Dioscorea Linn.

1. Stems twining to left:
    2. Leaves simple, cordate, 7 to 9-nerved ....**D. bulbifera**
    2. Leaves compound, 3 to 5-foliolate, less than 7-nerved ....**D. pentaphylla**
1. Stems twining to right:
    3. Leaves base hastate ....**D. belophylla**
    3. Leaves base ovate-cordate ....**D. glabra**

D. bulbifera Linn. Sp. Pl. 1033.1753; FUGP 2:324; Prain and Burkill in Ann. Roy. Gdns. Cal. 14(1):188.t.77.1936; Burkill in Fl. Males. ser.1.4(3):311.1951; P.C. Kanjilal, For. Fl. UP. 3:681.1966; Srivastava, FG. 316.1976 (Fig. 16).

D. sativa Thunb. Fl. Jap. 151.1784 (non Linn. 1753).

Large, glabrous, twining herbs. Bulbils tubercled, in the leave axil. Leaves simple, alternate, ovate-cordate, acute or acuminate, entire, 9-nerved, petiole 5-7 cm long; blade 8-15 x 6-10 cm. Flowers unisexual. Male flowers in slender axillary panicled spikes. Female flowers 3-4 together in spikes. Capsules quadrately oblong.

Local name : Ratalu,
Flowers and Fruits : September-December

Occurs commonly in open places and along the edges of the forest, twining on small trees. The tubers are edible. Singh 2972, 2982, 7371 (CDRI).

D. pentaphylla Linn. Sp. Pl. 1032. 1753; FBI 6:281; FUGP 2. 323; Prain & Burkill, loc.cit. 160. 1936 et Burkill in Fl. Males. 4(1): 315. 1951; P.C. Kanjilal, For. Fl. UP. 3: 680. 1966; Srivastava, FG. 317 1976.

Slender twining herbs. Leaves 3 to 5-foliolate; leaflets digitate, ovate-elliptic, 7-11 x 2.5-4 cm, base oblique, acuminate. Bulbils present. Male spikes 1 to 2-nate in long axillary or terminal panicles. Female spikes long, 2-nate. Capsules oblong, deflexed.

Flowers and Fruits : August-March

Occurs occasionally in the sal forests. The tubers are edible. Singh 4326, 7381 (CDRI).

D. belophylla Voigt, Hort. Suburb. Calc. 653.1845; FUGP 2:325; Prain & Burkill in Ann. Roy. Bot. Gdns. Cal. 14(2): 348. t.127. 1938; P.C. Kanjilal, For. Fl. UP. 3: 683. 1966; Saldanha & Nicolson Fl. Hass. Distt. 806.1976.

Slender climbing herbs. Leaves opposite and alternate, petiolate, blade 10 x 7.5 cm, cordate, 7-nerved, base hastate, acuminate. Male spikes axillary, 1 to 3-nate. Female spikes dropping from node of leafy branches. Capsules 1.5 x 2 cm, pendulous. Seeds winged.

Flowers and Fruits : September-January

Found frequently in the forests climbing on small trees. Singh 3046, 4329, 7778, 7887, 8481 (CDRI).

D. glabra Roxb. Fl. Ind. 3:803.1832; FBI 6:294 pro parte; FUGP 2:325; Burkill in Fl. Males. loc. cit. 311.1951; P.C. Kanjilal, loc. cit. 684.1966.

Slender climbing herbs. Leaves opposite and alternate, petiolate, blade 8-9 x 4.5 cm, ovate-cordate or elliptic, base ovate, 7-nerved, acuminate. Male spikes axillary in pairs; stamens 6. Female spikes elongate. Capsules 3-angled, transversely elliptic.

Flowers and Fruits : September-January

Found frequently in the forest areas. Singh 3089, 7440, 7522 (CDRI); Inayat 22827(a) (DD); G. Saran 26487 (LWG).

105. LILIACEAE

1. Plants woody; cladodes present ....**Asparagus**
1. Plants herbaceous; cladode absent:
    2. Flowers orange or scarlet; leaves ovate-lanceolate, tendrillar. ....**Gloriosa**
    2. Flowers white; leaves terete, hollow ....**Asphodelus**

Asparagus Linn.

1. Cladodes 2-6 together, linear, triquetrous ....**A. racemosus**
1. Cladodes 6-20 together, filiform ....**A. adscendens**

A. racemosus Willd. Sp. Pl. 2:152.1799; FBI 6:316; FUGP 2:327; P.C. Kanjilal, For. Fl. UP. 3:686.1966; Srivastava, FG. 318. 1976.

Woody, scandent or straggling much branched spinous shrubs with tuberous roots. Spines recurved. Cladodes in tufts of 2-6, falcate, 2-2.5 cm long, triquetrous, acuminate. Flowers white in 2 cm long solitary or fascicled racemes. Berries globose.

Local name : Satawar

Flowers : August-September; Fruits : November-January

Found frequently in the sal forests, straggling on small trees. The roots are used in ayurvedic medicine and tonic. Singh 3013, 7752 (CDRI); G. Saran 26240 (LWG).

A. adscendens Roxb. Fl. Ind. 2:153.1824; FBI 6:317; FUGP 2:328; P.C. Kanjilal, loc. cit. 687.1966.

Suberect, prickly shrubs with tuberous roots. Branches ash-grey, striate; spines 0.5 cm long, straight. Cladodes in dense tufts of 6-20, filiform, terete. Flowers white in 5 cm long racemes. Perianth segments spreading. Stamens shorter than the perianth.

Local Name : Satawari,
Flowers and Fruits : August-December

Found occasionally in sal forests in moist shady places. Singh 4277, 7581, 7984 (CDRI).

Gloriosa Linn.

G. superba Linn. Sp. Pl. 105.1753; FBI 6:358; FUGP 2:330; P.C. Kanjilal, loc. cit. 688.1966; Srivastava, FG. 318.1976.

Glabrous climbing herbs with tuberous roots. Leaves alternate, sessile, 10.5-15 x 2-2.5 cm, ovate-lanceolate, acuminate, tips spirally twisted in tendrils. Flowers solitary axillary on 8-9 cm long pedicels. Perianth 5-8 cm long, free, undulate, red and orange. Stamens spreading with dorsifixed anthers. Capsules green, smooth, linear-oblong.

Local name : Kalihari
Flowers and Fruits : August-November

Occurs occasionally in open situations amongst bushes. The tubers are used in rheumatism and gout. Singh 3006, 7375 (CDRI); C.L. Malhotra 22866 (LWG).

Asphodelus Linn.

A. tenuifolius Cav. in An. Cienc. Nat. 3:46.t.27.f.2.1801; FBI 6:358; FUGP 2:333; Srivastava, FG. 318.1976.

Stemless annual herbs. Leaves 15-30 x 3-5 cm, terete, hollow. Scapes many, 30-60 cm long, branched above. Flowers white, bracteate, in lax racemes. Perianth lobes oblong. Stamens black, acutely trigonous. Capsules globose. Seeds rugose, black.

Local name : Banpiyaji
Flowers and Fruits : January-June

Found commonly in cultivated fields and wastelands. Singh 7622 (CDRI).

106. SMILACACEAE

Smilax Roxb.

1. Umbels 2-3, alternate on a common peduncle:

   2. Petiole sheath auricled below the middle ....**S. ocreata**

   2. Petiole sheath not auricled, inconspicuous ....**S. ovalifolia**

1. Umbels many, ternate or whorled on an axillary peduncle......**S. perfoliata**

S. ocreata A.DC. Monogr. Phaner. 1:93.1878; T. Koyama in Adv. Front. Pl. Sci. 4:57.1963; M.A. Rau in Bull. Bot. Surv. India Suppl. No.2.1(10):80.1969.
S. roxburghiana Wall. ex Hook.f. in FBI 6:311.1892; FUGP 2:329.

Large prickly climbers with terete stems and angular branches. Leaves alternate, 10-20 x 5.5-11 cm, oblong-ovate, base rounded cordate or cuneate, 5 to 7-nerved; petiolar sheath auricled below the middle. Umbels many flowered, alternate on a common peduncle. Staminodes 3. Stigmas long, recurved. Berries globose, 0.5 cm in diameter.

Flowers and Fruits : July-December

Found frequently in the sal forests. Singh 7126, 7338 (CDRI).

S. ovalifolia Roxb. Fl. Ind. ed.2, 794.1832; Wight, Icon. t.809.1851; T. Koyama loc. cit. 48.1963.
S. macrophylla Roxb. Fl. Ind. 3: 793; FBI 6: 310. 1892 non Willd. 1805.

Large, prickly climbers with, terete striated branches. Leaves alternate, 18-20 x 12-13.5 cm, ovate-oblong, 5-nerved, base cordate or cuneate, cuspidate. Umbels 2-3, alternate, on a common peduncle. Berries green, globose, smooth.

Local name : Ramdatun
Flowers and Fruits : April-December

Found frequently in the forest areas climbing on trees. Singh 4378 (CDRI); Inayat 22807, 22808 (CAL).

S. perfoliata Lour. Fl. Cochinch. 622.1790; T. Koyama in Adv. Front. Pl. Sci. 4:1963; M.A. Rau in Bull. Bot. Surv. India Suppl. No. 2 10:80.1968.
S. prolifera Roxb. Fl. Ind. 3:795.1832; FBI 6:312; FUGP 2:330; Srivastava, FG. 319.1976.

Prickly, scandent shrubs. Leaves 9-15 x 5-5.5 cm, elliptic-lanceolate, retuse or shortly acuminate, 3 to 5-nerved, base cuneate or cordate; petiole conspicuously sheathed to half its length. Umbels many, whorled on an axillary peduncle. Berries oblong-globose, red when ripe.

Flowers and Fruits : March-September

Found frequently in the sal forests. Singh 7331, 7912 (CDRI); Inayat 22810 (DD).

107. PONTEDERIACEAE

1. Petioles swollen into a bladder; tepals forming a tube ....**Eichhornia**

1. Petioles not forming bladder; tepals free ....**Monochoria**

Eichhornia Kunth *nom. cons.*

E. crassipes (Mart.) Solms in DC. Mon. Phan. 4: 527. 1832; Subramanyam, Aq. Ang. 70. 1962; Srivastava, FG. 319. 1976; Raizada, Suppl. FUGP 279. 1976.

Pontederia crassipes Mart. Nov. Gen. Sp. Pl. 1:9,t.4.1823.

Floating aquatic herbs, rooting at the node. Leaves radical, rosulate, 9 x 8 cm, rhomboid or ovate; petiole swollen into bladders. Flowers violet, funnel-shaped in many flowered spikes. Perianth 6-lobed. Stamens 6, unequal. Ovaries 3-celled with many ovules.

Local name : Jalkumbhi
Flowers and Fruits : September-November

Found commonly in ponds and ditches. Singh 4370 (CDRI).

Monochoria Presl

M. vaginalis (Burm.f.) Presl, Reliq. Hoenk. 1:128.1827; FBI 6:363; FUGP 2:336; Backer in Fl. Male. ser.1,4:256.1951; Saldanha & Nicolson, Fl. Hass. Distt. 791.1976.

Pontederia vaginalis Burm.f. Fl. Ind. 80.1768.

Monochoria vaginalis var. *plantaginea* Solms in A. & C. DC, Mon. Phan. 4:524.1883; FUGP 2:337.

Erect, acuatic or marshy perennial herbs with creeping rootstocks. Leaves blade 5-12 x 2.5-8 cm, ovate-lanceolate to ovate-cordate, acuminate, base cordate; petiole 16-20 cm long Flowers blue in racemes. One stamen larger than the others. Ovaries ellipsoid; stigmas 3-lobed.

Flowers and Fruits : August-November

Occurs commonly in the marshy area and near margin of ponds. Singh 7483, 7748, 8489 (CDRI).

108. COMMELINACEAE

1. Fertile stamens 6:
   2. Leaf-sheaths inflated, subtending axillary flowers..**Amischophacelus**
   2. Leaf-sheaths short, flowers subtended by bracteole..........**Cyanotis**
1. Fertile stamens 2-3:
   3. Flowers arranged in cymes; perfect stamens 3 ....**Commelina**
   3. Flowers arranged in panicles or in clusters; perfect stamens 2 ....**Murdannia**

Amischophacelus R. Rao & Kammathy

A. axillaris (Linn.) R.Rao & Kammathy in JLS (Bot.) 59:306.1966; Raizada in Indian For. 94(6):434.1968; Srivastava, FG. 320.1976.
Commelina axillaris Linn. Sp. Pl. 42.1753.
Cyanotis axillaris (Linn.) Schult.f. Syst. 7:1154.1830; FBI 6:388; FUGP 2:344.

Slender, glabrous branched herbs. Leaves 4-9 x 0.5 cm, linear-lanceolate, acute; ocreate sheaths inflated. Flowers purple in axillary fascicled cymes subtended by the inflated sheath. Capsules oblong-ellipsoid.

Flowers and Fruits : August-December

Found commonly in damp shady places in sal forests. Singh 4272, 7541 (CDRI).

Cyanotis D.Don.

C. cristata (Linn.) D.Don, Prodr. Fl. Nep. 46.1825; FBI 6:385; FUGP 2:343.
Commelina cristata Linn. Sp. Pl. 42, 1753.

Slender, glabrous herbs. Leaves sessile, 3.5-5.5 x 1.2x1.7 cm, ovate-oblong, acute, base rounded, sheath short hairy. Flowers blue arranged in cymes enclosed by bracts and bracteoles. Bracts much longer than the cymes. Capsules trigonous, truncate.

Flowers and Fruits : July-November

Found frequently in damp and muddy places in forests and other wastelands. Singh 4216, 7819 (CDRI).

Commelina Linn.

1. Capsules 2 to 3-seeded:
   2. Leaves linear; spathes deeply cordate at the base ....**C. attenuata**
   2. Leaves elliptic-lanceolate; spathes acute, funnel-shaped..**C. paludosa**
1. Capsules 5-seeded:
   3. Leaves ovate or elliptic-ovate; seeds rugosa, pitted...**C. benghalensis**
   3. Leaves ovate-lanceolate; seeds smooth, cylindrical ......**C. hasskarlii**

C. attenuata Koen. ex Vahl, Enum. 2: 168. 1806; FBI 6: 372; FUGP 2: 340; Srivastava, FG. 321.1976.

Much branched, annual herbs. Leaves 11-12 x 0.5-0.9 cm, linear-lanceolate, acute; sheath 2.5 cm long, cylindric, ciliate. Spathes 5 cm long, narrowly ovate-lanceolate, base deeply cordate. Flowers blue. Capsules oblong, 2-seeded.

Flowers and Fruits : September-November

Found frequently in waste damp places. Singh 7411 (CDRI); G. Saran 22184 (LWG).

P. paludosa Blume, Enum. Pl. Jav. 1:2.1827; R. Rao in Notes R. Bot. Gdn. Edinb. 25:81.1964; Srivastava, FG. 322.1976.

C. obliqua Buch.-Ham. ex D.Don, Prodr. 45.1825 non Vahl (1805); FBI 6:372; FUGP 2:340.

Stout branched, annual herbs. Leaves 7-13 x 1-2.2 cm, ovate to elliptic-lanceolate, acuminate, base oblique; sheath 1.5-2 cm long, hairy. Flowers blue in racemes. Capsules obovoid, 3-loculed. Seeds 3, smooth.

Flowers and Fruits : September-November

Found frequently in moist shady places in sal forests. Singh 7513 (CDRI).

C. benghalensis Linn. Sp. Pl. 41.1753; FBI 6:370; FUGP 2:339; Srivastava, FG. 321.1976.

Branched annual herbs with underground stems having white, cleistogamous fertile flowers. Leaves ovate-oblong, 2.5-5.5 x 1.5-5 cm, acute, base rounded or cuneate, sheath 1 cm long, hairy at the mouth. Aerial flowers blue. Capsules 5-seeded. Seeds rugose, pitted.

Flowers and Fruits : August-November

Found commonly in waste places. The leaves are cooked as vegetable. Singh 7461 (CDRI); G. Saran 30136 (LWG).

C. hasskarlii Clarke, Comm. & Cyrt. Beng. 13.t.3.1874; FBI 6:370; FUGP 2:339; Maheshwari, Fl. Del. 340.1963.

Much branched annual herbs. Leaves 3.5-6.5 x 1-1.5 cm, ovate-lanceolate, acute; sheath 1.5 cm long, hairy. Spathes axillary, ovate-lanceolate, base cordate. Flowers in unequal cymes; the lower branch 1 to 2-flowered, upper 2 to 4-flowered. Capsules quadrate, apiculate. Seeds cylindric.

Flowers and Fruits : August-November

Occurs frequently in moist shady places in the sal forests. Singh 7412, 7539 (CDRI).

Murdannia Royle

1. Leaves radical; plant erect ....**M. scapiflora**

1. Leaves cauline; plant diffuse ....**M. nudiflora**

M. scapiflora Roxb. Fl. Ind. 1:175.1820; M.A. Rau in Bull. Bot. Surv. India 2.10:81.1969.
Aneilema scapiflorum Wight, Ic. f. 2073.1853; FBI 6:375; FUGP 2:341.

Erect annual herbs with elongated tuberous roots. Leaves radical, 18-20 x 1.5-2 cm, oblong-lanceolate, acute. Flowers blue on erect 20-35 cm long leafless scales with sheathing base. Capsules obovoid with 3-6 seeds in each locule.

Flowers and Fruits : August-November

Occurs occasionally in moist shady places in sal forests. Singh 4139 (CDRI); G. Saran 26262 (LWG).

M. nudiflora (Linn.) Brenan in Kew Bull. 7:189.1952; Backer & Bakh.f. Fl. Java 3:658.1968; Srivastava, FG. 322.1976.
Commelina nudiflora Linn. Sp. Pl. 41.1753; pro type parte.
Aneilema nudiflorum R.Br. Prodr. 271.1810; FBI 6:378; FUGP 2:342.

Diffuse, slender, branched herbs with fibrous roots. Leaves 5-12 x 0.5-0.7 cm, linear-lanceolate, acute or acuminate; sheath 1 cm. Flowers purplish in cymes at the end of the branches of terminal panicles. Capsules subglobose, 3-loculed with 2 to 7-seeds in each locule.

Flowers and Fruits : August-November

Occurs frequently on moist sandy soil in sal forests. Singh 4172, 4196, 7537, 8493 (CDRI).

109. JUNCACEAE

Juncus Linn.

1. Leaves septate; cymes terminal .... **J. prismatocarpus**

1. Leaves not septate; cymes scattered .... **J. bufonius**

J. prismatocarpus R.Br. Prodr. 259.1810; FBI 6:395; FUGP 2:346; Saldanha & Nicolson, Fl. Hass. Distt. 655.1976.

Tufted, perennial herbs. Leaves flat, 5-10 x 0.2 cm, transversely septate, shorter than the stem. Flowers greenish-brown in erect dense flowered cymes. Tepals glumaceous, linear, Capsules trigonous.

Flowers and Fruits : October-April

Occurs occasionally in the muddy places. Inayat 22831 (DD).

J. bufonius Linn. Sp. Pl. 328.1753; FBI 6:392; FUGP 2:346; Maheshwari, Fl. Del. 341.1963.

Slender, tufted, dichotomously branched herbs. Leaves radical, setaceous, grass-like. Flowers pale-green, solitary and in few-flowered scattered cymes. Tepals with membranous margin. Capsules obovoid embraced by the perianth.

Flowers and Fruits : October-March

Found frequently in moist and swampy areas. G. Saran 26298, 26374 (LWG).

110. ARECACEAE (nom. alt. PALMAE)

1. Unarmed erect shrubs .... **Phoenix**

1. Armed scandent shrubs .... **Calamus**

Phoenix Linn.

P. acaulis Buch.-Ham. ex Roxb. Fl. Ind. 3:783.1832; FBI 6:426; FUGP 2:354; P.C. Kanjilal, For. Fl. UP. 3:694.1966.

Erect dwarf shrubs with very short stem. Leaves pinnate; leaflets 28-50 x 1-1.5 cm, subopposite, rigid, acute. Spadix 15-25 cm long. Fruits elliptic-oblong, mucronate, blue-black when ripe.

Local name : Khajuri, Chindi
Flowers : December-January; Fruits : April-May

Occurs frequently in grasslands in open areas. The root stock is edible. Singh 2908, 7287 (CDRI).

Calamus Linn.

C. **tenuis** Roxb. Fl. Ind. 3:780.1832; FBI 6:447 FUGP 2:356; P.C. Kanjilal, loc. cit. 697.1966; Srivastava, FG. 324.1976.

Scandent armed shrubs. Leaves pinnate, 40-60 cm long, feathery; leaflets equidistant, lower 17 cm long, linear-lanceolate, acuminate; rachis armed with recurved prickles; upper leaflets 8-10 cm long. Spadix long, decompound; spathe tubular. Flowers unisexual; male flowers on scorpioid spikelets. Calyx 3-toothed, cupular.

Local name : Bent
Flowers and Fruits : September-December

Occurs frequently in marshy areas and along river banks. The stems are used for making basket & furniture. Singh 7713 (CDRI).

111. TYPHACEAE

Typha Linn.

T. **angustata** Bory & Chaub. Exp. Sc. Mon. 3(2):33.1832; FBI 6:489; Santapau in RBSI 16(1):328.1953; P.C. Kanjilal, For. Fl. UP. 3:702.1966.
T. elephantina Gr. 227.1838 (non Roxb.)
T. angustifolia auct. (non Linn. 1753).

Perennial marshy shrubs. Leaves 100-200 x 0.5-1 cm, subterete above the sheath, flattened at the apex. Spikes cylindrical. The distance between male and female spikes ranges from 5-15 cm. Female spikes pale-brown. Male flowers mixed with clavate pistillodes. Fruits an achene, dehiscing by lateral slits.

Local name : Pater, Gondi
Flowers and Fruits : May-November

Found commonly in marshy places. The leaves are used for making mat. Singh 7652, 7940 (CDRI).

112. ARACEAE

1. Plants floating aquatics .... **Pistia**
1. Plants terrestrials:
    2. Climbing shrubs:
        3. Petioles winged; fruiting spadix drooping .... **Scindapsus**
        3. Petioles deeply grooved; fruiting spadix erect..... **Rhaphidophora**
    2. Erect herbs:
        4. Leaves compound divided into 3 or more segments:
            5. Plants with leaves and inflorescence together; spadix with a barren terminal appendages.......... **Arisaema**
            5. Plants with leaves and inflorescence not together; spadix without a terminal appendages .... **Plesmonium**
        4. Leaves simple, ensiform .... **Acorus**

Pistia Linn.

P. stratiotes Linn. Sp. Pl. 963. 1753; FBI 6:497; FUGP 2:361; Subramanyam, Aq. Ang. 75. 1962; Srivastava, FG. 327. 1976.

Floating stoloniferous, stemless herbs. Leaves sessile in close spirals, 5 cm long, obovate, cuneate, pubescent, retuse at the apex. Spathes white, sessile; spadix with a single pistil adnate to the spathe and a few stamens gathered into a synandrium.

Flowers : August-November

Occurs frequently in still water of tanks and ditches. Singh 7830 (CDRI).

Scindapsus Schott.

S. officinalis (Roxb.) Schott in Melet. 21. 1832; FBI 6:541; FUGP 2:367; P.C. Kanjilal, For. Fl. UP. 3: 703. 1966. *Pothos officinalis* Roxb. Fl. Ind. 1:431.1820.

Climbing shrubs with deeply furrowed, corky bark. Stems rooting at the nodes. Leaves alternate, ovate or elliptic-ovate, 20-30 x 9-17 cm, caudate-acuminate, base rounded; petiole 9 cm long, winged. Flowers bisexual, all fertile. Spathes boat-shaped, thick, caducous. Spadix shorter than the spathe, sessile, dense flowered. Fruiting spadix 10-15 cm long, drooping; berries closely adpressed.

Local name ; Gajpipali, Hadjor
Flowers : April-May; Fruits : September-November

Occurs frequently in sal forests climbing on sal trees. The plant is used in bone fracture. Singh 7957 (CDRI); Mehrotra 162 (CDRI).

Rhaphidophora Hassk.

R. glauca (Wall.) Schott in Bonpland. 5:45.1857; FBI 6:547; P.C. Kanjilal, loc. cit. 3:704. 1966; Raizada, Suppl. FUGP 286.1976.

I have included this taxon on the authority of P.C. Kanjilal, loc. cit. who reports its occurrence from Gauriphanta forest areas of the Park.

Large evergreen climbers. Leaves 15-30 x 8-22 cm, obliquely ovate, pinnatipartite with 1-2 deep cut on either side of the midrib; petioles 10-30 cm long, deeply grooved. Fruiting spadix erect.

Arisaema Mart.

A. tortuosum (Wall.) Schott in melet. 17. 1832; FBI 6:520; FUGP 2:362.

Arum tortuosum Wall. Pl. As. Rar. 2:10.f.111.1830.

Erect tuberous herbs. Leaves 2-3 partite; leaflets 5-19, sessile, linear-lanceolate, caudate-acuminate; petiole green mottled with purple. Flowers unisexual. Spathes pale-green, 11-15 cm long; blade 7-10 cm. Spadix sigmoid, 18 cm long. Berries globose.

Local name : Jhagpapari
Flowers and Fruits : September-November

Occurs occasionally in open places. The stem after peeling is cooked as vegetable. Singh 3002, 4137, 7398 (CDRI).

Plesmonium Schott

P. margaritiferum Schott in Syn. Aroid. 34. 1856; FBI 6:518; FUGP 2:364.

Erect tuberous herbs. Leaves solitary, appearing long after the flowers, 3-partite; leaflets 10-15 cm long, linear, acuminate. Spathes 10 x 14 x 4-5 cm, cream coloured. Spadix stout as long as spathe. Male flowers occupying largest portion of spadix. Neuter flowers in the middle. Female flowers on the bottom.

Flowers and Fruits : August-October

Found frequently in marshy places. Singh 7745 (CDRI).

Acorus Linn.

A. calamus Linn. Sp. Pl. 324. 1753; FBI 6:555; P.C. Kanjilal, loc. cit. 704. 1966; Raizada, Suppl. 285.1976.

Perennial marshy herbs with aromatic rhizome. Leaves ensiform, 40-60 x 1-1.5 cm, with sheathing bases. peduncle and spathe leaflike. Spadix sessile, 4-5 cm long, dense flowered, cylindric. Flowers bisexual. Sepals 6, orbicular, concave, tips incurved. Stamens 6, filament linear, flat; anthers reniform.

Local name : Bach
Flowers : October-March

Found rarely in North Kheri near Tharu village in marshy area. The fresh rhizome are chewed in bronchitis, & asthma. It is used as insecticide. Singh 7250, 8406 (CDRI).

113. LEMNACEAE

1. Rootlets present:

    2. Fronds with single rootlet ....Lemna

    2. Fronds with many rootlets ....Spirodela

1. Rootlets absent ....Wolffia

Lemna Linn.

L. paucicostata Hegelm. Lemnac. 138. f. 8. 1868; FBI 6:556; FUGP 2:369; Subramanyam, Aq. Ang. 78; Srivastava, FG. 328. 1976.

Small scale like floating plants. Fronds oblong or obovate-oblong, nearly flat on both surfaces, oblique at base. Rootlet single with an appendaged root-sheath, and acute root cap.

Occurs frequently in ponds and ditches. The flowers could not be observed. Singh 8410 (CDRI).

Spirodela Schleid.

S. polyrhiza (Linn.) Schleid. in Linnaea 13: 392.1839; FUGP 2:368; Subramanyam, Aq. Ang. 77. 962; Srivastava, FG. 328. 1976.
Lemma polyrhiza Linn. Sp. Pl. 970. 1753; FBI 6:557.

Minute aquatic herbs. Fronds dark green above, purplish beneath, broad obovate or orbicular, flat above, slightly convex below, each frond with many rootlets.

Found commonly in still water of ponds and ditches. Flowers could not be observed in the area. Singh 841 (CDRI).

Wolffia Horkel ex Schleid. *nom. cons.*

W. arrhiza (Linn.) Horkel ex Wimmer, Fl. Schles. ed. 3: 140. 1857; FBI 6:557; Subramanyam, Aq. Ang. 79.1962; Srivastava, FG. 328.1976.
Lemna arrhiza Linn. Mant. 2:294.1767.

Smallest flowering plants. Fronds minute, subglobose or oblong, convex above. Rootlets absent.

Found occasionally in ponds and ditches. Flowers could not be observed in the area. Singh 84112 (CDRI).

114. ALISMATACEAE

1. Carpels inserted on large globose receptacle; stamens many.....**Sagittaria**
1. Carpels on small flat receptacle; stamens 6 ....**Caldesia**

Sagittaria Linn.

1. Leaves base hastate or sagittate, apex acute ....**S. trifolia**
1. Leaves base deep cordate, apex obtuse.....**S. guayanensis** subsp. **lappula**

S. trifolia Linn. Sp. Pl. 993. 1753; FBI 6:561; FUGP 2:371; Subramanyam, Aq. Ang. 86-87. 1962; Srivastava, FG. 329. 1976.
S. sagittifolia auct. non Linn. 1753; FBI 6: 561. 1893.

Erect aquatic herbs. Leaves radical, petiole 18-30 cm long; blade hastate, sagittate, acute, glabrous. Flowers whitish-purple, bracteate, in 1 to 3-flowered whorls of panicles; female flowers sessile; male flowers pedicellate. Stamens many. Achenes obliquely ovate; apiculate, compressed, winged.

Flowers and Fruits : September-February

Found frequently in water and muddy places in open areas. Singh 7353 (CDRI); P.C. Kanjilal 194 (LWG); G. Saran 26136 (LWG).

S. guayanesis H.B. & K. Nov. Gen. Sp. 1: 250, 1815; FBI 6: 561; FUGP 2: 371; Maheshwari, Fl. Del. 347. 1963.

subsp. **lappula** (D. Don) Bogin.

Floating or marshy herbs, Leavs radical, petiole 13 cm long; blade 4-5 cm long, ovate, deep, cordate, obtuse at the apex, membranaceous. Flowers white, in close whorls on stout pedicles. Achenes numerous, flat, winged, toothed.

Flowers and Fruits : September-December

Occurs occasionally in marshy places and in ponds. Singh 7484, 7695 (CDRI).

Caldesia Parl.

C. parnassiifolia (Linn.) Parl. Fl. Ital. 3:599. 1853; Hartog in Fl. Males. ser. 1:5:319.1957; (parnassifolia); Saldanha & Nicolson, Fl. Hass. Distt. 631.1976.

Alisma parnassifolia Bassi ex Linn. Syst. Nat. ed. 12, 3 (App): 230.1768 (parnassifolia).

A. reniforme D. Don, Prodr. Fl. Nep. 22. 1825; FBI 6:560; FUGP 2:370.

Erect aquatic stemless perennial herbs. Leaves radical, petiole 6-8 cm long; blade 6 x 3-5 cm, elliptic or suborbicular, cordate or reniform, 12-nerved, apex rounded. Flowers white in panicles arranged on 30-40 cm long scape. Sepals oblong, spreading in fruits. Achenes 20-30 in a single whorl, compressed with persistent style.

Flowers and Fruits : July-October

Found rarely in stagnant water of ponds in Bellrain range of the Park. Singh 7555 (CDRI).

115. BUTOMACEAE

Butomopsis Kunth

B. latifolia (D.Don) Kunth, Enum. Pl. 3: 165. 1841.
Butomus latifolius D.Don, Prodr. Fl. Nep. 22. 1825.
Butomopsis lanceolata Kunth, Enum. Pl. 3:165.1841; FBI 6:562; FUGP 2:372.
Tenagocharis latifolia (D.Don) Buchen. in Abh. Natur. Ver. Bremen 2:2, 3, 16.1868; van Steenis in Fl. Males. 5:118.t.1.1955.

Erect scapigerous marshy herbs with fibrous roots. Leaves radical, elliptic-lanceolate, 12-14 cm long, acute. Flowers white in umbels on 30-35 cm long scape. Sepals 3. Petals white, 3. Follicles 6-7, reticulately veined, beaked.

Flowers and Fruits : September-December

Occurs occasionally in rice fields in South Kheri Forest division and in open marshy places. Singh 4334, 7847 (CDRI).

116. APONOGETONACEAE

Aponogeton Linn.f.

A. undulatus Roxb. Fl. Ind. 2: 221. 1832. Bruggen in Blumea 18: 465. 1970; Bennet in Nom. Chang. Plants. 38. 1987.
A. crispus auct. non Thunb. Nov. Gen. 4:78.1781; FBI, 6:564; FUGP 2:374.

Floating, scapigerous aquatic herbs. Leaves submerged; blade linear-lanceolate, 7-15 cm long, petiole shorter than the blade. Flowers minute in lax spikes. Stamens 3. Carpels 3.

Flowers and Fruits : August-November

Found frequently in ponds and jhils. Singh 7553 (CDRI).

Note: A. crispus Thunb. is found in Ceylon and does not occur in India.

117. POTAMOGETONACEAE

Potamogeton Linn.

1. Upper or all leaves floating on surface, petiolate .... P. nodosus
1. All leaves submerged, sessile:
   2. Leaves linear-lanceolate, crisped .... P. crispus
   2. Leaves narrowly linear to filiform .... P. pectinatus

P. nodosus Poir. in Lamk. Encycl. Suppl. 4:538.1816; Saldanha & Nicolson, Fl. Hass. Distt. 638.1976.
P. indicus Roxb. Hort. Beng. 12. 1814, nom. nud. et Fl. Ind. 1:452.1820; FBI 6:565; FUGP 2:374.

Submerged aquatic herbs with creeping rootstocks. Leaves petiolate, 6-9 x 3-4 cm, elliptic-lanceolate, entire, many nerved; stipules scarious. Flowers white, numerous in dense spikes.

Flowers and Fruits : November-December

Occurs occasionally in ponds and in muddy places. Singh 7268 (CDRI).

P. crispus Linn. Sp. Pl. 126.1753; FBI 6:566; FUGP 2:315; Srivastava, FG. 330. 1976.

Slender submerged aquatic herbs. Stems compressed. Leaves sessile, 3-5 x 0.5 cm, linear or linear-oblong, crisped, serrulate, 3-nerved, translucent; stipules hyaline, axillary. Flowers white in 1-1.5 cm long spikes.

Flowers and Fruits : December-April

Occurs frequently in the ponds. Singh 7245 (CDRI); G. Saran 26157 (LWG).

P. pectinatus Linn. Sp. Pl. 127.1753; FBI 6:567; FUGP 2:375; Subramanyam, Aq. Ang. 95.1962; Srivastava, FG. 330.1976.

Profusely distichously branched aquatic herbs. Leaves narrowly linear to filiform; stipules adnate to the leaf sheath with free tips. Flowers white in interrupted spikes.

Flowers and Fruits : October-April

Found commonly in ponds and tanks. Singh 7246, 8460 (CDRI); G. Saran 26160 (LWG).

118. ERIOCAULACEAE

Eriocaulon Linn.

E. cinereum R. Br. Prodr. 254.1810; Santapau in RBSI 16(1):294.1967; Srivastava, FG. 337.1976.
E. sieboldianum Sieb. & Zucc ex Steud. Syn. Pl. Cyp. 2:272.1855; FBI 6:577; FUGP 2:378.

Stemless, tufted marshy herbs. Leaves 2-5 cm long, narrowly linear, glabrous. Peduncle many upto 13 cm long. Heads whitish-purple, globose or ovoid, 0.3 cm across, bracteate. Male flowers with corolla provided with an apical glands. Stamens 6. Female flower's sepals hyaline; petals eglandular. Stigmas 3, filiform.

Flowers and Fruits : September-January

Occurs commonly in open marshy places and in rice fields. Singh 7489, 7854 (CDRI).

119. CYPERACEAE

1. Nut enclosed in an utricle .... **Carex**

1. Nut not enclosed in an utricle:

    2. Flowers bisexual:

3. Flowering glumes distichous ....Cyperus

3. Flowering glumes spirally arranged or sometimes the lowest subdistichous:

4. Styles bases swollen with a constriction above the point of insertion on the nut:

5. Leaves absent; hypogynous bristles present...Eleocharis

5. Leaves present; hypogynous bristles absent:

6. Styles deciduous, leaving a tumor on the nut ....Bulbostylis

6. Styles persistent, if falling not leaving a tumor on the nut ....Fimbristylis

4. Styles bases not swollen, continuous with the nut..Scirpus

2. Flowers unisexual, nut white polished ....Scleria

Carex Linn.

1. Stigmas 2; utricle compressed ....C. cernua

1. Stigmas 3; utricle not compressed ....C. fedia

C. cernua Boott, III. 4.171, t.578.1876; FBI 708; FUGP 2:417; M.A. Rau in Bull. Bot. Surv. India 2(10):84.1969.

Erect perennial herbs with stoloniferous rhizomes. Stems 30-40 cm long, triquetrous, basal sheath brown. Leaves 30-40 x 0.5-0.6 cm, flat, scabrous. Spikes 5-8, dense, cylindric, 3-5 cm long; the female at the top; male at the base. Bracts foliaceous, much exceeding the inflorescence. Stigmas 2. Utricle flattened, obovate, compressed, shortly beaked.

Flowers and Fruits: February-April

Occurs occasionally in marshy places. Inayat 22886 (DD); Sis Ram 46488 (DD).

C. fedia Nees in Wt. Contrib. 129. 1834; Srivastava, FG. 332. 1976.

C. wallichiana Presc. ex Nees in Wt. Contrib. 129.1834; FBI 6:747; FUGP 2:418.

Erect glabrous herbs with a stout rhizome. Stems triquetrous, smooth, leafy; basal sheath dark-brown. Leaves 60 x 0.5 cm, longer than the stem, flat, coriaceous, scabrid. Spikes 5-6 on lax, 25-40 cm long panicles; lower 2-3 female; upper 2-3 spikes male. Stigmas 3. Utricle hairy, longitudinally nerved, not compressed.

Flowers and Frutis : September-April

Found commonly in marshy places and near ponds etc. Singh 2819, 7172, 7660 (CDRI).

Cyperus Linn. sensu lato

1. Stigmas 2; nut compressed or flattened:
    2. Nut flattened:
        3. Glumes obovate, 1.2 mm long, retuse .... C. pumilus
        3. Glumes ovate-lanceolate, 2 mm long, acuminate. C. michelianus subsp. pygmaeus
    2. Nut laterally compressed:
        4. Spikes 3, collected into a compact head..................... C. triceps
        4. Spikes in solitary head:
            5. Fruiting glumes with a crest on the keel........... C. kyllingia
            5. Fruiting glumes not crested........................... C. brevifolia
1. Stigmas 3, nuts triquetrous:
    6. Glumes few; rachilla of spikelets deciduous:
        7. Umbels simple .... C. cyperoides
        7. Umbels compound .... C. compactus
    6. Glumes numerous; rachilla of spikelets persistent:
        8. Stigmas minute, tapering .... C. platystylis
        8. Stigmas filiform:

9. Slender annuals:

10. Spikelets clustered:

11. Spikelets in dense globose heads:

12. Glumes ovate-orbicular ....**C. difformis**

12. Glumes ovate-oblong ....**C. silletensis**

11. Spikelets digitate:

13. Nuts rounded ....**C. tenuispica**

13. Nuts angular ....**C. haspan**

10. Spikelets spicate:

14. Glumes cuspidate or aristate; spikes compressed:

15. Spikelets 3 mm wide ....**C. compressus**

15. Spikelets 1-2 mm wide ....**C. cuspidatus**

14. Glumes truncate; spikes cylindrical:

16. Spikelets golden in lax spikes ....**C. iria**

16. Spikelets greenish-yellow in umbels ....**C. alulatus**

9. Perennials with thick rootstock or slender rhizome:

17. Spikelets clustered in a solitary head ....**C. niveus**

17. Spikelets distinctly spicate:

18. Bracts half the length of the inflorescence.........**C. corymbosus**

18. Bracts not as above:

19. Spikelets numerous in elongated spikes:

20. Secondary rays absent or occasional:

21. Spikelets yellowish-brown; rachis glabrous........ ....**C. digitatus**

21. Spikelets red-brown; rachis hairy.........**C. pilosus**

19. Spikelets 3-8 in short spikes ....**C. rotundus**

20. Secondary rays present:

22. Spikes ovoid or obovoid:

23. Spikelets acuminate ....**C. distans**

23. Spikelets acute ....**C. pangorei**

22. Spikes cylindrical, peduncled ....**C. exaltatus**

C. pumilus Linn. Cent. Pl. 2:6.1756; Kuekenth. in Pfreich. 101.375, f.44 A-E; Kern in Fl. Males. ser. 1.(3):650.1974; Srivastava, FG. 340.1976.
Pycreus nitens Nees in Nov. Act. Nat. Cur. 19.Suppl. 1: 53. 1843; FBI 6:591.

Small glabrous tufted annual herbs with fibrous roots. Leaves linear, acuminate; midrib porminent. Inflorescence an umbel of spikes on slender rays; primary bracts foliar, much exceeding the inflorescence. Spikelets 10-30, linear, compressed, many flowered, 2 ranked. Glumes ovate, 1.2 mm long, boat-shaped. Nuts ellipsoid.

Flowers and Fruits : August-October

Occurs occasionally in open moist places. Singh 8570 (CDRI); C.L. Malhotra 23578 (BSD).

C. michelianus (Linn.) Link subsp. **pygmaeus** (Rottb.) Aschers. & Graebn. Syn. 2(2):273.1903; Kuekenth. loc. cit. 312, f. 35, f-G; Srivastava, FG. 339.1976.
Scirpus michelianus Linn. Sp. Pl. 52. 1753.
Cyperus pygmaeus Rottb. Descr. & Icon. 20, t. 14.f. 4-5.1773.

Small tufted erect annuals with many close stems. Leaves narrowly linear, acuminate, equalling stems. Spikelets numerous, many flowered in dense, solitary terminal heads. Bracts leaf-like, 3-5, unequal. Glumes ovate-lanceolate, distichous, acute or acuminate. Nuts ellipsoid, plano-convex, tipped by style base.

Flowers and Fruits : October-July

Found frequently near ponds and in marshy places. Inayat 23898 (DD); C.L. Malhotra 31536 (BSD).

C. triceps (Rottb.) Endl. Cat. Hort. Acad. Vindob. 1:94, 1842; Kuekenth. loc. cit. 578; Maheshwari, Fl. Del. 353. 1963; Kern in Fl. Males. loc. cit. 659.1974.

Kyllinga triceps Rottb. Descr. & Icon. 14.t.4, f.6.1773; FBI 6:587; FUGP 2:396.

Small glabrous tufted herbs, 8-20 cm tall. Leaves linear, half or nearly as long as the stem. Bracts 3-4, leaflike. Spikes 3, ovoid-oblong, white, aggregated into a compact head. Spikelets 1-flowered. Nuts oblong, laterally compressed.

Flowers and Fruits : September-December

Occurs occasionally in the moist shady places. Singh 8568 (CDRI).

C. kyllingia Endl. Cat. Hort. Acad. Vindob. 1:94.1842; Kuekenth. loc. cit. 606; Kern in Fl. Males. loc. cit. 659.1974; Srivastava, FG. 339.1976.
Kyllinga monocephala Rottb. Descr. & Icon. 13, t.4, fig. 4.1773; FBI 6:588; FUGP 2:397.

Erect glabrous herbs up to 30 cm tall with creeping rhizomes. Stems slender with pale-brown sheath. Leaves linear, as long as the stem, acuminate. Spikes subglobose, solitary, white. Spikelets obliquely lanceolate, elliptic, 1-flowered. Third glume boat-shaped, acuminate, with 3 nerves on each side of the winged keel. Nuts obovate, compressed.

Flowers and Fruits : August-March

Occurs frequently in marshy places. Singh 2955, 4181, 7418, 7774 (CDRI).

C. brevifolius (Rottb.) Hassk. Cat. Hort. Bogor. 24.1844; Kuekenth. loc. cit. 600; Kern in Fl. Males. loc. cit. 656. 1974; Srivastava, FG. 335.1976.
Kyllinga brevifolia Rottb. Descr. & Icon. 13, t.4.f.3.1773; FBI 6:588; FUGP 2:396.

Erect glabrous herbs with slender horizontal rhizomes, 15-30 cm tall. Stems covered with brown sheath at the base. Leaves linear, erect, shorter or equalling the stem. Spikes oblong, cylindrical, white. Spikelets ovate-lanceolate, 1-flowered. Third glume boat shaped, acuminate, 2-nerved on either side of the non winged keel. Nuts obovate-ellipsoid, compressed laterally.

Flowers and Fruits : June-October

Found frequently in moist places. Singh 8569 (CDRI); G. Saran 26389, 26514 (LWG).

C. cyperoides (Linn.) O. Kuntze, Rev. Gen. Pl. 3.2:133.1898; Kuekenth. loc. cit. 514; Kern in Reinwardtia 6(1):64.1961 et Backer & Bakh.f. Fl. Java 2:474.1968 et Fl. Males. loc. cit. 642.1974; Raizada, Suppl. FUGP 296.1976.

Scirpus cyperoides Linn. Mant. 2:181.1771.

Mariscus sieberianus Nees in Linnaea 9:289.1835; FBI 6:622.

Perennial herbs with short rhizomes; stolon absent. Stems many, triquetrous, rigid, 20-90 cm long, Leaves flaccid, flat, gradually long, acuminate. Spikes arranged in terminal whorls. Glumes membranous, ovate-oblong. Nuts triquetrous, linear, slightly curved.

Flowers and Fruits : September-November

Found frequently in moist sandy and swampy places. Singh 3017, 7916, 8470, 8522 (CDRI).

C. compactus Retz. Obs. Bot. 5:10.1789; Kuekenth. loc. cit. 423; Kern in Reinwardtia 6(1):63.1961 et Fl. Males. loc. cit. 638.1974; Srivastava, FG. 335.1976.

Mariscus dilutus Nees in Wight Contrib. 90.1834; FUGP 2:398.

M. microcephalus Presl, Rel. Hoenk. 1:182.1830; FBI 6:624.

Erect perennial glabrous herbs with short and stout rhizomes. Stems obscurely trigonous. Leaves as long as stem, linear, acuminate. Spikelets numerous, narrowly linear, 1.5 cm long, terete, reddish-brown. Glumes imbricate. Stamens 3. Nuts 3-angled.

Flowers and Fruits : August-November

Occurs frequently in moist and marshy places. Singh 2953, 4271, 7451, 7558, 8448 (CDRI).

C. platystylis R.Br. Prodr. 214.1810; FBI 6:598; Kuekenth. loc. cit. 185.f.21; Kern in Reinwardtia 2.112.f.8.1952 et in Backer & Bakh.f. Fl. Java 3:472 et Fl. Males 1.7(3):618.1974; Raizada, Suppl. FUGP 302.1976; Srivastava, FG. 340.1976.

Robust stoloniferous herbs. Stems flattened below, trigonous upwards. Leaves and bracts large. Spikes numerous, close together, arranged in umbels. Spikelets bright-brown, 0.4-1.3 cm long. Glumes ovate-oblong, concave. Nuts unequally trigonous, ellipsoid.

Flowers and Fruits : July-February

Found occasionally in marshy places and near the banks of rivers. Sis Ram 46433, 46434 (DD).

C. difformis Linn. Cent. Pl. 2:6.1756; FBI 6:599; FUGP 2:382; Kuekenth. loc. cit. 77; Kern in Backer & Bakh.f. Fl. Java 3: 475. 1968 et Fl. Males. loc. cit. 629.1974; Srivastava, FG. 337.1976.

Tufted annuals with purplish fibrous roots. Leaves shorter than the stem, smooth, flaccid, acuminate. Bracts 2-3, unequal, leaflike, exceeding rays. Inflorescence of many globose, dense clusters of spikelets. Spikelets linear-lanceolate. Glumes ovate-orbicular, imbricate. Nuts elliptical to obovoid, pale-brown.

Flowers and Fruits : August-December

Occurs occasionally in marshy places and other moist places. Singh 8425 (CDRI).

C. silletensis Nees in Wight, Contrib. 79.1834; FBI 6:600.

Slender erect perennial herbs without stolons. Leaves 26-30 cm long, narrowly linear, acute. Spikes globose, dense, pale-brown. Spikelets linear, many flowered. Glumes ovate-oblong, subtriangular at tip. Nuts oblong-obovoid.

Flowers and Fruits : August-November

Found frequently in marshy places. Singh 4362, 7893, 8510 (CDRI); Sis Ram 46431 (DD); C.L. Malhotra 22883 (BSD).

C. tenuispica Steud. Syn. Pl. Glum. 2:11.1855; Kuekenth. loc. cit. 245; Kern in Backer & Bakh.f. Fl. Java 3:476.1968 et Fl. Males. loc. cit. 625.1974; Srivastava, FG. 341.1976.
C. flavidus auct. non Retz.

Erect glabrous herbs with numerous fibrous roots. Stems weak, 10-20 cm long. Leaves linear, acute, shorter or longer than stems. Spikelets in dense clusters, 3-6 mm long, linear-lanceolate. Glumes oblong, purple-brown. Nuts globosely obovoid.

Flowers and Fruits : October-November

Found occasionally in rice fields and other marshy places. Singh 7491 (CDRI).

C. haspan Linn. Sp. Pl. 45.1753; FBI 6:600; Kuekenth. loc. cit. 247; Kern in Reinwardtia 6:56.1961 et in Backer & Bakh.f. Fl. Java 3:476.1968 et in Fl. Males. loc. cit. 624.1974; Raizada, Suppl. FUGP 298.1976.

Perennial herbs with slender and short rhizomes. Stems slender, weak, triquetrous, 3-winged, 10-40 cm long. Leaves flat, smooth. Spikelets 0.5-0.7 cm, elliptic in clusters of 4-10, linear-lanceolate, acute. Glumes elliptic, 1 mm long. Nuts trigonous, stipitate, smooth to verrucose.

Flowers and Fruits : September-November

Occurs occasionally in marshy places in the Park. Singh 7889 (CDRI).

C. **compressus** Linn. Sp. Pl. 46.1753; FBI 6:605; FUGP 2:384; Kern in Backer & Bakh.f. Fl. Java 3:480.1968 et Fl. Males. loc. cit. 617. 1974; Srivastava, FG. 336.1976.

Erect tufted annuals with fibrous roots. Leaves narrowly linear, acuminate with purplish sheath. Bracts leaf-like, erect or spreading. Inflorescence umbellate; spikelets digitately clustered, much compressed, silvery. Glumes ovate, 3-4 mm long, winged. Nuts obovate, triquetrous, dark brown.

Flowers and Fruits : August-November

Found frequently in moist waste places and in rice fields. Singh 8432 (CDRI).

C. **cuspidatus** H.B. & K., Nov. Gen. Sp. 1:204.1816; FBI 6:598; Kuekenth. loc. cit. 101.261; Kern in Fl. Males. ser. 1.7(3):629.1974.
C. uncinatus auct. non. Poir. 1806.

Small tufted annuals. Stems 6 cm high. Leaves equalling stem, setaceous. Bracts spreading longer than the rays. Spikelets digitately clustered, linear, 5-6 mm long, compressed. Glumes oblong, 1.5 mm long, green. Nuts oblong or obovate, reddish-brown.

Flowers and Fruits : September-October

Found frequently in damp places in open grounds. Singh 8525 (CDRI).

C. iria Linn. Sp. Pl. 45.1753; FBI 6:606; Kuekenth. loc. cit. 150; FUGP 2:385; Kern in Backer & Bakh.f. Fl. Java 3:478.1968 et Fl. Males. loc. cit. 616.1974; Srivastava, FG. 338.1976.

Erect glabrous annuals, 15-60 cm tall. Stems triangular and longitudinally striated, with brown papery sheath. Leaves linear, acuminate. Spikes cylindrical, lax arranged in compound umbels. Spikelets golden yellow, linear oblong, compressed, 6 to 12-flowered. Glumes obovate, obtuse. Nuts obovate-elliptic with concave smooth sides.

Flowers and Fruits : August-November

Found commonly in rice fields and in other marshy places. Singh 2976, 7472, 8446 (CDRI).

C. alulatus Kern in Reinwardtia 1(4):463-466, f. 1.1952; Som Deva & Naithani in Indian For. 100(10): 659. 1974; Raizada, Suppl. FUGP 293.1976.

Erect annual herbs. Stems triquetrous, 15-16 cm tall. Leaves 20-28 x 0.4-0.6 cm, linear, acuminate. Spikes 3-3.5 cm long on an umbellate inflorescence. Spikelets greenish-yellow, 0.3-0.7 x 0.2 cm. Glumes ovate, imbricate, keel 7-nerved, winged with a serrulate margin in the upper part. Nutlets obovoid, reddish-brown.

Flowers and Fruits : September-November

Found commonly in rice fields and in other marshy places. Singh 2977 (CDRI).

C. niveus Retz. Obs. 5:12.1788; FBI 6:601; FUGP 2:383; Srivastava, FG. 339. 1976.

Erect glabrous perennial herbs with short woody rhizomes and slender roots. Leaves equalling or shorter than the stem, narrowly linear, acute. Spikelets 3-8 in one head, oblong-elliptic, much compressed, whitish. Glumes ovate-lanceolate, distichous. Nuts obovoid, triquetrous, dark-brown.

Flowers and Fruits : July-October

Found occasionally in moist open places in sandy soils. Singh 7387 (CDRI); Inayat 22855 (DD).

C. corymbosus Rottb. Descr. & Icon. 42, t.7, f. 4. 1773; FBI 6:612; FUGP 2;391; Srivastava, FG. 336.1976.

Tall glabrous herbs with creeping rhizomes. Leaves reduced or a short acute blade. Bracts 3, erect, shorter than the rays. Umbels compound, primary rays slender; secondary rays terminated by corymbs of spikelets. Spikelets 4-16, light brown. Glumes erect, ovate-oblong, pale-reddish-brown. Nuts narrowly obovoid.

Flowers and Fruits : August-November

Found frequently in marshy places. Singh 8505 (CDRI).

C. digitatus Roxb. Fl. Ind. 1:205; 1820; FBI 6:618; FUGP 2:387; Kern in Fl. Males. ser. 1.7(3):601.1974; Srivastava, FG. 337.1976.

Tall glabrous perennial herbs with horizontal rhizomes. Stems trigonous with smooth angles. Leaves flat, coriaceous, as long as stem. Bracts longer than the rays. Umbels long, simple or compound. Spikelets sub-terete, acute, 12 to 20-flowered, spreading horizontally, yellowish-brown. Glumes linear, apiculate. Nuts ellipsoid, trigonous, yellowish.

Flowers and Fruits : August-December

Found frequently in moist shady and marshy places. Singh 8472 (CDRI).

C. pilosus Vahl, Enum. Pl. 2:354.1805; FBI 6:609; FUGP 2:389; Kuekenth. loc. cit. 92; Kern in Backer & Bakh.f. Fl. Java 3:477.1968 et in Fl. Males. loc. cit. 611.1974.

Erect perennial herbs with several slender stolons. Stems up to 60 cm long, acutely triquetrous, clothed with loose membranous sheath at the base. Leaves acuminate. Bracts leafy much exceeding rays. Spikes clustered, cylindrical; lateral spikes spreading or reflexed with rachilla, hispid-pilose. Spikelets flat, distant, reddish-brown. Glumes distant, ovate, 2-2.5 mm long, obtuse and apiculate with sharp keel, 2 to 3-nerved. Nuts obovate-elliptic.

Flowers and Fruits : August-December

Occurs occasionally in muddy places and near rice fields. Singh 7557, 7763 (CDRI).

C. rotundus Linn. Sp. Pl. 45. 1753; FBI 6:614; FUGP 2:388; Kuekenth. loc. cit. 107.f.13; Kern in Reinwardtia 6(1):53. 1961 et in Backer & Bakh.f. Fl. Java 3:479.1968 et in Fl. Males. loc. cit. 604.1974; Srivastava, FG. 341.1976.

Erect glabrous perennial herbs. Stems leafy at the base, arising from a tuber. Leaves linear, acuminate, shorter than the stem. Bracts leaf-like, unequal, three, exceeding the umbel. Spikes arranged in an umbel. Spikelets 1-3.5 cm long, 20 to 40-flowered, reddish-brown. Glumes ovate to elliptic, subacute, imbricate with hyaline margins. Nuts oblong-ovate.

Local name : Motha
Flowers and Fruits : July-December

Found commonly in lawns and other moist grasslands in open areas. The tubers are used in rheumatism. Singh 4388 (CDRI).

C. distans Linn.f. Suppl. 103.1781; FBI 6:607; FUGP 2:390; Kern in Fl. Males. loc. cit. 610.1974.

Erect tall herbs with rhizomes and stolons covered with dark brown imbricate scales. Leaves as long as stem or slightly shorter than the stem. Bracts leafy, longer than the rays. Spikelets 2.5 cm long, narrowly linear, 10 to 20-flowered. Glumes oblong-elliptic, obtuse. Nuts ellipsoid, trigonous, dark brown.

Flowers and Fruits : August-October

Found frequently along forest roads in moist muddy places. Singh 8459, 8526 (CDRI).

C. pangorei Rottb. Oescr. Pl. Rar. Programm. 18.1772; Kuekenth. loc. cit. 101.76; Srivastava, FG. 340.1976. *C. tegetum* Roxb. Fl. Ind. 1.208.1820; FBI 6:613; FUGP 2:389.

Tall glabrous perennial herbs with stout, creeping rhizomes. Leaves 20-30 cm long, linear, acuminate, shorter than the stem. Bracts leafy, equalling or much exceeding rays. Umbels compound. Spikelets linear, acute, red-brown, lax flowered. Glumes 2-3 mm long, elliptic-oblong, obtuse. Nuts oblong, brown.

Flowers and Fruits : August-January

Occurs frequently in moist and muddy places. Singh 8437 (CDRI).

C. exaltatus Retz. Obs. 5:11.1789; FBI 6:617; FUGP 2:387; Kuekenth. loc. cit. 64. f.9. A.f; Kern in Fl. Males. loc. cit. 602.1974; Srivastava, FG. 337.1976.

Large glabrous perennial herbs upto 1 m tall. Stems triquetrous with dark-purple tough sheath. Leaves linear, acuminate. Spikes narrowly cylindrical in compound umbels. Spikelets elliptic-lanceolate, suberect, 8 to 16-flowered, much compressed. Glumes imbricate, ovate-oblong, mucronate. Nuts ovate to elliptic, apiculate.

Flowers and Fruits : August-December

Flowers commonly in marshy places and dried ditches. Singh 2951, 7452, 7755 (CDRI); Sis Ram 46424 (DD).

Eleocharis R.Br.

1. Spikelets 2-4 cm long; stems transversely septate .... **E. dulcis**

1. Spikelets 0.3-2 cm long; stems not septate:

2. Rhizomes absent; spikelets 0.3-0.5 cm long ....**E. atropurpurea**

2. Rhizomes present; spikelets 1.5 cm long.................................**E. palustris**

E. dulcis (Burm.f.) Henschel, Vita Rumph. 186.1833; Kern in Backer & Bakh.f. Fl. Java 3:460.1968 et in Fl. Males. ser.1.7 (3): 529. 1974; Srivastava, FG. 343.1976.
Andropogon dulcis Burm.f. Fl. Ind. 219.1768.
Heleocharis plantaginea Roem. & Shult. Syst. Veg. 2: 150. 1817; FBI 6: 625; FUGP 2:401.

Robust densely tufted, stoloniferous perennial herbs. Stems terete, spongy, 40-60 cm, septate. Spikelets cylindrical, 2-4.5 cm long, terete, yellowish-green. Glumes imbricate, oblong, obtuse with a hyaline margin. Bristles 6-8, hispid. Nuts obovoid, smooth, yellowish-brown, tipped by persistent conical style base.

Local name : Naryee
Flowers and Fruits : September-December

Found frequently in marshy area and shallow water of ponds and ditches. Singh 8426 (CDRI).

E. atropurpurea (Retz.) Kunth, Enum. 2:151.1837; FBI 6:627; FUGP 2:400; Kern in Backer & Bakh.f. Fl. Java 3:461. 1968; Srivastava, FG. 342.1976 (Fig. 17).
Scirpus atropurpurea Retz. Obs. 5:14.1788.

Slender filiform tufted annuals upto 12 cm. Leaves absent. Sheath apex membranous, apiculate. Spikelets solitary, terminal, ovoid-ellipsoid upto 0.3 cm long, brown. Glumes ovate-oblong, obtuse, loosely imbricate with a hyaline margin. Bristles 4, white. Nuts black, biconvex, smooth.

Local name : Naryee
Flowers and Fruits : September-December

Occurs commonly in marshy places and as a weed of rice fields. Singh 7488 (CDRI).

E. palustris R.Br. Prodr. 224.1810; FBI 6:628; FUGP 2:400; Srivastava, FG. 343.1976.

Erect tufted herbs with creeping rhizomes. Stems terete, longitudinally striate. Leaves absent. Spikelets solitary, terminal, 1.5-2 cm long, ellipsoid, cylindrical, yellowish-brown, wider than the stem. Glumes imbricate, ovate-lanceolate or elliptic. Bristles brown, retrorsely scabrid. Nuts obovoid, biconvex, tipped by the broad style base.

Flowers and Fruits : October-April

Occurs commonly in marshy places. Singh 7664, 8428 (CDRI); Inayat 22882 (DD); Sis Ram 46409 (DD).

Bulbostylis Kunth. *nom. cons.*

B. barbata (Rottb.) Clarke in FBI 6:651.1893; FUGP 2:408; Kern in Reinwardtia 6:51.1961 et Fl. Males. loc. cit. 539.1974; Srivastava, FG. 332.1976.
Scirpus barbatus Rottb. Descr. & Icon. 52, t.17, f.4.1773.

Slender tufted annual herbs with fibrous roots. Stems filiform upto 20 cm long. Leaves filiform; sheath with long fine hairs in the mouth. Spikelets in small, dense, terminal heads, brown. Glumes ovate-lanceolate, boat shaped, red-brown. Nuts obovoid, trigonous.

Flowers and Fruits : September-October

Found frequently in open sandy soil and in cultivated fields. Singh 7450 (CDRI).

Fimbristylis Vahl *nom. cons.*

1. Spikelets 1-2:
    2. Lower glumes distichous; nuts terete .... **F. ovata**
    2. Lower glumes spiral; nuts ellipsoid, obovoid .... **F. schoenoides**
1. Spikelets many:
    3. Stigmas 3; nuts trigonous:
        4. Spikelets discrete:

5. Spikelets subglobose ....**F. littoralis**

5. Spikelets terete-lanceolate....................**F. tenera var. oxylepis**

4. Spikelets or some of them in clusters of 2-5 ....**F. falcata**

3. Stigmas 2; nuts biconvex:

6. Nuts smooth:

7. Gynophore prominent ....**F. podocarpa**

7. Gynophore not distinguishable from the nut ....**F. rigidula**

6. Nuts longitudinally striate:

8. Spikelets 1-1.5 mm wide; glumes 1.25 mm long..**F. bisumbellata**

8. Spikelets 2-4 mm wide; glumes 2.5 mm long........**F. dichotoma**

F. ovata (Burm.f.) Kern in Blumea 15(1):126.1967 et in Fl. Males. ser.1.7(3):565.1974; Srivastava, FG. 347.1976. *Carex ovata* Burm.f. Fl. Ind. 194.1768. *Fimbristylis monostachya* Hassk. Pl. Jav. Rar. 61.1806; FBI 6:649; FUGP 2:405.

Shortly rhizomatous, perennial tufted herbs. Stems 12-30 cm long, slender. Leaves linear, shorter than the stem, acute. Spikelets solitary, 5 mm-1.2 cm long, narrowly ovate, depressed, acute. Glumes triangular-ovate, acute with pale scarious sides. Nuts 2.5 mm long, terete or somewhat ridged.

Flowers and Fruits : August-December

Occurs frequently in damp sandy soils. Singh 8541 (CDRI).

F. schoenoides (Retz.) Vahl, Enum. Pl. 2:286.1805; FBI 6:634; FUGP 2:402; Kern in Fl. Males. loc. cit. 573.1974; Srivastava, FG. 346.1976. *Scirpus schoenoides* Retz. Obs. 5:14.1789.

Tufted, perennials upto 30 cm tall. Leaves narrow, filiform; sheath with membranous margins. Inflorescence of 1-2 spikelets. Spikelets broadly ovate to orbicular, 6.0 x 4.0 mm. Glumes ovate of suborbicular, shortly mucronate, rounded on the back. Nuts elliptic, flattened, obovoid.

Flowers and Fruits : August-October

Occurs occasionally in moist places. Singh 7511 (CDRI).

F. littoralis Gaud. in Freycinet, Voy. Bot. 413.1826; Kern in Fl. Males. loc. cit. 551.1974; Srivastava, FG. 345.1976. *F. miliacea* Vahl, Enum. Pl. 2:287.1805; FBI 6:644; FUGP 2:406; Kern in Backer & Bakh.f. Fl. Java 3:463.1968.

Glabrous leafy annuals upto 50 cm tall. Leaves shorter than the stem, tapering to a filiform point, margin smooth. Umbels decompound, rays unequal, filiform, bearing many scattered spikelets. Spikelets subglobose, reddish-brown. Glumes closely imbricate, ovate, obtuse, rounded on back. Nuts obovoid, obtusely trigonous.

Flowers and Fruits : September-December

Found commonly in rice fields and other moist places. Singh 8455 (CDRI).

F. tenera Roem. & Schutt. var. **oxylepis** Clarke in FBI 6: 642. 1893; FUGP 2: 406; Maheshwari, Fl. Del. 360. 1963.

Erect slender tufted annuals, 15-30 cm tall. Leaves shorter than the stem. Umbels lax, subcompound, consisting of a single spikelet and 2-3 pedicellate spikelets. Spikelets lanceolate, terete, acute. Glumes ovate, brown, 2.5 mm long, acute. Nuts obovoid, verrucose.

Flowers and Fruits : August-October

Found frequently near ponds and in rice fields. Singh 8550 (CDRI).

F. falcata (Vahl) Kunth, Enum. 2:239.1837; Kern in Blumea 8:113.1955 et in Fl. Males. loc. cit. 557. 1974. *Scirpus falcatus* Vahl, Enum. Pl. 2:275.1805. *Fimbristylis junciformis* Kunth, Enum. 2:239.1837; FBI 6:647; FUGP 2:407.

Perennials with a short, creeping horizontal rhizome. Stems angular, striate. Leaves spreading, crowded at the base, pointed, shorter than the stem. Bracts foliar. Spikelets elliptic, clustered and solitary in compound umbels. Glumes obtuse, red-brown with a white margin. Nuts trigonous, obovoid, pale.

Flowers and Fruits : August-October

Found occasionally in moist sandy places. Singh 4118 (CDRI).

F. podocarpa Nees in Wight Contrib. 98.1834; pro parte typica; FBI 6:638; FUGP 2:404; Saldanha & Nicolson, Fl. Hass. Distt. 680.1976.

Tufted perennial herbs 30-50 cm tall. Leaves ligulate, half the length of the stem; sheath hairy at the top. Spikelets ovate, acute, terete in simple or compound umbels. Glumes broadly ovate or orbicular. Nuts orbicular, flattened, about 1.3 x 1 mm with a conspicuous gynophore.

Flowers and Fruits : August-February

Occurs frequently in moist sandy places. Singh 7706, 8441, 8445 (CDRI).

F. rigidula Nees in Wight Contrib. 99. 1834; FBI 6:640; FUGP 2:404; Kern in Fl. Males. loc. cit. 582.1974.

Erect perennial herbs with horizontal woody rhizomes. Stems 12-40 cm long, thickened at the base. Leaves shorter than the stem, subobtuse. Spikelets obtusely ellipsoid or subglobose, 0.7-1 cm long. Glumes concave, broadly ovate, dark-brown. Nuts biconvex, brown.

Flowers and Fruits : September-December

Found occasionally in moist places. Inayat 22880, 23895 (DD).

F. bisumbellata (Forsk.) Bub. Dodec, 30. 1850; Kern in Rein wardtia 6(1):47.1961 et in Fl. Males. loc. cit. 579.1974; Saldanha & Nicolson, Fl. Hass. Distt. 676.1976.
Scirpus bisumbellata Forsk. Fl. Aeg.-Arab. 1:15. 1775.
Fimbristylis dichotoma auct. (non Vahl 1805).

Tufted fibrous rooted annual herbs. Stems striate, slender. Leaves as long as stem or shorter, scabrous on the margins; sheath membranous, hairy. Spikelets numerous, oblong. Rachilla much elongating. Glumes ovate-oblong, 1.2 mm long, membranous, red-brown with green apiculate keel. Nuts obovate.

Flowers and Fruits : February-April

Found commonly in damp sandy soils and along river banks. Singh 7174, 7241, 7709, 8469, 8509 (CDRI).

F. dichotoma (Linn.) Vahl, Enum. Fl. 2:287.1805; Kern in Fl. Males. loc. cit. 575.1974; Saldanha & Nicolson, Fl. Hass. Distt. 677.1976.
Scirpus dichotomus Linn. Sp. Pl. 1:50.1753.
Fimbristylis diphylla (Retz.) Vahl, Enum. Pl. 2:289. 1805; FBI 6:636; FUGP 2:402; Srivastava, FG.345.1976.

Tufted perennials with fibrous roots. Stems 12-40 cm long, slender, striated. Leaves linear with spinulose tip; sheath hairy, brown spotted. Spikelets scattered, ovate, 0.5 cm long, red-brown, acute. Glumes 2.5 mm long, concave, broadly ovate, acute or apiculate. Nuts obovate or elliptical, 1 mm long, compressed.

Flowers and Fruits : July-November

Occurs frequently in muddy places and in rice fields. Singh 3045, 4159, 7552, 8439 (CDRI); C.L. Malhotra 22884 (BSD).

Scirpus Linn.

1. Spikelets sessile, 1-many in simple clusters or heads:
    2. Clusters terminal on the stem:

3. Bristles present .... S. affinis

3. Bristles absent .... S. michelianus

2. Clusters lateral on the stem:

4. Spikelets born near the base of the septate stem..... S. articulatus

4. Spikelets born below the top of the triquetrous stem .... S. mucronatus

1. Spikelets stalked, many in umbels:

5. Umbels terminal, simple or compound .... S. tuberosus

5. Umbels panicled decompound .... S. grossus

S. affinis Roth, in Roem. & Schult. Syst. 2:140.1817 & Nov. Pl. Sp. 30.1821; Maheshwari, Fl. Del. 362.1963.
S. maritimus var. *affinis* Clarke in FBI 6:659.1893; FUGP 2:412.

Erect spongy-rooted sedges with creeping rhizomes. Stems triquetrous, leafy upto 40 cm tall. Leaves blade 20 cm long with a prominent sheath on the stems. Bracts 1-3, unequal. Spikelets 1-5, sessile, ovoid-lanceolate, 2 cm long, light brown. Nuts lenticular, dark brown, ovate, apiculate.

Flowers and Fruits : February-May

Found frequently in marshy places. Singh 7323, 8392 (CDRI); Inayat 22872 (DD).

S. michelianus Linn. Sp. Pl. 52.1753; FBI 6:662; FUGP 2:411.

Erect tufted fibrous rooted annuals upto 15 cm tall. Leaves as long as the stem, narrow, grass like; sheath membranous, striate. Bracts many, leaflike. Spikelets numerous in dense terminal heads. Glumes elliptic, margins hyaline. Bristle absent. Nuts fusiform, pale-brown.

Flowers and Fruits : September-November

Occurs occasionally in marshy places. Inayat 22864 (DD).

S. articulatus Linn. Sp. Pl. 47.1753; FBI 6:556; FUGP 2:410; Kern in Backer & Bakh.f. Fl. Java 3:458.1968 et in Fl. Males. loc. cit. 513.1974; Srivastava, FG. 347.1976.

Erect amphibious sedges with fibrous roots. Stems terete, fistular, transversely septate when dry. Leaves absent or the sheath with a membranous acute tips. Spikelets pale-brown in a e. Bristles absent. Nuts obovoid, triquetrous, apiculate, black.

Flowers and Fruits : September-November

Found frequently in marshy and muddy places. Singh 4295, 7351, 7802 (CDRI); Sis Ram 46400 (DD).

S. mucronatus Linn. Sp. Pl. 50.1753; FBI 6:657; FUGP 2:409; Kern in Reinwardtia 6:33.1961 et in Backer & Bakh.f. Fl. Java 3:459. 1968 et in Fl. Males. ser.1. 7(3):510.1974; Srivastava, FG. 348.1976.

Erect perennial herbs, 30-50 cm tall. Stem stout, acutely triquetrous with a loose sheath. Leaf blade absent. Spikelets 4-10 in single lateral clusters near the top of the stem. Glumes 4 mm long, broadly ovate, acute, finely multiseriate. Bristles 5-6, unequal, hispid. Nuts obovoid, unequally trigonous, black, smooth.

Flowers and Fruits : September-November

Found occasionally in muddy places. Singh 8338 (CDRI); Inayat 22870 (DD).

S. tuberosus Desf. Pl. Allant. 1:50.1798; Srivastava, FG. 349.1976.
S. maritimus Linn. Sp. Pl. 51.1753 pro parte, FBI 6:658; FUGP 2:411.

Erect glabrous sedges with creeping rhizomes and woody tubers. Stem stout, trigonous. Leaves grasslike, as long as the stem. Umbels simple or compound; rays unequal. Spikelets 3-8 or solitary on each ray, dark-brown, ovate-oblong, elliptic. Glumes ovate, bifid at the apex. Bristles 3-6, unequal scabrid. Nuts planoconvex, smooth, brown.

Flowers and Fruits : September-December

Found occasionally in marshy localities in the area. Inayat 22872, 22866 (DD); G. Saran 26460, 26410 (LWG).

S. grossus Linn.f. Suppl. Pl. 104.1781; FBI 6:659; Kern in Fl. Males. ser.1 vol.7(3):498.1974; Srivastava, FG. 348.1976.

Large sedges with the stolons frequently bearing black tubers. Stem sharply triquetrous. Leaves few, radical, long, acuminate, coriaceous. Spikelets dark-brown, ovoid in panicles. Glumes brown, scarious, hypogynous bristles usually 6. Nuts obovoid, trigonous.

Local name : Ata, Ater
Flowers and Fruits : September-December

Found occasionally in ponds and other muddy places. The leaves are used for making mats. G. Saran 26151 (LWG).

Scleria Berg.

S. laevis Retz. Obs. Bot. 4:13.1786; M.A. Rau in Bull. Bot. Surv. India 2 (10:87.1969; Kern in Fl. Males. ser. 1.7(3): 732. 1974.

S. hebecarpa Nees in Linnaea 9:303.1835; FBI 6:689; FUGP 2:416.

Erect perennial sedges upto 1 m tall. Stems trigonous, leafy. Leaves 15-30 cm long, acuminate; sheath 3-winged, scabrid. Spikes 5-8, dense, cylindric, 3-7 cm long, the terminal usually female, other male at the base arranged in panicles. Bracts foliaceous, much exceeding the inflorescence. Fruiting glumes 3, broadly ovate, cuspidate. Nuts ovoid, white, polished, apiculate.

Flowers and Fruits : August-November

Occurs frequently on the edge of the sal forests in moist shady places. Singh 3044, 7543, 8466, 8514 (CDRI).

120. POACEAE (nom. alt. GRAMINEAE)

1. Culms woody with culm sheath (reduced leaves) ....**Dendrocalamus**

1. Culms not woody; leaves not reduced to culm sheath:

    2. Spikelets of two florets, the upper perfect and the lower staminate or neutral and usually falling off entirely together with glumes at maturity:

        3. Male and female spikelets in separate inflorescence:

            4. Plant glabrous or slightly hairy; caryopsis enclosed in a stony bract..**Coix**

            4. Plant hairy; caryopsis enclosed in a thick lower glume....... ....**Chionachne**

        3. Spikelets all hermaphrodite with male or barren or hermaphrodite spikelets mixed in the same inflorescence:

            5. Spikelets often paired; with one sessile and other pedicelled, glumes as long as the spikelets; upper lemma usually awned:

                6. All spikelets alike:

                    7. Racemes in more or less compound panicles or racemosely arranged on an elongated common axis:

                        8. Spikelets all pedicelled ....**Imperata**

                        8. Spikelets one sessile, the other pedicelled:

                            9. Panicles narrow, villous to tomentose; glumes coriaceous.................**Narenga**

                            9. Panicles wide, silky hairy; glumes membranous ....**Saccharum**

                    7. Racemes on a short common axis or digitate or sometimes solitary:

                        10. Spikelets 2-flowered ....**Eulaliopsis**

                        10. Spikelets 1-flowered:

11. Lower glumes dorsally flattened or shallowly concave between the keels; upper lemma bidentate at the apex ....**Eulalia**

11. Lower glumes convex on the back; upper lemma stipiform ....**Pseudopogonatherum**

6. Sessile and pedicelled spikelets dissimilar;

12. Joints and pedicels trigonous, rounded or flat:

13. Upper lemm awned:

14. Racemes many noded, not enclosed in a spathe like sheath ....**Ischaemum**

14. Racemes 1-noded, partially enclosed into boat-shape, spathe ....**Apluda**

13. Upper lemma not awned:

15. Sessile spikelets globose ....**Hackelochloa**

15. Sessile spikelets not globose:

16. Racemes compressed; spikelets of each pair similar ....**Hemarthria**

16. Racemes cylindrical; spikelets of each pair dissimilar:

17. Pedicelled spikelets distinct and more or less of same size ....**Rottboellia**

17. Spikelets all sessile; pedicelled rudimentary ....**Mnesithea**

12. Joints and pedicels slender, rarely thickened upwards:

18. Racemes in whorled panicles, never spatheate:

19. Lower glumes flattened on the back:

20. Panicles dense, contracted ....**Pseudosorghum**

20. Panicles lax, open ....**Sorghum**

19. Lower glumes not flattened on the back:

21. Spikelets triad, one sessile and two pedicelled ....**Chrysopogon**

21. Spikelets biad, one sessile and other pedicelled ....**Vetiveria**

18. Racemes digitate, subdigitate or in pairs or solitary terminal, usually spatheate:

22. Lemma awned from the back ....**Arthraxon**

22. Lemma awned from the tip or sinus:

23. Margins of lower glumes of sessile spikelets sharply infolded, two keeled; awn glabrous:

24. Lemma awned from the tip:

25. Racemes in compound espatheate panicles...**Capillipedium**

25. Racemes not in compound espatheate panicles:

26. Lower involucral glumes with pits..........**Bothriochloa**

26. Lower involucral glumes without pits .....**Dichanthium**

24. Lemma awned from a sinus:

27. Plant aromatic; racemes binnate, supported by spathe arranged in panicles ....**Cymbopogon**

27. Plant not aromatic; racemes solitary terminal on the culms and branches ....**Schizachyrium**

23. Margins of lower glumes inturned and rounded at the sides; awn usually hairy:

28. Racemes surrounded at the base by an involucre of homogamous, sessile, non-deciduous spikelets..........**Themeda**

28. Racemes without an involucre ....**Heteropogon**

5. Spikelets solitary or paired, more or less similar; glumes usually membranous, the lower usually smaller or sometimes suppressed; upper lemma usually awnless:

29. Spikelets in open or contracted panicles:

30. Spikelets in open panicles:

31. Spikelets not supported by bristles:

32. Spikelets gibbous, laterally compressed...**Cyrtococcum**

32. Spikelets not gibbous ....**Panicum**

30. Spikelets in contracted spike like panicle ....**Hymenachne**

29. Spikelets in secundate spiciform racemes or false spikes:

33. Upper lemma crustaceous or coriaceous:

34. Spikelets adaxial;

35. Glumes awned:

36. Blades lanceolate to ovate ....**Oplismenus**

36. Blades linear, narrow ....**Echinochloa**

35. Glumes awnless:

37. Lower glumes well developed:

38. Aquatic plants ....**Paspalidium**

38. Terrestrial or marshy plants ....**Urochloa**

37. Lower glumes absent ....**Paspalum**

34. Spikelets abaxial ....**Brachiaria**

33. Upper lemma thinly cartilaginous, usually with flat hyaline margin ....**Digitaria**

31. Spikelets supported by bristles:

39. Bristles caducous, forming an involucre; upper lemma smooth ....**Pennisetum**

39. Bristles persistent, solitary; upper lemma transversely rugose ....**Setaria**

2. Spikelets always with a single hermaphrodite floret; if two-flowered, with one of the florets barren, then lemmas as firm as glumes:

40. Spikelets arranged in open or contracted spike-like panicles:

41. Spikelets with 2 or more fertile florets:

42. Glumes shorter than the lowest florets:

43. Lemmas 5-many nerved ....**Poa**

43. Lemmas 1-3-nerved:

44. Spikelets in open, contracted or spikelike panicle ....**Eragrostis**

44. Spikelets sessile in digitate or racemosely arranged spikelike racemes:

45. Rachis ending in spikelets:

46. Spikelets falling entire of straight spikes which are numerous and crowded into long, narrow dense panicles ....**Desmostachya**

46. Spikelets breaking up at maturity; spikes few to many...**Eleusine**

42. Glumes longer than the lowest florets:

47. Lemmas and rachila glabrous ....**Avena**

47. Lemmas and rachila joints with long silky hairs ....**Phragmites**

41. Spikelets with one fertile floret:

48. Glumes minute or suppressed; fertile lemma and palea very similar in shape and texture:

49. Blade ovate-oblong or ovate-lanceolate with inflated sheath ....**Hygroryza**

49. Blade linear, acute or acuminate ....**Oryza**

48. Glumes well developed; fertile lemma and palea dissimilar:

50. Spikelets with 3-florets ....**Phalaris**

50. Spikelets with 1 to 2-florets:

51. Spikelets with 2-florets ....**Arundinella**

51. Spikelets with 1-floret:

52. Lemmas 1 to 3-nerved ....**Perotis**

52. Lemmas 5-nerved ....**Polypogon**

40. Spikelets arranged on one side of the tough rachis:

53. Spikelets awned ....**Chloris**

53. Spikelets awnless ....**Cynodon**

45. Rachis ending in a sharp point ....**Dactyloctenium**

Dendrocalamus Nees

D. strictus (Roxb.) Nees in Linnaea 9:476.1834.
Bambusa stricta Roxb. Pl. Cor. t. 80. 1798; FBI 7:404.

Small trees. Culms variable, dark green, nodes swollen, lower often rooting; culm sheath triangular, with brown shining hairs on the back; ligule narrow. Leaves linear-lanceolate, rough, hairy; tip twisted, margins scabrous; petiole minute, seath hairy; auricle ciliate; ligule serrate.

Local name : Bans

Occurs occasionally in the Park near Tharu villages. The stem is used for making baskets. This species has not been collected in flowering. Singh 7684 (CDRI).

Coix Linn.

C. lacryma-jobi Linn. Sp. Pl. 972.1753; Raizada *et al.,* in Indian For. Rec. 4(7):176.1957; Bor, Grasses India 264.1960; FBI 7:100; Srivastava, FG. 360.1976.

Stout annual grass. Culms branching at the lower nodes, glabrous, polished. Leaves linear-lanceolate from a cordate base, acuminate, margins spinuous, serrate; ligule a narrow membrane. Peduncles flattened or trigonous, 2.5-7 cm long, usually in large numbers from the axil of one leaf together with oblong bracts. Male inflorescence glabrous, erect or nodding. Spikelets sessile and one pedicelled. Glumes lanceolate. Caryopsis ovoid, enclosed in involucre.

Flowers and Fruits : September-December

Found rarely near village side in moist places. The seeds are used for making ornaments. Singh 4359 (CDRI).

Chionachne R.Br.

C. koenigii (Spreng.) Thw., Enum. Pl. Zeyl. 357.1864; Raizada *et al.* in Indian For. Rec. 4(7):177.1957; Bor, Grasses India 262.

Coix koenigii Spreng. Syst. 1:239.1825.

Polytoca barbata (Roxb.) Stapf in Hook.f. FBI 7: 102. 1896.

Tall annual or perennial grasses. Stems upto 2 m high, slightly hairy with the leaves and sheath bristly with tubercled hairs. Leaves 30-60 x 1.5-2 cm, acuminate, with thickened margin. Male spikelets glabrous or softly pilose. Female spikelets enclosed in spathiform bract, solitary. Caryopsis enclosed in thickened lower glumes.

Flowers and Fruits : August-November

Found commonly in open areas in grasslands. Singh 4193, 4291, 7524, 7760, 7914 (CDRI).

Imperata Cyrill.

I. cylindrica (Linn.) Beauv. Agrost. 165,t.5,f.1. Expl. Planch, 5.1812; Raizada *et al.*, loc.cit. 182.1957; Bor, Grasses India 169; Srivastava, FG. 369.1976.

Lagurus cylindricus Linn. Syst. 878.1759.
Imperata arundinacea Cyrill. Pl. Rar. Neap. 2:26, t.11.1792; FBI 7:106.

Perennial tufted grasses with culms densely beared at nodes. Leaves linear, flat, 30-40 x 2.5 mm, acuminate, scabrid on the margins. Inflorescence silvery white, dense, cylindric, spikelike panicle. Spikelets lanceolate, in pairs. Lower glumes 3.5-5 mm long, 3 to 8-nerved. Lower lemma smaller than the glumes.

Local name : Bharuee
Flowers and Fruits : June-October

Common in grasslands and in open fields. Singh 7672, 8420 (CDRI); G. Saran 26568, 26179 (LWG).

Narenga Bor

N. porphyrocoma (Hance) Bor in Indian For. 66:267.1940; Raizada *et al.*, loc.cit. 184; Bor, Grasses India 198.1960.
Eriochrysis porphyrocoma Hance in J. Bot. 14:294.1876.
Saccharum narenga Ham. in Wall. Cat. No. 8856; FBI 7:267.1896.

Erect tall perennial grasses. Culms 2-2.5 m tall, slender, rounded, appressed silky below the panicles and on the nodes. Leaves 30-60 cm long, flat, scabrid; ligules rounded, ciliolate. Panicles narrow, dense, linear, woolly, purplish, 30-40 cm long. Sessile spikelets oblong, brown; callus small with rigid hairs. Glumes coriaceous.

Flowers and Fruits : October-November

Found occasionally in open areas and in grasslands. Singh 4294, 7844 (CDRI).

Saccharum Linn.

1. Peduncles hairy below the panicle; rachis very fragile ....**S. spontaneum**
1. Peduncles glabrous; rachis less fragile ....**S. bengalense**

S. spontaneum Linn. Mant. 183.1771; Raizada *et al.,* loc.cit. 185; Bor, Grasses India 214; Srivastava, FG. 376.1976.

Tall perennial grasses. Leaves 40-100 cm long, narrow, linear, acuminate, coriaceous, often blotched reddish-black inside. Inflorescence a silvery panicle, lanceolate; branches whorled. Spikelets paired, lanceolate, sessile and pedicelled, similar, 1-flowered, hermaphrodite.

Local name : Kans
Flowers and Fruits : September-January

Found commonly along river banks, in swampy places and along the bunds of fields etc. The plants is used for making basket. Singh 3028, 7500 (CDRI).

S. bengalense Retz. Obs. Bot. 5.16.1789; Bor, Grasses India 210.1960.
S. munja Roxb. Fl. India 1:246.1820.
Erianthus munja (Roxb.) Jesw. in Rev. Bot. Appliq. 581. 1929; Raizada, *et al.,* loc. cit. 186.1957.

Tall perennial grasses. Leaves flat, 20-60 cm long, acuminate, scabrid on mrgins, glaucous. Spikelets in pairs, one sessile and other pedicelled, clothed with soft creamy or purplish woolly hairs.

Local name : Munj
Flowers and Fruits : September-October

Found frequently in marshy places and along railway lines and in other open areas. The leaf sheath is used for making basket. Singh 4330 (CDRI) G. Saran 25783 (LWG).

Eulaliopsis Honda

E. binata (Retz.) Hubbard in Hook. Ic. Pl. Sub. tab. 3262.1935; Raizada *et al.,* loc. cit. 191; Bor, Grasses India 158.1960.
Andropogon binatus Retz. Obs. Bot. 5:21.1789.

Perennial grasses. Culms tufted, erect, slender, glabrous, woolly at the base. Leaves flat, glabrous; mouth of sheath ciliate, margin scaberulous; ligule a ridge of short hairs. Racemes sessile, unequal, 2-4 cm long, dense, dark yellow. Spikelets lanceolate; callus very short, obtuse with rusty hairs.

Local name : Baib.
Flowers and Fruits : October-February

Occurs occasionally in open dry places near sal forests. The leaves are used for making ropes. Singh 7618 (CDRI); G. Saran 26443 (LWG).

Eulalia Kunth

E. leschenaultiana (Decne) Ohwi in Bull. Tokyo Sci. Mus. No. 18:2.1947; Raizada, *et al.,* loc.cit. 194; Bor, Grasses India 155.1960.

Andropogon leschenaultiana Decne, Herb. Timor. Descr. 29:1835.

Eulalia cummingii (Nees) A. Camus in Lecomte Fl. Gen. de L' Indo-Chine 7:250.1922.

Erect perennial grasses. Culms slender, pubescent, terete, 5-6 noded. Leaves 5-15 cm long, linear, acuminate, margins scabrid. Inflorescence of digitate or fascicled racemes, covered with dense golden-yellow or fulvous hairs. Spikelets oblong, 2.5-3 mm long. Upper lemma deeply bifid, awn 1.2-1.5 mm long, slender.

Flowers and Fruits : September-November

Occurs occasionally in moist places. Singh 7879 (CDRI).

Pseudopogonatherum A. Camus

P. contortum (Brongn.) A. Camus in Lecomte, Fl. Gen. de L' Indo-Chine 7:255. 1922; Raizada *et al.,* loc. cit. 194; Bor, Grasses India 204. 1960.

Pogonatherum contortum Brongn. in Duperr. Voy. Bot. Coq. 90 t.17.1831.

Pollinia articulata Trin. in Bull. Sc. Acad. Petersb. 71.1836; FBI 7:109.

Erect annual grasses. Culms slender, densely tufted, glabrous with short internodes. Leaves linear-lanceolate, acuminate, hairy on the inside; ligule a tuft of hairs. Racemes on a common axis, alternate or opposite, silky white, 3.5-6 cm long. Spikelets all pedicelled, lanceolate. Lower glumes convex on the back. Upper lemma hermaphrodite, membranous, linear, 1-nerved.

Flowers and Fruits : October-December

Found frequently in grasslands. Singh 7865 (CDRI).

Ischaemum Linn.

I. rugosum Salisb. Icon. Strip. Rar. 1.t.1.1791; FBI 7:127; Raizada et al. loc. cit. 200; Bor, Grasses India 184; Srivastava, FG. 369.1976.

Erect annual grasses, 30-60 cm tall. Leaves linear-lanceolate, acuminate, 5-20 cm long, glabrous or sparsely hairy; sheaths inflated. Racemes usually two, erect, pale-yellow, arising together from the top of the culms, rachis trigonous, ciliate on the dorsal angle. Spikelets 4-5 cm long, linear-oblong, the sessile and pedicelled closely appressed together.

Flowers and Fruits : September-December

Occurs commonly in wet marshy places near rice fields. Singh 7945, 8423 (CDRI).

Apluda Linn.

A. mutica Linn. Sp. Pl. 82.1753; Raizada, *et al.*, loc. cit. 203; Bor, Grasses India 93; Srivastava, FG. 356.1976.

A. varia Hack. var. *aristata* Hack. in DC. Mon. Phan. 6:196.1889; FBI 7:150.

Erect leafy perennial grasses. Culms many moded, smooth, glabrous. Leaves 20-30 x 1.5 cm, linear-lanceolate, apex setaceous. Inflorescence a leafy panicles of reduced racemes, enclosed in a boat-shaped shortly awned spathe. Spikelets in trios, one sessile, two pedicelled. Sessile hermaphrodite, upper floral glumes hyaline, cleft half way down, awned in the cleft.

Flowers and Fruits : September-February

Found commonly in open grasslands and along railway lines. The Plant is used for thatching. Singh 4188, 7787, 8534 (CDRI).

Hackelochloa O. Kuntze

H. granularis (Linn.) O. Kuntze, Rev. Gen. Pl. 2.776.1891; Raizada, *et al.,* loc. cit. 205; Bor, Grasses India 159; Srivastava, Fg. 367.1976.

Cenchrus granularis Linn. Mant. 2, App. 575.1771.

Manisuris granularis Linn. F. Nov. Gram. Gen. 40.1779; FBI 7:159.

Erect annual much branched grasses. Leaves 5-15 cm long, lanceolate, base cordate, acute, hirsute, soft, margins ciliate. Spikelets in pairs, sessile and pedicelled; all pedicelled spikelets facing one direction, all sessile the other; pedicels completely fused with the adjacent rachis forming a cavity. Caryopsis subglobose.

Flowers and Fruits : September-December

Found frequently in sal forests in moist places and in open grasslands. Singh 7397, 7788, 8520 (CDRI).

Hemarthria R.Br.

H. compressa (Linn.f.) R.Br. Prodr. 207.1810; Raizada *et al.,* loc. cit. 207; Bor, Grasses India 160; Srivastava, FG.

367.1976.
Rottboellia compressa Linn.f. Suppl. 114.1781; FBI 7:153 in part.

Hygrophilous perennial grasses. Culms erect from a decumbent, rooting base, compressed. Leaves 6-12 x 0.3-0.5 cm, linear, acute. Raceme erect, compressed, dark green; joints of the rachis thickened, glabrous. Sessile spikelets 4-4.5 mm long, linear, lanceolate. Pedicelled spikelets hermaphrodite or male.

Flowers and Fruits : July-September

Occurs commonly in marshy areas and in rice fields. Singh 8440, 8467 (CDRI).

Rottboellia Linn.f.

R. exaltata Linn.f. Nov. Gram. Gen. 40.t.1.1779 et in Suppl. 114.1781; FBI 7:156; Raizada *et al.*, cot. cit. 208; Bor, Grasses India 206; FBI 7:156; Srivastava, FG. 374.1976.

Erect annual grasses. Culms branched, solid, leafy, smooth. Leaves linear-lanceolate, acuminate, scabrid or hispid; sheath loose, often covered with bulbous-based hairs in the upper half. Racemes 6-13 cm long, yellowish-green. Sessile spikelets 6-8 mm long, ovate-lanceolate, glabrous. Glumes 4. Pedicelled spikelets imperfect.

Flowers and Fruits : August-November

Found frequently along the water streams. Singh 7951 (CDRI).

Mnesithea Kunth

M. laevis (Retz.) Kunth, Rev. Gram. 1:154.1829; Raizada *et al.*, loc. cit. 209; Bor, Grasses India 196.
Rottboelia laevis Retz. Obs. 3:11.1783.
R. perforata Roxb. Pl. Corom. 2, 43. t.182. 1798; FBI 7:158.

Perennial branched grasses with culms upto 90 cm long, 4-5 noded. Leaves 7-30 cm long, linear, base narrow, glabrous except the base. Spikelets 8-16 cm long, joints semicylindric. Spikelets 3-5 mm long. Lower florets neuter, upper bisexual. Caryopsis narrow-oblong.

Flowers and Fruits : July-December

Found commonly in grasslands and in moist beds of rice fields. Singh 7707 (CDRI).

Pseudosorghum A. Camus

P. fasciculare (Roxb.) A. Camus in Bull. Mus. Hist. Nat. Paris 26:662.1920; Raizada *et al.*, loc. cit. 213; Bor, Grasses India 205.
Andropogon fascicularis Roxb. Fl. Ind. 265.1832; FBI 7:177.

Erect annual grasses. Culms slender, rounded, glabrous. Leaves 20-25 cm long, linear or sublanceolate-linear, acuminate. Panicles 3-8 cm long, contracted, dense, green or purplish. Sessile spikelets lanceolate. Pedicelled spikelets male or neuter.

Flowers and Fruits : September-December

Found commonly in grasslands and along railway lines. Singh 7876, 7881, 7974 (CDRI).

Sorghum Moench

1. Primary branches of panicles undivided .... **S. nitidum**
1. Primary branches of panicles branched .... **S. halepense**

S. nitidum (Vahl) Pers. Syn. Pl. 1:101.1805; Raizada *et al.*, loc. cit. 214; Bor, Grasses India 245.1960.
Holcus nitidum Vahl, Symb. Bot. 2:102.1791.
Andropogon serratus Thunb. Fl. Jap. 41. 1784; FBI 7:185.

Erect tufted grasses upto 1 m high. Culms densely villous at the nodes. Leaves linear-lanceolate, setaceously acuminate, with a prominent white midrib, sparsely hairy on both surfaces. Inflorescence an elongate panicle, oblong; lower branches whorled, upper alternate. Spikelets 2-nate; differing in shape and sex.

Flowers and Fruits : September-December

Found commonly in open grasslands and along railway lines. Singh 4179, 4206, 4325, 7832, 7878 (CDRI).

S. halepense (Linn.) Pers. Syn. Pl. 1:101.1805; Raizada *et al.,* loc. cit. 215; Bor, Grasses India 222; Srivastava, FG. 378.1976.
Holcus halepensis Linn. Sp. Pl. 1047.1753.
Andropogon halepensis Brot. Fl. Lus. 1:89.1804; FBI 7:182.

Erect stout perennial grasses. Culms erect smooth, glabrous and polished. Leaves linear-lanceolate, setaceously acuminate. Inflorescence a decompound panicles, open, lax, brown or purplish in colour. Spikelets in groups at the end of the branches, ovoid-lanceolate. Sessile spikelets elliptic, acute, 4-5 mm long, hermaphrodite. Pedicelled one male or neuter.

Local name : Ghamoy.
Flowers and Fruits : September-December

Found occasionally in cultivated fields and along railway lines. Singh 8533 (CDRI).

Chrysopogon Trin. nom. cons.

C. aciculatus (Retz.) Trin. Fund. Agrost. 188.1820; Raizada *et al.,* loc. cit. 218; Bor, Grasses India 115; Srivastava, FG. 360. 1976.
Andropogon aciculatus Retz. Obs. 5:22.1788; FBI 7:188.

Erect or ascending perennial grasses with creeping rhizomes. Leaves 2-14 cm x 3-6 mm, densely tufted, linear, margins spinulose, sheath terete. Panicles 2-9 cm long, pale-purplish with spreading, filiform branches. Spikelets in three, awned, one sessile, hermaphrodite; the two lateral male or neuter, pedicellate; peduncles hairy and fulvously bearded at the top. Caryopsis linear, laterally compressed.

Flowers and Fruits : August-November

Frequent in open wastelands. Singh 2984 (CDRI).

Vetiveria Bory

V. zizanioides (Linn.) Nash in Small, Fl. Southeast U.S. 67.1903; Raizada *et al.,* loc. cit. 220; Bor, Grasses India 258; Srivastava, FG. 380.1976.
Phalaris zizanioides Linn. Mant. 183.1771.
Andropogon squarrosus Hook.f. in FBI 7:186.1896 (non Linn.f).

Densely tufted perennial grasses upto 3 m high with aromatic roots and rhizomes. Leaves linear, rigid, 25-85 cm x 6-18 mm, usually glabrous, ligule reduced to a scarious rim. Panicles oblong, erect, 18-40 cm long, composed of whorls of compound racemes. Spikelets 2-nate, one sessile and one pedicelled. Sessile spikelets 3.5-5 mm long, linear-lanceolate, lower glumes muriculate all over back.

Local name : Seenk, Khas
Flowers and Fruits : July-November

Found commonly in sal forests and other open marshy places. The roots are aromatic. The plant is used in making brooms. Singh 2983, 7550, 7764 (CDRI).

Arthraxon P. Beauv.

A. prionodes (Steud.) Dandy in Andrews, Fl. Sudan 3:399. 1956.1956; Bor, Grasses India 101 et in Rech. f. Fl. Iran. 70:535.1970.

Andropogon prionodes Steud. Syn. Pl. Glum. 1:383.1954.

A. serrulatus Hochst. in Flora 39: 188.1856.

A. lanceolatus Hochst. in Flora 39:188.1856; FBI 7:143 (in part); Raizada *et al.,* loc. cit. 224.1957.

Perennial grasses with short rhizomes; culms covered with red silky cataphylls. Leaves 2-3.5 cm long, ovate-lanceolate, acuminate, base cordate-amplexicaul with cartilaginous margins. Racemes 2 to 6-nate, green or purplish-violet, upto 6 cm long. Sessile spikelets upto 7 mm long, glabrous. Pedicelled spikelets male, upto 6 mm long.

Flowers and Fruits : October-February

Occurs occasionally in the open areas. G. Saran 26286 (LWG).

Capillipedium Stapf

C. assimile (Steud.) A. Camus in Lecomte, Fl. Gen. de Indo-Chine 7:314.1922; Raizada *et al.,* loc. cit. 226; Bor, Grasses India 110. et in Rech. f. Fl. Iran 70:536.1970.

Andropogon assimilis Steud. in Zoll. Syst. Verz. 58.1854; FBI 7:179.

Perennial suffruticose grasses. Culms decumbent and rooting at the base, branched. Leaves 7-15 cm x 2.5-5 mm, linear-lanceolate, narrowed at base, glabrous or hispidulous with spinously toothed margins. Inflorescence decompound panicles, 6-11 cm long, slender, triangular. Sessile spikelets 2-3 mm long, linear-oblong, all hermaphrodite, pale-green. Lower lemma empty, nerveless; upper fertile; awn upto 10 mm long. Pedicelled spikelets male or neuter; lower lemma empty.

Flowers and Fruits : August-December

Occurs commonly along railway lines. The plant is used for thatching material. Singh 4182, 7979 (CDRI).

Bothriochloa O. Kuntze

1. Lower racemes shorter than the rachis ....**B. intermedia**
1. Lower racemes longer than the rachis:
   2. Lower glumes of sessile spikelets hairy below the middle..**B. pertusa**
   2. Lower glumes of sessile spikelets glabrous below the middle ....**B. kuntzeana**

B. intermedia (R.Br.) A Camus in Ann. Soc. Linn. Lyon n.s. 76.164.1931; Bor, Grasses India 108.1960.
Andropogon intermedius R.Br. Prodr. 202.1810; FBI 7:175.
Dichanthium bladhii (Retz.) Clayton.

Tall erect grasses. Leaves narrowly linear, sheath terete. Sessile spikelets 3-4.5 mm long, oblong, lanceolate, acute. Lower lemma of the sessile spikelets nearly as long as glume. Racemes 3-7 cm long. Sessile spikelets not pitted.

Flowers and Fruits : August-December

Found commonly in grasslands. The plant is used for thatching. Singh 4191, 7686, 7937 (CDRI).

B. pertusa (Linn.) A. Camus in Ann. Soc. Linn. Lyon n.s.76.164.1931; Raizada *et al.,* loc. cit. 230; Bor, Grasses India 109; Srivastava, FG. 358.1976.
Holcus pertusus Linn. Mant. 301.1771.
Andropogon pertusus Willd. Sp. Pl. 4:922.1806; FBI 7:173.

Perennial grasses with slender, suberect or ascending branches. Leaves linear, 15 cm x 3-4.5 mm, acuminate, contracted at the base. Racemes subdigitate, 2-8, slender, flexuous, purplish. Spikelets in pairs, hermaphrodite sessile, pedicelled. Glumes of sessile spikelets with a pit.

Local name : Sandhur
Flowers and Fruits : September-January

Found commonly in grasslands and in dried up rice fields, wastelands etc. Singh 8540, 8551 (CDRI).

B. kuntzeana (Hack.) Henr. in Blumea 3:456.1940; Raizada *et al.,* loc.cit. 231; Bor, Grasses India 108.
Andropogon kuntzeanus Hack. Mon. Androp. 478.1889; FBI 7:175.
Dichanthium kuntzeana (Hack.) Jain *et* Deshpande in Bull. bot. Sur. India. 20 No 1-4, 136, 1978.

Tall perennial grasses with nodes densely bearded. Leaves 16-30 cm x 3-6 mm, narrowly linear, glabrous, glaucous, with scaberulous margins, the upper cauline leaves short, rigid. Racemes 12-15, corymbosely fasciculate, slender, purplish. Lower glumes of sessile spikelets ovate-lanceolate, 7 to 9-nerved, pitted. Upper glumes glabrous, thin, lanceolate. Pedicelled spikelets male.

Flowers and Fruits : September-January

Occurs occasionally in open areas and in grasslands. Singh 7978 (CDRI).

Dichanthium Willem.

D. annulatum (Forsk.) Stapf in Prain, Fl. Trop. Afr. 9:178.1917; Raizada *et al.,* loc. cit. 233; Bor, Grasses India 133; Srivastava, FG. 362.1976.
Andropogon annulatus Forsk. Fl. Aegypt.-Arab. 173. 1775; FBI 7:196.

Perennial densely tufted leafy grasses. Culms erect or ascending upto 1 m tall. Leaves linear, setaceous-acuminate, rigid, glaucous. Racemes greenish or purplish, 3 or often more, subdigitately fascicled; joints and pedicels silky hairy. Spikelets in pairs. Sessile spikelets hermaphrodite, upto 3.5 mm long, oblong, concave, awned. Pedicelled spikelets male or neuter, exceeding the sessile spikelets. Caryopsis obovate-oblong.

Flowers and Fruits : August-January

Occurs commonly in lawns and other wastelands. Singh 7238, 7600, 7619, 8332 (CDRI).

Cymbopogon Spreng.

1. Sessile spikelets awnless .... **C. osmastonii**

1. Sessile spikelets awned:
    2. Panicles short in dense fascicles pairs; joints and pedicels villous with long hairs .... **C. jwarancusa**
    2. Panicles large, lax, slender, joints and pedicels without villous hairs .... **C. flexuosus var. microstachys**

C. osmastonii Parker in Fedde, Repert. Sp. Nov. 31:126.1932; Raizada *et al.*, loc. cit. 240; Bor, Grasses India 130.1960.

Perennial grasses. Culms upto 1.5 m high erect, terete, glabrous. Leaves flat, linear, setaceous-acuminate, upto 18 cm long, rounded or subcordate at the base. Inflorescence a narrow panicle; spatheoles contain a pair of racemes. Sessile spikelets awnless, hermaphrodite, 2.5-3 mm long, greenish or purplish. Lower glumes of sessile spikelets with a slit-like furrow in the lower half. Pedicelled spikelets 3 mm long, male, golden brown or purplish.

Flowers and Fruits : September-December

Occurs occasionally in the Dudhwa range along forest roads, mixed with other grasses. This grass is endemic to the Park. Singh 7877 (CDRI).

C. jwarancusa (Jones) Schult. Mant. 2:458.1824; Raizada *et al.*, loc. cit. 243; Bor, Grasses India 128.1960.
Andropogon jwarancusa Jones in Asiat. Res. 4:109.1795.

Perennial aromatic grasses with a stout rhizomatous rootstock. Leaves flat, linear, 20-25 cm x 3-5 mm, glabrous on both surfaces, coriaceous; basal sheath when old slipping from the culms and becoming curled in a spiral. Inflorescence a panicle of paired racemes seated on a peduncle, supported by spathe. Sessile spikelets 5 mm long, linear, awned, hermaphrodite. Pedicelled spikelets oblong, elliptic, acuminate, male.

Local name : Jaracus
Flowers and Fruits : August-December

Found rarely in the area in open places in sal forests. The decoction of the plant is given in fever. Singh 7181 (CDRI).

C. flexuosus (Nees) Wats. var. **microstachys** (Hook.f.) Bor in JBNHS 52:162.1954; Raizada *et al.,* loc. cit. 247; Bor, Grasses India 127.1960.

Erect tall aromatic slender perennial grasses. Leaves linear, acuminate; sheath very narrow, tubular, erect. Panicles erect long, slender, bearing few distant solitary or subsolitary erect filiform branches with few small deflexed spikes. Spikelets lanceolate, 4-5 mm long, awned.

Flowers and Fruits : September-December

Found frequently in grasslands. Singh 4315 (CDRI).

Schizachyrium Nees

S. brevifolium (Sw.) Nees Agrost. Bras. 332. 1829; Raizada *et al.,* loc. cit. 235; Bor, Grasses India 215.1960.
Andropogon brevifolium Sw. Prodr. Veg. Ind. Occ. 26.1788; FBI 7:165.

Slender weak divaricately branched annual grasses. Leaves linear, constricted at the base, 2-3 x 2-5 mm, acute; turning red. Racemes borne on filiform, clavate, slender peduncles. Sessile spikelets linear-lanceolate, acuminate, pale-green. Pedicelled spikelets reduced to a minute glume, provided with bristle like awn.

Flowers and Fruits : September-December

Found occasionally in open areas. Singh 7839 (CDRI).

Themeda Forsk.

T. **arundinacea** (Roxb.) Ridley in TLS 2:401.1893; Raizada *et al.*, loc. cit. 252; Bor, Grasses India 250.1960 (Plate v1).
Anthistiria arundinacea Roxb. Fl. Ind. 1:251.1820.
A. gigantea Cav. subsp. *arundinacea* Hack. Mon. Androp. 674.1889; FBI 7:217.

Tall tufted perennial grasses. Culms upto 3 m tall, compressed, smooth and polished. Leaves linear, alternate towards the base, apex subulate, scabrid on the margins; ligule a membranous ring. Inflorescence large consisting of numerous compound racemes arising from the spatheoles. Involucral spikelets lanceolate, covered with long spreading hairs. Sessile spikelets dark-brown, smooth. Pedicelled spikelet seated on the pedicels.

Local name : Ulla
Flowers and Fruits : August-December

Occurs commonly in low-lying areas along the sal forests and in grasslands. The grass is used for thatching. Singh 4211, 8554 (CDRI).

Heteropogon Pers.

H. **contortus** (Linn.) Beauv. ex R & S. Syst. Veg. 2:836.1817; Raizada et al. loc. cit. 247; Bor, Grasses India 162.1960; Srivastava, FG. 368.1976.
Andropogon contortus Linn. Sp. Pl. 1045.1753; FBI 7:199.

Erect or decumbent branched perennial grasses. Leaves 11-20 cm x 2-3.5 mm, linear, abruptly acuminate, flat, sheath compressed and keeled, glabrous. Racemes terminal, spiciform, bristly hairy, 4-7 cm long. Spikelets in pairs; lower homogamous, awnless, male or neuter, 2-6 pairs, covered with bulbous based hairs. Sessile spikelets 6-7 mm long. Lower glumes terete; upper floral glume reduced to an awn, shedding along the grain, with sharp pointe callus. Pedicelled spikelets 8.5-11 mm long, hispid at the back. Caryopsis sublinear, terete.

Flowers and Fruits : September-December

Common in grasslands and in open wastelands. Singh 4274, 7838 (CDRI).

Cyrtococcum Stapf

C. patens (Linn.) A. Camus, in Bull. Mus. Nat. Hist. Paris 27:118.1921; Bor, Grasses India 292. 1960; Raizada & Jain in Indian For. Rec. 5(3):176.1964.
Panicum patens Linn. Sp. Pl. 86.1753; FBI 7:57.

Erect perennial grasses upto 50 cm tall, rooting at the base. Leaves linear-lanceolate, acuminate; sheath ciliate on margins; ligules membranous short, obtuse. Spikelets round, green, arranged in lax spreading panicles. Glumes herbaceous, membranous. Lower lemma 1.8 mm long, sub-elliptic; upper lemma hermaphrodite, cordate.

Flowers and Fruits : August-November

Found frequently in low-lying areas and other moist places. Singh 7536 (CDRI).

Panicum Linn.

1. Spikelets 1.5 mm long .... **P. austroasiaticum**
1. Spikelets 2-4 mm long:
   2. Lower glumes cuspidate-acuminate .... **P. trypheron**
   2. Lower glumes cup shaped .... **P. paludosum**

P. austro-asiaticum Ohwi in Acta. Phytotax. & Geobot. 2:45.1942; Bor, Grasses India 324.1960; Raizada & Jain

in Indian For. Rec. 5(3):155.1964.
P. humile Nees ex Steud. Syn. Pl. Gram. 84. 1854; FBI 7:48.

Slender glabrous annual tufted grasses. Leaves 2-12 cm x 2.5 mm, liner, flat, glabrous, tapering to a slender point. Panicle 10 cm long, open, lax, branches filiform. Spikelets 1.5 mm long, acute. Lower glume obliquely lanceolate, acute.

Flowers and Fruits : September-December

Found frequently in moist places. Singh 8524 (CDRI).

P. trypheron Schult. Syst. Veg. Mant. 2:244.1824; FBI 7:47; Bor, Grasses India 331; Raizada & Jain loc. cit. 154.1964.

Annual, tufted, geniculately ascending grasses with culms upto 80 cm tall. Leaves 6.5-22 cm x 2-3 mm, narrow, linear, acuminate, hairy. Panicles 8-20 cm long, open, branches filiform. Spikelets 2-3 mm long, gaping widely at anthesis. Lower glumes cuspidate-acuminate, upto 2.5 mm long; upper glumes broadly ovate, 9-nerved, subcuspidate.

Flowers and Fruits : August-December

Occurs frequently in cultivated fields and in grasslands. Singh 7722 (CDRI).

P. paludosum Roxb. Hort. Beng. 8:1814, nom. nud. et Fl. Ind. 1:307.1820; Bor, Grasses India 329.1960; Raizada & Jain, loc.cit. 157.1964 (Fig.19).

Perennial aquatic grasses. Culms rooting at the node, spongy, upto 60 cm long. Leaves linear, ensiform, 12-33 cm x 7-15 mm, base rounded, subcordate. Panicles 12-33 cm, whorled. Spikelets 3-3.5 mm long, ovate-lanceolate, acuminate. Lower glumes 1-nerved; upper glumes 9 to 11-nerved, orbicular, apiculate. Lower florets barren; upper florets male.

Flowers and Fruits : September-December

Found commonly in marshy places, still water and banks of tank, ponds throughout the area. Singh 7730, 8473, 8563 (CDRI).

Hymenachne P. Beauv.

H. pseudointerrupta C. Muell. in Bot. Z. 19:333.1861; Bor, Grasses India 313.1960.

Erect leafy stout grasses with long spongy creeping rhizomes. Leaves linear, base cordate, acuminate, 15-40 cm x 15-20 mm, ligules short, truncate. Panicles dense, narrow upto 30 cm long, interrupted at the base. Spikelets 4-6 mm long, narrow, lanceolate-acuminate. Lower glumes 2-3 mm long, 1-nerved, keeled, cuspidate; upper glumes 5 mm long, 5-nerved. Caryopsis oblong, top contracted.

Flowers and Fruits : September-December

Found occasionally in swampy areas in South Kheri Forest division. Singh 7874 (CDRI).

Oplismenus P. Beauv.

1. Awns smooth .... **O. compositus**
1. Awns barbellatie.. **O. burmannii**

O. compositus (Linn.) Beauv. Ess. Agrost. 54:168.1812; Bor, Grasses India 317.1960; Raizada & Jain, loc. cit. 180; Srivastava, FG. 370.1976.
Panicum compositus Linn. Sp. Pl. 57. 1753.

Perennial slender grasses. Culms terete, rooting at the nodes. Leaves lanceolate to linear-lanceolate, acuminate, hairy below. Panicles consisting of a series of sub-secund, spiciform racemes. Spikelets oblong-lanceolate, acuminate, produced into a long awn.

Flowers and Fruits : August-January

Found frequently forming a dense herbaceous-undergrowth in moist shady grounds of the orchard. Singh 7822 (CDRI).

O. burmannii (Retz.) Beauv. Agrost 54: 168. 1812. Bor., loc. cit. 317. 1960, Raizada & Jain loc. cit 181; Srivastava FG. 370. 1976 (Fig. 21).

Panicum burmannii Retz. Obs. 3: 10. 1783.

Prostrate or decumbent, slender annual grasses rooting at the nodes. Leaves 2.5-5 cm x 4-6 mm, lanceolate to ovate, sparsely hairy. Panicles secund racemes up to 17 cm long. Spikelets oblong-lanceolate, awned, one of a pair often imperfect. Involucral glumes and lower floral glumes awned. Caryopsis lanceolate-oblong, convex on ventral surface.

Flowers and Fruits : September-November

Common in moist shady places in orchard and in forests. Singh 7771, 7821 (CDRI).

Echinochloa P.Beauv.

1. Ligules present ....**E. stagnina**
1. Ligules absent:
    2. Lower lemma and upper glumes acute, awnless ....**E. colonum**
    2. Lower lemma and upper glumes acuminate, awned.........**E. crusgalli**

E. stagnina (Retz.) P. Beauv. Ess. Agrost. 53; 161.1812; Bor, Grasses India 311.1960; Raizada & Jain loc. cit. 172.1964.

Panicum stagninum Retz. Obs. Bot. 5:17.1789.

P. crusgalli Linn. Sp. Pl. 56.1753; FBI 7:30 in part.

Perennial aquatic grasses with floating spongy culms. Leaves 12-40 x 2 cm, linear; ligules a fringe of tough hairs in lower leaves. Panicles secund, 12 cm long. Spikelets 5-7 mm long, ovate-oblong. Conspicuously hispid. Lower lemma awned. Upper lemma cuspidate. Caryopsis elliptic, 1.6-2 mm long.

Flowers and Fruits : August-December

Common in stagnant ditches and banks of ponds etc. Singh 7824, 8501 (CDRI).

E. colonum (Linn.) Link, Hort. Berol. 2:209.1833; Bor, Grasses India 308.1960; Raizada & Jain loc. cit. 170; Srivastava, FG. 363.1976.
Panicum colonum Linn. Syst. 870.1759; FBI 7:32.

Annual subaquatic grasses. Culms erect or ascending, branched. Leaves linear-lanceolate, flat, glabrous. Spikes arranged in erect panicles. Spikelets crowded, globose, arranged in two rows on the same side of the rachis. Caryopsis elliptic.

Local name : Sanwa
Flowers and Fruits : July-November

Common in rice fields and other marshy places. The seeds are edible. Singh 7523, 8417 (CDRI) G. Saran 26390, 26303 (LWG).

E. crusgalli (Linn.) Beauv. Agrost. 53.1812; Bor, Grasses India 310.1960; Raizada & Jain loc. cit. 171; Srivastava, FG. 363.1976.
Panicum crusgalli Linn. Sp. Pl. 56.1753; FBI 7:30 in part.

Annual grasses. Culms decumbent or shortly creeping below, branched. Leaves linear, 7-24 cm x 7-21 mm, acute, margins cartilaginous. Panicles 7-18 mm long, lobed. Spikelets 3-4 mm long, crowded ovate-elliptic, cuspidate, greenish or tinged with purple. Lower florets barren. Upper florets hermaphrodite, polished. Caryopsis broadly elliptic.

Local name : Sanwa
Flowers and Fruits : August-November

Common in moist places, near ponds, ditches and in rice fields. The seeds are edible. Singh 7761 (CDRI).

Paspalidium Stapf

1. Terrestrial annuals; upper lemma granular ....**P. flavidum**
1. Aquatic perennial; upper lemma rugulose ....**P. punctatum**

P. flavidum (Retz.) A. Camus in Lecomte, Fl. Gen Indo-Chine 7:419.1922; Bor, Grasses India 333.1960; Raizada & Jain loc. cit. 167; Srivastava, FG. 372.1976.
Panicum flavidum Retz. Obs. 4:1786; FBI 7:28.

Tufted annual grasses with culms upto 1 m tall, branching from the base. Leaves 6-14 cm x 3.5 mm, linear-lanceolate, acuminate. Inflorescence a panicle of 6-9, false, sessile, erect distanct spike, appressed to the main axis. Spikelets two rowed, ovate-elliptic. Lower glumes upto 1.7 mm long, convex, 3-nerved. Upper glumes 1.25-2.5 mm long, convex, membranous.

Flowers and Fruits : July-November

Occurs commonly in grasslands and other wastelands. Singh 7866, 8424, 8442, 8508 (CDRI).

P. punctatum (Burm.f.) A. Camus, Fl. Gen. de L' Indo-Chine 7:419.1922; Bor, Grasses India 333.1960; Raizada & Jain loc. cit. 169.1964 (Fig.20).
Panicum punctatum Burm.f. Fl. Ind. 26.1768; FBI 7:29.

Perennial aquatic grasses with culms rooting at the nodes. Leaves 8-27 cm x 4-7 mm, linear, acute or acuminate, rigid. Panicle of spike-like racemes, alternate upto 4.5 cm long, appressed to the rachis. Spikelets many, imbricate, 1.7-2.6 mm. Lower florets neuter. Upper florets hermaphrodite. Caryopsis orbicular, compressed.

Flowers and Fruits : August-December

Common in moist places and in rice fields. Singh 8475 (CDRI).

Urochloa P. Beauv.

U. panicoides Beauv. Ess. Agrost. 53. t.11, Fig.1.1812; Bor, Grasses India 372.1960; Raizada & Jain loc. cit. 177.1964.
Panicum javanicum Hook.f. in FBI 7:35.1896.

Annual decumbent grasses, rooting at the nodes with culms 25-50 cm long. Leaves lanceolate, 6-16 x 0.6-2.2 cm, semiamplexicaul at the base, margins crisped-undulate. Inflorescence of 4-7, erect or spreading, spiciform, secund racemes upto 6 cm long. Spikelets 4.5 mm long, ovate-elliptic to oblong-ovate. Involucral glumes dissimilar. Caryopsis upto 2 mm long, compressed.

Flowers and Fruits : August-November

Commonly found in moist places and in pasture grounds. Singh 8429 (CDRI).

Paspalum Linn.

1. Spikelets elliptic .... **P. distichum**
1. Spikelets rotundate-elliptic .... **P. commersonii**

P. distichum Linn. Syst. Nat. ed. 10.2:855.1759; FBI 7:12; Bor, Grasses India 338; Raizada & Jain loc. cit. 198.1964; Srivastava, FG. 372.1976.

Perennial aquatic grasses with creeping stolons. Leaves linear-lanceolate, 3-11 cm long, glabrous. Racemes 2, terminal, erect, divaricate, 2.5-6 cm long. Spikelets 2.5-3 mm long, broadly elliptic, acute, appressed on the rachis. Upper glumes fertile, pubescent. Upper lemma granulate, enclosing the membranous palea. Caryopsis 2.3-2.6 mm long, elliptic.

Flowers and Fruits : June-September

Commonly found on moist places in ponds and ditches. Singh 7417, 8427 (CDRI).

P. commersonii Lamk. Encycl. 1:175. t.43.f. 1. 1791; Bor, Grasses India 335.1960.

Erect shortly decumbent grasses. Leaves linear-lanceolate, 10-30 cm x 2.5-6 mm, acuminate. Inflorescence panicle. Spikelets biseriate, rotundate-elliptic, 2-2.8 mm long.

Flowers and Fruits : September-January

Occurs as a weed in rice fields. Singh 2988 (CDRI).

Brachiaria Griseb.

1. Spikelets 1.25-1.8 mm long, glabrous ....B. reptans
1. Spikelets 2.5-4 mm long:
    2. Spikelets crowded:
        3. Spikelets sub-sessile, glabrous ....B. distachya
        3. Spikelets long pedicelled, pubescent ....B. ramosa
    2. Spikelets distant, lanceolate ....B. kurzii

B. reptans (Linn.) Gard. & C.E. Hubbard in Hook. Icon. Pl. 3363.1938; Bor, Grasses India 285.1960; Raizada & Jain loc. cit. 162.1964; Srivastava, FG. 359.1976.
Panicum reptans Linn. Syst. 870.1759.
P. prostratum Lamk. Tab. Encycl. 1:171.1791; FBI 7:33.

Annual creeping semi-aquatic grasses with culms upto 40 cm tall, rooting at the nodes. Leaves linear-lanceolate, 2.5-5.5 cm x -.4-1.3 mm, amplexicaul, acuminate; sheath ciliate. Racemes trigonous, 1.2-3.5 cm long. Spikelets 1.2-3 mm long, paired, ellipsoid, glabrous. Pedicels with long white bristles. Lower glumes reduced to a semi-lunar scale, nerveless. Upper glumes 2 x 0.5-0.7 mm, ovate, 5 to 7-nerved.

Flowers and Fruits : September-January

Found frequently in the moist places and near rice fields. Singh 8549 (CDRI).

B. distachya (Linn.) Stapf in Prain, Fl. Trop. Afr. 9:565.1919; Bor, Grasses India 281.1960; Raizada & Jain loc. cit. 164.1964.
Panicum distachyon Linn. Mant. Alt. 183.1771; FBI 7:37.

Annual creeping or erect, branched grasses, with culms upto 40 cm tall, rooting at the base. Leaves 1.5-5 cm x 3.7 mm, linear-lanceolate, acuminate, rounded at the base, glabrous. Panicles 2-5.5 cm long, consisting of 2-3 simple racemes. Spikelets 2-3 mm long, elliptic-obovate. Caryopsis 2 x 1.5 mm, oblong.

Flowers and Fruits : August-December

Found frequently in grasslands. Singh 8444 (CDRI).

B. ramosa (Linn.) Stapf in Prain, Fl. Trop. Afr. 9:542.1919; Bor, Grasses India 284.1960; Raizada & Jain loc. cit. 164.1964; Srivastava, FG. 359.1976.
Panicum ramosum Linn. Mant. 29.1767; FBI 7:36 in part.

Erect or ascending, annual grasses with culms upto 50 cm tall, branched, rooting at the base. Leaves 5-13 cm long, linear-lanceolate, base rounded. Panicles 4-16 cm long; rachis terminating into a spikelets. Spikelets alternate, ovoid, 3-4 mm long, often in pairs, one sessile, other pedicelled. Caryopsis elliptic.

Flowers and Fruits : July-November
Found frequently in waste moist places. Singh 8418 (CDRI).

B. kurzii (Hook.f.) A. Camus in Lecomte Fl. Gen. de L' Indo-Chine 7:438.1922; Bor, Grasses India 283.1960; Raizada & Jain loc. cit. 165.1964.
Panicum kurzii Hook.f. in FBI 7:38.1896.

Erect or decumbent, annual grasses, rooting at the nodes. Leaves 4-6 cm x 6-14 mm, lanceolate, acuminate, cordate at the base, margins crisped, long ciliate at the base. Inflorescence 5-6.5 cm long, composed of false spikes. Spikelets 2.5 mm long, ovate-lanceolate, subsessile. Lower glumes ovate, sub-obtuse, 3 to 5-nerved.

Flowers and Fruits : August-November

Occurs occasionally in moist places. Singh 8465 (CDRI).

Digitaria Heist. ex Fabr.

D. adscendens (H.B. & K.) Henr. in Blumea 1:92.1934; Bor, Grasses India 298.1960; Raizada & Jain loc. cit. 192; Srivastava, FG. 362.1976.
Panicum adscendens H.B. & K. Nov. Gen. & Sp. 1:97.1815.

Erect or decumbent annual grasses. Leaves linear to linear-lanceolate, rounded at the base, acuminate. Inflorescence of 2-4 subsessile, digitate racemes. Spikelets 3 mm long, lanceolate-elliptic, acute, appressed to the rachis, binate. Floral glumes densely bearded with soft, spreading hairs.

Flowers and Fruits : July-November

Found frequently in the open areas. Singh 8553 (CDRI).

Pennisetum Rich.

P. polystachyon (Linn.) Schult. Mant. Syst. Veg. 2:146.1824; Bor, Grasses India 346.1960; Raizada & Jain oc. cit. 210.1964.
Panicum polystachyon Linn. Syst. Nat. ed. 10:870.1759.

Perennial grasses. Leaves 30-45 cm x 3-11 mm, linear-lanceolate, acute, narrowed at the base. Spikes 12-14 cm long, purple; involucre densely ciliate around spikelets. Spikelets solitary, sessile, oblong-acute, 4-5 mm long. Lower glumes linear-lanceolate, hyaline, nerveless. Caryopsis elliptic-oblong.

Flowers and Fruits : September-February

Occurs occasionally in grasslands. Singh 7645 (CDRI).

Setaria P. Beauv.

1. Bristles retrorsely barbed .... **S. verticillata**

1. Bristles antrorsely barbed:

2. Inflorescence a cylindrical spike ....S. glauca

2. Inflorescence a narrow panicle ....S. tomentosa

S. verticillata (Linn.) Beauv. Ess. Agrost. 51:178.1812; FBI 7:80; Bor, Grasses India 365.1960; Raizada & Jain loc. cit. 203. 1964.
Panicum verticillatum Linn. Sp. Pl. ed. 2:82.1762.

Slender erect leafy annual grasses. Leaves linear-lanceolate, 5-20 x .5-1.5 cm, acute, scabrid on both surfaces. Panicles cylindrical, 3-15 cm long, coarsely bristly. Spikelets ellipsoid, obtuse, glabrous, 2.5 mm long. Lower glumes, broadly ovate, acute, hyaline; upper glumes membranous, ovate, subacute.

Flowers and Fruits : August-November

Found frequently in the area in moist waste places. Singh 3067 (CDRI).

S. glauca (Linn.) Beauv. Agrost. 51.1812; FBI 7:78 in part; Bor, Grasses India 360.1960; Raizada & Jain loc. cit. 204 Srivastava, FG. 376.1976.
Panicum glaucum Linn. Sp. Pl. 56.1753.
Pennisetum glaucum (L.) R.Br. Prod. 1. 195.1810.

Tufted annual grasses with culms upto 90 cm tall. Leaves 6-25 cm x 4-9 mm, linear-lanceolate, acute, glabrous or sparsely hairy. Racemes spiciform, dense cylindric, yellow. Spikelets 3.5-4.5 mm, turgid, 2-flowered, broadly oblong. Lower glumes 3-nerved, ovate-lanceolate. Upper glume broadly lanceolate.

Local name: Bharuee.
Flowers and Fruits : August-October

Commonly found in wastelands and in open grasslands. Singh 3029, 7786, 8414, 8431, 8519 (CDRI).

S. tomentosa (Roxb.) Kunth, Rev. Gram. 1:47.1829; Bor, Grasses India 365.1960; Raizada & Jain loc. cit. 205;

Srivastava, FG. 377.1976.
Panicum tomentosum Roxb. Fl. Ind. 1:303.1820.
Setaria intermedia R. & S. Syst. 2:489.1817; FBI 7:79.

Weak flaccid annual grasses. Leaves 7-18 x 0.4-2 cm, narrowly linear-lanceolate, sparsely hairy, scabrid at margins. Panicles 4-22 cm long, conico-cylindrical, lax. Spikelets straw coloured, upto 2 mm long, broadly ovate to elliptic. Lower glume half the length of spikelets. Upper glume dorsally curved, not keeled. Caryopsis elliptic, planoconvex, upto 1.5 mm long.

Flowers and Fruits : August-December
Common in moist grasslands. Singh 7572 (CDRI).

Poa Linn.

P. annua Linn. Sp. Pl. 68.1753; FBI 7:345; Bor, Grasses India 555.1960.

Tufted glabrous suberect grasses. Leaves linear, flat, flaccid, margins scabrid. Panicles upto 9 cm long, ovate or lax, smooth. Spikelets 3-6 x 2 mm, oblong, ovate-lanceolate, 3 to 7-flowered. Involucral glumes unequal. Floral glumes oblong, ciliate below along the keel.

Flowers and Fruits : December-February

Occurs occasionally on moist grounds. Singh 7890 (CDRI).

Eragrostis P. Beauv.

1. Rachilla of spikelets jointed; spikelets breaking up from above downwards:
   2. Paleas ciliolate on the keels .... **E. tenella**
   2. Paleas scaberulous or smooth on the keels .... **E. japonica**
1. Rachilla of spikelets tough, persistent; spikelets falling away from its base upward:
   3. Spikelets 1.5 mm wide .... **E. atrovirens**
   3. Spikelets more than 2 mm wide .... **E. unioloides**

E. tenella (Linn.) P. Beauv. ex Roem. & Schult. Syst. Veg. 2:576.1817; Bor, Grasses India 513.1960; Srivastava, FG. 366.1976.
Poa tenella Linn. Sp. Pl. 69.1753.

Erect tufted slender annual grasses with culms 15-30 cm tall. Leaves linear, tapering to a fine point; sheath long, ciliate near the mouth. Panicles plumose, green or purplish, loose, open or contracted. Spikelets 1.5-4 mm, oblong. Palea ciliate on the keels. Caryopsis ovoid, pale-brown.

Flowers and Fruits : August-February

Occurs commonly in moist places. Singh 4385, 8421, 8430 (CDRI) G. Saran 26167 (LWG).

E. japonica (Thunb.) Trin. in Mem. Acad. Petersb. ser. 6.1:405.1831; Bor, Grasses India 509.1960.
Poa japonica Thunb. Fl. Jap. 51.1784.

Erect slender annual grasses. Leaves linear-lanceolate, acute. panicles oblong or linear; whorled. Spikelets loose, open, few flowered. Lemma less than 1 mm long.

Flowers and Fruits : September-February

Occurs commonly in moist places in wastelands and in grasslands. Singh 4187, 7818, 7845, 7892 (CDRI).

E. atrovirens (Desf.) Trin. ex Steud. Nom. Bot. ed. 2.1:562.1840; Bor, Grasses India 503.1960.
Poa atrovirens Desf. Fl. Atlant. 1:75,t.14.1798.

Erect perennial grasses upto 90 cm tall. Leaves linear, glabrous, acuminate. Inflorescence a contracted panicle. Spikelets 5-9 x 2-2.5 mm; pedicel 1.5 mm long. Lemmas 1.5-1.8 mm long, elliptic-oblong, acute. Caryopsis oblong.

Flowers and Fruits : August-November

Found frequently in grasslands near moist places. Singh 8434, 8476 (CDRI).

E. unioloides (Retz.) Nees ex Steud. Syn. Pl. Glum. 264.1855; Bor, Grasses India 515.1960; Srivastava, FG. 366.1976.
Poa unioloides Retz. Obs. 5:19.1789.
Eragrostis amabilis W. & A. ex Nees in Hook. & Arn. Bot. Beach Voy. 251. 1830-1840; FBI 7:317.

Slender erect or ascending tufted annual grasses. Leaves linear, base cordate, acute; sheath smooth, glabrous. Panicles ovate to oblong. Spikelets ovate to oblong, compressed, shortly pedicelled, green to reddish-purple, more than 2 mm wide. Glumes ovate-lanceolate, 1-nerved. Caryopsis obovoid, or ellipsoid, laterally compressed.

Flowers and Fruits : September-November

Occurs commonly in moist, shady places. Singh 4214, 4297, 7376, 7540, 7811, 7891 (CDRI).

Desmostachya Stapf

D. bipinnata (Linn.) Stapf in Dyer, Fl. Cap. 7:632.1900; Bor, Grasses India 491.1960; Srivastava, FG. 362.1976.
Briza bipinnata Linn. Syst. Nat. ed. 10.2:875.1759.
Eragrostris cynosuroides Beauv. Agrost. 162.1812; FBI 7:324.

Erect tufted branched perennial grasses upto 150 cm tall. Leaves linear-lanceolate, rigid, 40 x 1 cm, with filiform tips, margins hispid. Panicles interrupted, racemiform, linear, greyish-green. Spikelets sessile, secund, biseriate, deflexed. Glumes unequal.

Local name : Kusha
Flowers and Fruits : June-November

Found frequently in grasslands. The leaves are used in religious ceremonies. Singh 2979, 4108, 8552 (CDRI).

Eleusine Gaertn.

E. indica (Linn.) Gaertn. Fruct. 1:8.1788; Bor, Grasses India 493.1960; Srivastava, FG. 364.1976.

Cynosurus indicus Linn. Sp. Pl. 72.1753.
Eleusine tristachya Lamk. Illus. 1:203. t.203.1791; FBI 7:263.

Erect tufted annual grasses. Culms upto 65 cm tall, glabrous, compressed. Leaves distichous, flat; sheath ciliate. Spikes 2-7. Spikelets biseriate, 3 to 6-flowered, glabrous, upto 5 mm wide. Glumes unequal. Lemma ovate-oblong. Caryopsis obtusely trigonous, oblong, reddish-brown.

Flowers and Fruits : August-November

Occurs as a weed in cultivated fields. Singh 8433 (CDRI).

Avena Linn.

A. fatua Linn. Sp. Pl. 80.1753; Bor, Grasses India 434.1960; FBI 7:275.

Erect annual grasses upto 1 m tall. Culms stout fistular, nodes hairy. Leaves linear-lanceolate, scaberulous; sheath smooth. Spikelets in large, terminal, open panicles. Glumes 20-25 mm long. Lemmas awned from the back, often densely hairy except towards the apex. Rachilla readily disarticulating between all the florets at maturity.

Local name : Jaraee
Flowers and Fruits : February-April

Occurs occasionally in cultivated wheat fields. Singh 7650 (CDRI).

Phragmites Trin.

P. karka Trin. ex Steud. Nom. 2:324.1841; Bor, Grasses India 416.1960. Saldanha (Fl. Hass. Distt.
P. maxima (Forsk.) Blatt. & McC. Bomb. Gr. 202.1935; Srivastava, FG. 374.1976.
Arundo maxima Forsk. Fl. Aegypt.-Arab. 24.1775.
Arundo phragmites Linn. Sp. Pl. 81.1753.

Tall erect perennial grasses upto 3 m tall. Culms hollow, terete, smooth, covered with leaf sheaths. Leaves linear, acuminate, coriaceous. Inflorescence large decompound panicles, lax, silky hairy at the base. Spikelets hermaphrodite, several flowered. Glumes glabrous. Caryopsis oblong.

Local name - Narkul, Narkat
Flowers and Fruits : December-February

Found commonly in marshy places and near streams. The stems are used for making mats. Singh 2818, 2997, 4229, 7240, 7498 (CDRI).

Hygroryza Nees

H. aristata (Retz.) Nees ex W. & A. in Edinb. New Phil. J. 15.380.1833; Bor, Grasses India 597.1960; Srivastava, FG. 368.1976.
Pharus aristatus Retz. Obs. Bot. 5:23.1789.

Aquatic floating grasses; stems spongy with feathery whorled roots. Leaves ovate-oblong, obtuse, 6-7 x 2 cm, base rounded; ligule membranous. Panicles 3-6 cm long with spreading branches. Spikelets narrow, sessile or pedicelled, bisexual, 1-flowered. Floral glumes lanceolate, awned.

Flowers and Fruits : October-February

Occurs commonly in lakes and ponds in the forest areas. Singh 7454 (CDRI).

Oryza Linn.

1. Spikelets persistent ....**O. sativa**
1. Spikelets caducous ....**O. rufipogon**

O. sativa Linn. Sp. Pl. 333. 1753; Bor, Grasses India 605.1960; Srivastava, FG. 371.1976.

Erect tufted annual grasses. Leaves linear, acuminate, striate, 1-nerved; ligule upto 18 mm long. Inflorescence a lax, or erect, panicle. Spikelets 7-10 mm long. Floral glumes hispid-ciliate, dorsally spinescently ciliate. Caryopsis oblong, angular.

Local name : Dhan, Chaval
Flowers and Fruits : September-November

Cultivated in the area for its edible grains. Singh 7758 (CDRI).

O. rufipogon Griff. Notul. 3:5.1851; Bor, Grasses India 605.1960.

Erect perennial grasses. Leaves long, linear, acuminate, flat; ligule of the lower leaves upto 20 mm long. Inflorescence a lax panicle. Spikelets caducous. Mature florets with 2-3 grains.

Local name : Pasaee
Flowers and Fruits : October-November

Occurs commonly mixed with paddy fields in marshy places. The seeds are used for worshipping gods and are edible. Singh 7746, 8487 (CDRI).

Phalaris Linn.

P. minor Retz. Obs. Bot. 3:8.1783; FBI 7:221; Bor, Grasses India 616.1960; Srivastava, FG. 373.1976.

Erect annual grasses upto 1 mm tall. Culms branched, nodes swollen. Leaves linear-lanceolate, 15-25 cm long, finely pointed, glabrous. Panicles erect, spikelike, contracted. Spikelets laterally compressed, flattened, 1-flowered and with 1 or 2 reduced imperfect glumes below the floral glumes. Floral glume half the length of empty glumes.

Local name : Genhu mama, Senhu.
Flowers and Fruits : January-April

Found frequently as a weed in wheat crop. Singh 7645 (CDRI).

Arundinella Reddi

A. bengalensis (Spreng.) Druce in Rep. Bot. Exch. Club Brit. Isle 605.1916; Bor, Grasses India 421.1960.
Panicum bengalensis Spreng. Syst. Veg. 1.311.1825.

Erect stout grasses upto 1 m tall. Culms smooth, glabrous. Leaves 25-36 cm x 2.5-4.5 mm, lanceolate. Inflorescence erect, elongated panicles, upto 27 cm long. Spikelets 3 to 4-seriate, ovoid, 1.5-2 mm long, dark purple. Lower glumes 2-3.3 mm long. Upper glumes upto 3 mm long. Lower florets male. Lemma 5-nerved. Caryopsis 1-1.2 mm long.

Flower and Fruits : July-October

Found frequently in moist shady places. Singh 8539 (CDRI).

Perotis Ait.

P. indica (Linn.) O. Kuntze, Rev. Gen. Pl. 2:787.1891; Bor, Grasses India 611.1960; Srivastava, FG. 373.1976.
Anthoxanthum indicum Linn. Sp. Pl. 28.1753.

Prostrate and ascending tufted annual grasses. Stems leafy, terminating in inflorescence. Leaves 2.5-6 x 0.5-0.9 cm, ovate-cordate or lanceolate, base amplexicaul. Racemes slender, purplish-pink, dense. Spikelets hermaphrodite, 1-flowered, linear. Involucral glumes long-awned.

Flowers and Fruits : August-October

Found frequently in open waste sandy places. Singh 7432, 7720 (CDRI).

Polypogon Desf.

P. monspeliensis (Linn.) Desf. Fl. Atlant. 1:67.1798; Bor, Grasses India 403.1960; Srivastava, FG. 374.1976.
Alopecurus monspeliensis Linn. Sp. Pl. 61.1753.

Erect tufted annual grasses from a geniculate base. Leaves 2.5-6 x 0.5 cm, ovate-lanceolate, scabrid. Panicle spikelike, dense, light green, silky. Spikelets pubescent, shortly pedicelled. Involucral glumes, ciliate, awned. Floral glumes ovate-truncate, awned. Caryopsis obovoid.

Flowers and Frutis : January-May

Occurs commonly in moist sandy places and in sal forests. Singh 2813, 7117,7243, 7617, 8306 (CDRI).

Chloris Sw.

1. Empty lemmas above the fertile florets single represented by five arista ....**C. dolichostachya**
1. Empty lemmas above the florets 2 or 3 ....**C. barbata**

C. dolichostachya Lagasca, Gen. Spec. Pl. 5. 1816; Bor, Grasses India 466.1960.
C. imcompleta Roth, Nov. Pl. Sp. 60.1821; FBI 7:290.

Erect or procumbent branched annual grasses. Leaves linear-lanceolate, acute; ligules hairy. Spikes 5, digitate, spreading, forming terminal whorls. Spikelets 2.5-3 mm long, 2-rowed, imbricate, awned, lanceolate, closely appressed. Lower glumes short, linear. Upper glumes awned, glabrous.

Flowers and Fruits : August-November

Found frequently in moist shady places. Singh 4151 (CDRI).

C. barbata (Linn.) Sw. Fl. Ind. Occ. 1:200.1797; FBI 7:297; Bor, Grasses India 465.1960; Srivastava, FG. 360.1976.

Perennial tufted grasses. Culms erect or ascending from the geniculate, creeping base, compressed and branched. Leaves narrowly linear; sheath keeled and compressed. spikes 5-20, digitately arranged in terminal fascicles, 4-8 cm long. Spikelets secund, 3-awned. Involucral glumes lanceolate. Lowest floret hermaphrodite; glumes elliptic, densely hairy on the margins, awned.

Flowers and Fruits : August-November

Occurs commonly in open grasslands in sandy places. Singh 4359 (CDRI).

Cynodon Rich. ex Pers. *nom. cons.*

C. dactylon (Linn.) Pers. Syn. 1:85.1805; FBI 7:288; Bor, Grasses India 469.1960; Srivastava, FG. 361.1976.
Panicum dactylon Linn. Sp. Pl. 58.1753.

Perennial grasses with stem prostrate, creeping and rooting at the nodes. Culms forming matted tufts. Leaves linear-lanceolate, acuminate. Spikes 4-5, digitate, green or purplish. Spikelets compressed, sessile, appressed, imbricate.

Local name : Dubghas
Flowers and Fruits : Most part of the year

Occurs commonly in wet and dry places, used as lawn grass. The plant is used in religious ceremonies. Singh 8381 (CDRI).

Dactyloctenium Willd.

D. aegypticum (Linn.) Beauv. Agrost. 72.1812; Bor, Grasses India 489.1960; Srivastava, FG. 361.1976.
Cymosus aegypticum Linn. Sp. Pl. 72.1753.
Eleusine aegyptiaca Dasf. Fl. Atlant. 1:85.1798; FBI 7:295.

Erect or decumbent annual grasses. Leaves narrow-linear, flat, distichous, acute. Spikes 3-10 digitate. Spikelets densely crowded, 3 to 6-flowered, 3-4 x 2.5 mm. Lower glumes upto 2 mm long, ovate, acute. Upper glumes oblong. Lower lemma upto 3.5 mm long, mucronate-awned. Caryopsis laterally flattened.

Flowers and Fruits : June-November

Occurs as a weed in waste sandy places and cultivated fields. Singh 4386 (CDRI).

19. SELECTED BIBLIOGRAPHY

Anonymous, 1883-85. Index Kewensis Plantarum Phanerogamarum 2 Vols. (15 Suppl., 1886-1970). Oxford.

------- 1962. Recommendation made by the members of the summer school in Botany, Kodaikanal, 15th June 1962. *Bull. bot. Surv. India* 4: 543-577.

------- 1967. Notes on Flora of Rajnikunj (Gokarna Forest). Bull. Deptt. Med. Plants No.1., Thapathali, Kathmandu.

------- 1969. Flora of Nagarjun.*ibid.* No.4. Thapathali, Kathmandu.

------- 1974. Supplement to the Flora of Phulchoki and Godawari, Bull. Deptt. Med. Plants No. 2; Thapathali, Kathmandu.

------ 1927. Gazetteer of Kheri District, U.P. Allahabad.

Airy Shaw, H.K. 1972. The Euphorbiaceae of Siam. *Kew Bull.* 26(2):191-363.

Ali, S.I. 1958. Revision of genus *Indigofera* Linn. from West Pakistan and N.W. Himalayas. *Bot. Notiser* 111: 543-577.

Almeida, M.R. 1968. Notes on *Boerhavia* Linn. *J. Bombay nat. Hist. Soc.* 65:266-267.

Austin, D.F. 1977. *Ipomoea carnea* Jacq. vs. *Ipomoea fistulosa* Mart. *Taxon* 26:235-238.

Babu, C.R. 1977. Herbaceous Flora of Dehra Dun, U.P. CSIR, New Delhi.

Bailey, L.H. 1949. Manual of Cultivated Plants. New York.

Bahadur, K.N. and Gaur, R.C. 1976. On the nomenclature of two Indian *Acacias. Acta Bot. Indica* 4: 66-67.

Balakrishan, N.P. 1966. Nomenclatural notes on some flowering plants of India. *J. Bombay nat. Hist. Soc.* 63: 327-331.

Balakrishan, N.P. 1970. Nomenclatural notes on some flowering plants II. *J. Bombay nat. Hist. Soc.* 67: 57-66.

Banerjee, S.P. 1962. Identity of *Trichodesma amplexicaule* DC. *Bull. bot. Soc. Bengal* 16(1&2): 9-11.

Banerji, M.L. 1965. Contributions to the flora of Eastern Nepal. *Rec. bot. Surv. India* 19(2):1-90.

Bennete, S.S.R. 1987 Name changes in Flowering Plants India and adjacent areas Dehra Dun.

Bhakuni, D.S. *et al.* 1968. Screening of Indian Plants for biological activity part II. *J. Exp. Biology* 7:250-262.

Bhandari, M.M. 1977. Flora of the Indian Desert., Jodhpur.

Bhattacharyya, U.C. 1963. Contribution to the flora of Mirzapur 1. Some new records for the district and for the Upper Gangetic Plain, *Bull. bot. Surv. India.* 5:59-62.

Bor, N.L. 1947. Common Grasses of the United Provinces. *Indian For. Rec.* (n.s. Bot.) 2:1-220.

------- 1960. The Grasses of Burma, Ceylon, India and Pakistan (excluding Bambuseae). London.

Borssum, W.J.V. 1966-67. Malesian Malvaceae revised. *Blumea* 14(1):1-251.

Brizicky, G.K. 1962. Taxonomic and nomenclatural notes on *Zanthoxylum* and *Glycosmis* (Rutaceae) *J. Arn. Arb.* 43:80-93.

Burtt, B.L. 1965. Notes, chiefly nomenclatural on Gentianaceae of Pakistan and India. *Notes R. bot. Gdn. Edinb.* 26(3):269-278.

------- 1973. *Polygala arvensis, chinensis* and *glomerata. Notes R. bot. Gdn., Edinb.* 32(3):403-404.

Burkill, I.H. 1916. Notes on the Terai forest between the Gandak the Teesta. *J.* and *Proc. Asiat. Soc. Bengal* 12:267-272.

Cain, S.A. 1944. Foundations of Plant Geography. New York.

------- 1950. Life-forms and Phytoclimate. *Bot. Rev.* 16:1-32.

Chakravarty, H.L. 1959. Monograph on Indian Cucurbitaceae. Taxonomy and distribution. *Rec. bot. Surv. India* 17(1):1-234.

Champion, H.G. 1961. A preliminary survey of the forest types of India. *Indian For. Rec.* n.s. vol.1.no.1:1-195. Silviculture.

Chandra, G. 1970-71 to 1979-80. Working Plan of South Kheri Forest Division, Uttar Pradesh (Cyclostyled).

Channer, F.F.R. 1908. The Forest of Terai, Bhabhar. Govt. estate in United Province. *Indian For.* 34:393-406.

Chatterjee, D. 1939. Studies on the endemic flora of India and Burma. *J.R. Asiat. Soc. Bengal* 5(21):19-67.

Chatterjee, D. 1962. Floristic pattern in Indian vegetation. Proc. Summer School of Botany, Darjeeling (held in 1960) 32-42.

Chatterjee and **Bhardwaja, R.C.** 1955. Revision of the Scrophulariaceae of the Upper Gangetic Plain and adjoining areas. *Bull. bot. Soc. Bengal* 9(2):127-153.

Chopra, R.N. *et al.*, 1956. Glossary of Indian Medicinal Plants. CSIR, New Delhi.

------- 1959. Supplement to the Glossary of Indian Medicinal Plants, CSIR, New Delhi.

Constance, L. 1957. Plant taxonomy in an age of experiment. *Am. J. Bot.* 44:88-92.

------- 1964. Systematic Botany, An unending synthesis. *Taxon* 13:257-273.

Corner, E.J.H. 1960. Taxonomic notes on *Ficus* Linn. Asia and Australia *Gdns. Bull. Singapore* 17:368-485.

------- 1965. Check list of *Ficus* Linn. Asia and Australia with keys to identification. *ibid.* 21:1-186.

------- 1977. A revised handbook to the Flora of Ceylon. 1(2):111-196.

Croizat, L. 1952. Manual of Phytogeography of an account of plant dispersal throughout the world. Hague.

Cronquist, A. 1968. The Relevance of National Herbaria to Modern Taxonomic Researches in the United States of America. In Heywood, V.H. (ed). Modern Method in Plant Taxonomy, London 15-22.

Davis, P.H. and Heywood, V.H. 1963. Principles of Angiosperm Taxonomy. London.

Deva, S. 1968. A Critique on Duthie's Flora of Upper Gangetic Plain. *Bull. bot. Surv. India* 10:178-182.

Dhar, M.L. *et al.*, 1968. Screening of Indian plants for biological activity part V. *J. Exp. Biology* 6:232.

Dhawan, B.N. *et al.* 1977. Screening of Indian Plants for biological activity part VI. *J. Exp. Biology* 15:208-219.

Dixit, S.N. *et al.* 1966a. Additions to the rainy season weeds of Gorakhpur. *Proc. nat. Acad. Sci. India* (B) 36:149-156.

Duthie, J.F. 1899. A list of the trees and shrubs of forest of Pilibhit, Northern Oudh and Gorakhpur. Allahabad.

------- 1903-29. Flora of Upper Gangetic Plain and of the adjacent Siwalik & Sub-Himalayan Tracts. Calcutta. 2 Vol. (repr. ed. 1960).

Exell, A.W. 1944. Catalogue of the Vascular Plants of S. Tome. London.

Gamble, J.S. 1920. The Indian species of *Mimosa* Linn. *Kew Bull.* 1-6.

------- 1915-36. Flora of Presidency of Madras. Calcutta. (repr. ed. 1957. 3 Vol.).

Good, R. 1953. The Geography of the Flowering Plants. London.

Gupta, R.K. 1969. Common grasses of Gorakhpur. *Indian For.* 95:324-329.

Haines, H.H. 1921-25. Botany of Bihar and Orissa. 3 vols. London.

Heine, H. 1962. Four representatives of the Saharo-Sindian Element in the flora of Senegal and Mauritania. *Kew Bull.* 16(2). 16(2):203-207.

Henry, A.N. 1967. Nomenclatural changes in Indian plants. *Bull. bot. Surv. India.* 9:290.

Hiroshi Hara, 1966. The Flora of Eastern Himalayas First report. University of Tokyo, Japan.

------- 1971. The Flora of Eastern Himalayas Second report. University of Tokyo, Japan.

Hirayoshi Ohashi. 1975. The Flora of Eastern Himalayas Third report. University of Tokyo, Japan.

Hooker, J.D. *et al.* 1872-97. The Flora of British India, London. 7 Vols.

------- 1904. A Sketch of the Flora of British India, London.

Husain, W. 1969. Some new records for plants in the Upper Gangetic Plain. *Indian For.* 93:582-584.

Hutchinson, J. 1973. The Families of Flowering Plants. Third ed. Oxford.

Jacobs, M. 1960. Capparidaceae. Flora Malesiana ser. 1, 6:61-105.

Jain, S.K. 1965. Medicinal plant lore of the tribals of Bastar. *Econ. Bot.* 19(3):236-250.

Jain, S.K. 1967. Phytogeographical consideration of the flora of Mt. Abu. *Bull. bot. Surv. India* 9:(1)4:68-79.

Jain, S.K. and Doli Das. 1973. Notes on the nomenclature of some grasses. *Indian For.* 99: 570-578.

Jeffrey, C. 1962. Notes on Cucurbitaceae, including proposed new classification of the family. *Kew Bull.* 15(3):337-371.

Kakkar, K.V. 1963-73. Working plan for the North Kheri Forest Division. Part I & II. (Cyclostyled).

Kammathy, R.B. and Rao, H. 1961. Notes on Indian Commelinaceae II. Cytological observation. *Bull. bot. Surv. India* 3: 167-169.

Kanjilal, P.C. 1933. A Forest Flora of Pilibhit, Oudh, Gorakhpur and Bundelkhand. Allahabad.

Kanjilal, P.C. 1966. A Forest Flora for the Plains of Uttar Pradesh, Part II & III. Lucknow.

Kapoor, S.L. 1962. On the botany of Lucknow District. *J. Bombay nat. Hist. Soc.* 59:862-896.

Kazmi, S.M.A. 1971. A Revision of the Boraginaceae of West Pakistan and Kashmir. *J. Arn. Arb.* 52: 110-136; 334-363; 486-522; 666-690.

Kern, J.H. 1974. Cyperaceae, Flora Malesiana 7(3) ser. 1. 435-753.

Kihara, H. ed. 1955. Fauna and Flora of Nepal Himalayas, Vol. 1. Tokyo.

Kitamura, S. 1960. Flora of Afghanistan, Kyoto University, Japan.

Knaap-van, Meeuwen, M.S. *et al.* 1961. Preliminary revision of some genera of Malayasian Papilionaceae. *Reinwardtia* 6:85-108.

Kobuski, C.E. 1959. A revised key to the Chinese species of *Jasminum. J. Arn. Arb.* 40:385-390.

Legris, P. and Meher-Homji, V.H. 1971. Floristic element in vegetation of India. Symposium on Recent Advances in Tropical Ecology.

Maheshwari, J.K. 1957. Some name changes in the flora of India. *J. Bombay nat. Hist. Soc.* 54:804-805.

------- 1961. On the identity and nomenclature of some Indian plants. *Bull. bot. Surv. India* 3:91-94.

------- 1962. Studies on the naturalized flora of India. Proc. Summer School in Botany, Darjeeling, pp. 156-170.

------- 1963. The Flora of Delhi, CSIR, New Delhi.

Maheshwari, J.K. 1966. Illustrations to the Flora of Delhi, CSIR, New Delhi.

Maheshwari, J.K., Singh, K.K. and S. Saha 1981. The Ethnobotany of the Tharus of Kheri, district Uttar Pradesh & EBIS. NBRI, Lucknow.

Maheshwari, J.K. 1975. Floristics as a field for research and teaching. *The Botanica* (Silver Jubilee Volume) pp. 63-69.

Malhotra, C.L. 1966. New distributional records of plants from the Upper Gangetic Plain. *Bull. bot. Surv. India* 8:77.

Marriot, R.G. 1917. The Sal forest of the Gorakhpur division. *Indian For.* 43:442-456.

Meher-Homji, V.H. 1962. The phytoclimates of India in relation to the Vegetational criteria. *Bull. bot. Surv. India* 4(1-4):105-112.

------- 1970. Some phytogeographic aspect of Rajasthan. *Vegetatio* 21:299-330.

------- 1972. Himalayan plants on South Indian Hills. *Sci. Cult.* 38:8-12.

------- 1975. On the montane species of Kodaikanal, South India. *Phytocoenologia* 2:28-39.

------- 1977. Tropical deciduous forest of Peninsular India. *Feddes Repert.* Band 88(1-2)113-134.

Melville, R. 1958-59. Notes on *Alternanthera. Kew Bull.* 13:171-175.

Misra, R. 1944. The Vegetation of Rajghat ravines. *J. Indian bot. Soc.* 23:113-121.

Mooney, H.F. 1950. A Supplement to the Botany of Bihar and Orissa, Ranchi.

Mukerjee, S.K. 1940. A revision of the Labiatae of Indian Empire. *Rec. bot. Surv. India* 14(i):1-228.

------- 1950. A revision of the Indo-Burmese *Polygala. Bull. bot. Soc. Bengal* 12:29-50.

------- 1968. The genus *Alangium* in India. *Bull. bot. Surv. India* 10(3&4):330-331.

------- 1970. Botanical Survey of India. Its objectives and working. Abstract of Symposium I "Modern Trend in Taxonomy", Delhi University, Delhi.

Murty, Y.S. and Singh, V. 1961. Flora of Hastinapur. *Agra Univ. J. Res.* (Sci.) 10(2):193-242.

Murty, Y.S. and Singh, V. 1966. Some little known plant from Upper Gangetic Plain. *Sci. Cult.* 32:597-598.

Narayanaswami, V. and Seshagiri & Rao, R. 1950 A preliminary notes on the Indo-Burmese species of *Grewia* Linn. *J. Indian bot. Soc.* 29: 177-190.

Nayar, S.L. 1964. Medicinal plants of commercial importance found wild in Uttar Pradesh and their distribution. *J. Bombay nat. Hist. Soc.* 61: 651-661.

Ownbey, G.B. 1958. Monograph of the genus *Argemone* Linn. for North America and the West Indies. *Mem. Torr. bot. Club* 21(i):1-159.

Panigrahi, G. and Ram Saran, 1967. Contribution to the flora of Gorakhpur Forest Division, Uttar Pradesh. *Bull. bot. Surv. India.* 7:249-261.

------- *et al.* 1969. Contribution to the Botany of the Terai forest of the Bahraich District of Uttar Pradesh. *Bull. bot. Surv. India* 11(1&2):89-114.

Panigrahi, G. 1975. The genus *Adenostemma* (Commpositae) in the Indian region. *Kew Bull.* 30(4):647-655.

Panigrahi, G. 1976. Taxonomic notes on certain taxa of Asiatic Angiosperms. *Phytologia* 32(6):473-479.

Pennell, F.W. 1943. The Scrophulariaceae of Western Himalayas. *Acad. nat. Sci. Philad. Mon.* no. 5:1-163.

Raizada, M.B. 1931. Contribution to Duthie's Flora of Upper Gangetic Plain from the neighbourhood of Dehra Dun. *J. Indian bot. Soc.* 10:155-158.

------- 1932. New Indian species of forest importance. *Indian For.* 58:133-140.

------- 1936. Recently introduced or otherwise imperfectly known plants from the Upper Gangetic Plain. *ibid.* 14:339-348; 14:9-67.

------- 1942. Name changes in important Indian plants. *Indian For.* 68: 206-207.

------- 1950-51. Noteworthy plants from the Upper Gangetic Plain. *Indian For.* 76:489-497; *Indian For. Rec.* 4(i):65-72.

------- 1954. Grasses of Upper Gangetic Plain and some aspect of their ecology. *Indian For.* 80:24-46.

------- 1958. New plant records for the Upper Gangetic Plain. *Proc. nat. Inst. Sci. India* 24:198-208.

------- 1959. Name changes in common Indian plants. *Indian For* 84:467-538.

------- 1966. Nomenclatural changes in Indian plants. *Indian For.* 92:299-339.

------- 1968a. Some name changes in the flora of India. *Indian For.* 94:432-462.

------- 1968b. What is there in a name, why change? *ibid.* 94:37-46.

Raizada, M.B. 1976. Supplement to Duthie's Flora of Upper Gangetic Plain and of the adjacent Siwalik and Sub-Himalayan Tracts. 1-355.

------- *et al.* 1957. Grasses of the Upper Gangetic Plain. Part I Maydeae and Andropogoneae. *Indian Forest Rec.* 4(1):171-277.

Raizada, M.B. and Jain, S.K. 1964 Grasses of the Upper Gangetic Plain. Panicoideae Part II Paniceae. *Indian Forest. Rec.* 5(3):151-226.

Rajagopal, T. 1965. New records of species for "Flora of Allahabad" I *Proc. nat. Acad. Sci. India* 36: 57-84.

Randeria, A.J. 1960 The composite genus *Blumea.* A taxonomic revision. *Blumea* 10: 176-317, 28 plates.

Rao, R.S. 1958 History and importance of Indian Herbaria, *J. Indian bot. Soc.* "Modern Trend in Taxonomy", Delhi University, Delhi.

------- 1970 Morphology and Taxonomy of Angiosperms. Abstract of Symposium "Modern Trend in Taxonomy" Delhi University, Delhi.

Rao, R.S. and Kammathy, R.V. 1962 Notes on Indian Commelinaceae I. *J. Bombay nat. Hist. Soc.* 59:58-70.

Rau, M.A.1969 Flora of Upper Gangetic Plain and of the adjacent Siwalik and Sub-Himalayan Tracts. Check List. *Bull. bot. Surv. India* Suppl. no.2. 10:1-86.

Rhodes, D.G. 1975 A revision of genus *Cissampelos. Phytologia* 30 (6):446.

Saldanha, C.J. and Nicolson, D.H. 1976 (Editor) Flora of Hassan District, Karnataka, New Delhi.

Santapau, H. 1953 The Flora of Khandala on the Western Ghats of India. *Rec. bot. Surv. India* 16(1):1-396.

Santapau, H. 1956a. The Botanical Exploration of India. *J. Indian bot. Soc.* 35: 1-6.

------- 1956b The Flora of Purandhar. New Delhi.

------- 1958a Floristic studies in india. *Mem. Indian bot. Soc.* 1:117-121.

------- 1958b History of Botanical Researches in India, Burma and Ceylon Part. II. Systematic Botany of Angiosperms.

------- 1961 Critical notes on the identity and nomenclature of some Indian Plants. *Bull. bot. Surv. India* 3.1:111-121.

------- 1962 The Flora of Saurashtra Part I. Rajkot.

------- 1967 Changes in plant names. *Bull. bot. Soc. Bengal* 21:59-64.

Santapau, H. and Irani, N.A. 1960 The Asclepiadaceae & Periplocaceae of Bombay, *Bot. Mem. Univ. Bombay.* 4:1-97.

------- and Wagh, S.K. Nomenclatural changes in Indian plants. *Bull. bot. Surv. India* 5(2):105-109.

Saxena, H.O. 1967 New plant record for the Upper Gangetic Plain. *Indian For.* 93:329.

Schnell, R. 1955 Vegetation et Flora de la Region montagneuse du Nimba. *Mem. Inst. franc. Afrique Noira* n.22.Dakar.

Shah, G.L. 1962 Nomenclatural notes on some Bombay plants. Part I. *J. Bombay nat. Hist. Soc.* 59(1):320-321.

------- 1963 Nomenclatural notes on some Bombay plants. Part II. *J. Bombay nat. Hist. Soc.* 60(1):296-298.

Shah, G.L. 1969 Nomenclatural notes on some Bombay plants. Part III. *J. Bombay nat. Hist. Soc.* 66:231-233.

Singh, V. 1971 Addition to Duthie's Flora of Upper Gangetic Plain. *J. Bombay nat. Hist. Soc.* 68:2:339-346.

Srivastava, J.D. 1938 Flora of Allahabad 2 parts. Univ. Allahabad. *Student Bot. Soc.* Allahabad.

Srivastava, T.N. 1976 Flora Gorakhpurensis. 1-477. New Delhi.

Stafleu, F.A. 1963 Dates of Botanical publication 1788-1792.

------- *et al.* 1972 International Code of Botanical Nomenclature, Utrecht-Netherland.

Steenis, C.G.G.J. van 1948-74 Flora Malesiana Vol.4, ser. 1. 1948-54; Vol.5, ser.1.1955-58; Vol.6, ser.1.1960-67; Vol.7, ser.1.part 1-3. 1971-74.

------- 1962 The mountain flora of Malesian tropics, Endeavour 21:183-193.

------- 1965 Concise plant geography. *Fl. Java* 2:1-72.

Stewart, R.R. 1972 An Annotated Catalogue of the Vascular Plants of W. Pakistan and Kashmir. Karachi pp. 1-1028.

Subramanyam, K. 1962 Aquatic Angiosperms. CSIR, New Delhi.

------- and **Sreemadhavan, C.P.** 1970 The paramount role of Herbaria in modern Taxonomic Researches. Abstract of Symposium I "Modern Trend in Taxonomy" Delhi University, Delhi.

Thothathri, K. 1965 Studies in Leguminosae 5. Taxonomic and nomenclatural notes on Indo-Burmese species of *Bauhinia* Linn. *Bull. bot. Soc. Bengal* 19(2):130-134.

Varma, S.C. 1936 Some ecological aspect of the Upper Gangetic Flora. *J. Indian bot. Soc.* 15:247-67.

